UP AND DOWN CALIFORNIA

THE FIELD PARTY OF 1864

GARDINER   COTTER   BREWER   KING

UP AND DOWN
CALIFORNIA
IN 1860–1864

THE JOURNAL OF WILLIAM H. BREWER

FOURTH EDITION, WITH MAPS

EDITED BY FRANCIS P. FARQUHAR
WITH A NEW FOREWORD BY WILLIAM BRIGHT

UNIVERSITY OF CALIFORNIA PRESS
BERKELEY · LOS ANGELES · LONDON

UNIVERSITY OF CALIFORNIA PRESS
BERKELEY AND LOS ANGELES, CALIFORNIA

UNIVERSITY OF CALIFORNIA PRESS, LTD.
LONDON, ENGLAND

© 2003 by the Regents of the University of California

Fourth edition, with maps

LIBRARY OF CONGRESS CATALOGING-IN-PUBLICATION DATA

Brewer, William H., 1828-1910.
    Up and down California in 1860-1864 : the journal of William H. Brewer / edited by Francis P. Farquhar ; with a new foreword by William Bright.—4th ed.
       p. cm.
    Includes index.
    ISBN 978-0-520-23865-7 (pbk. : alk. paper)
    1. California—Description and travel. 2. California—Surveys. 3. California—History, Local. 4. Geological surveys—California. 5. Scientific expeditions—California. 6. Natural history—California. 7. Brewer, William Henry, 1828-1910—Journeys—California. 8. Brewer, William Henry, 1828-1910—Diaries. 9. Brewer, William Henry, 1828-1910—Correspondence. I. Farquhar, Francis Peloubet, 1887- II. Title.

F864.B75 2003
979.4′04′092—dc21
                                          2002032224

16 15 14 13
10 9 8 7 6 5

CONTENTS

Foreword	vii
Maps	ix
Itinerary	xiii
Preface to the Third Edition	xxix
Introduction	xxxi

Illustrations follow page 168

BOOK I: 1860–1861

I.	To California via Panama	3
II.	Los Angeles and Environs	11
III.	More of Southern California	29
IV.	Starting Northward	43
V.	Santa Barbara	55
VI.	The Coast Road	73
VII.	Salinas Valley and Monterey	91

BOOK II: 1861

I.	An Interlude	117
II.	New Idria	135
III.	New Almaden	149
IV.	Approaching the Bay	169
V.	The Mount Diablo Range	191
VI.	Napa Valley and the Geysers	213

BOOK III: 1862

I.	The Rainy Season	241
II.	Tamalpais and Diablo	255
III.	The Diablo Range South	275
IV.	Up the Sacramento River	291

V.	Mount Shasta	309
VI.	West and East of the Sacramento	325
VII.	Closing the Year—A Miscellany	347

BOOK IV: 1863

I.	In and About San Francisco	365
II.	Tejon—Tehachapi—Walker's Pass	375
III.	The Big Trees—Yosemite—Tuolumne Meadows	397
IV.	Mono Lake—Aurora—Sonora Pass	415
V.	To Carson Pass and Lake Tahoe	429
VI.	The Northern Mines and Lassen's Peak	451
VII.	Siskiyou	471
VIII.	Crescent City and San Francisco	489

BOOK V: 1864

I.	San Joaquin Valley—Giant Sequoias	505
II.	The High Sierra of Kings River	517
III.	Owens Valley and the San Joaquin Sierra	533
IV.	The Washoe Mines	551
V.	Homeward Bound—Nicaragua	561

Index   571

FOREWORD

William H. Brewer, the author of this book, was born in New York State in 1828. After studying science at Yale University, as well as in Europe, he came to California in 1860 as a member of a geological survey team headed by Josiah D. Whitney, after whom Mount Whitney, the tallest peak in California, is named. During Brewer's four years of fieldwork, he related his travels and described the young state of California in long letters sent to his brother "back East"—and what letters those were! Few of us now, in the twenty-first century, can boast of receiving such vivid correspondence.

California in the 1860s was a lively place. It had once been the home of a hundred Indian tribes; it had been a remote outpost of the Spanish colonial empire, then a neglected province of Mexico, then the scene of the frenzied gold rush of 1849, during which the state was flooded with immigrants from the eastern United States, from Europe, and indeed from China. By 1860, fortunes were still being made and lost overnight, but agriculture and commerce were beginning to take root. The Civil War, which was dividing the eastern states, had its repercussions in California as well: the state was nominally part of the Union, but prominent political figures openly supported the Confederacy. Everything was in flux, and Brewer saw it all—with humor and sympathy, but also with the cool eye of a scientist.

When Brewer's account of his travels was first published in 1930, it became an instant classic of Californiana, not only as a picture of the state's varied geography, but also as a compelling and eye-opening portrait of what life was like in the 1860s in places such as Los Angeles, San Francisco, and Angels Camp—a delight both for the professional historian and for the general reader. Revised editions were published in 1949 and 1966; and now the University of California Press has asked me to assist in the preparation of yet another edition, specifically by charting the course of Brewer's travels on a map of California. It had been years since I first read *Up and Down California*, but it was a pleasure to take it up again—this time with gazetteer, place-name dictionary, and topographic maps close at hand as I traveled with Brewer by ship, by stagecoach, on horseback or mule back, and on foot over the length and breadth of the state. For anyone interested in the Old West, picking up this book amounts to embarking on an adventure in time travel.

William Bright
August 2001

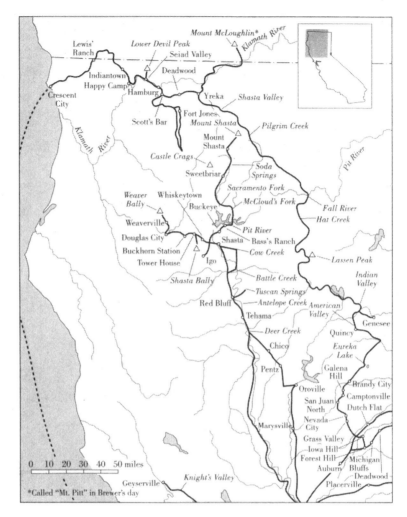

MAP 1. Brewer's route, Northern California. Solid lines represent the route traveled over land; dotted lines represent the route traveled over water.

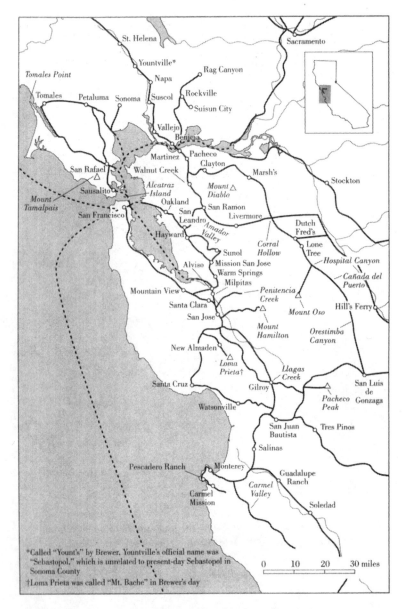

MAP 2. Brewer's route, San Francisco Bay–Monterey Bay area.

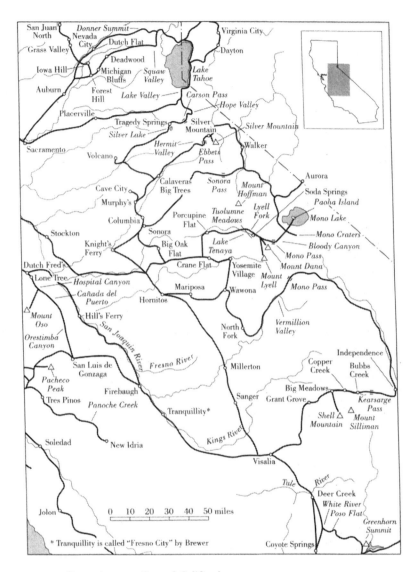

MAP 3. Brewer's route, Central California.

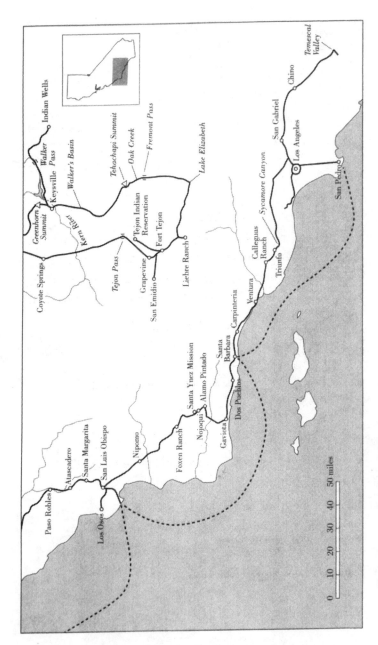

MAP 4. Brewer's route, Southern California.

ITINERARY

Following is the itinerary Brewer and the survey team followed over the four years covered in *Up and Down California*. The dates are sometimes approximate; corresponding page numbers follow the entries.

1860	Nov.	15	San Francisco, 7
		24	Departed from San Francisco by steamer, 11
		25	Steamer arrives at San Luis Obispo, 11
		26	Steamer stops at Santa Barbara, arrives at San Pedro, 12; stagecoach to Los Angeles
	Dec.	4	Excursion to San Gabriel, 14
		17	Santa Monica Mountains (location not specified), 15
		24	Back to camp in Los Angeles, 20
		28	San Gabriel, 28
1861	Jan.	7	Chino (San Bernardino Co.), 32
		20	Temescal Valley (Riverside Co.), 32
	Feb.	3	San Gabriel, 33
		5	Los Angeles, 40
		17	Sycamore Canyon (Ventura Co.), 43
		23	Triunfo (Ventura Co.), 45, 47
		27	Calleguas Ranch (Ventura Co.), 49
	Mar.	1	San Buenaventura (= Ventura), 49
		5	Carpinteria (Santa Barbara Co.), 48
		7	Santa Barbara, 56

April	2	Dos Pueblos (Santa Barbara Co.), 73
	4	Gaviota to Nojoqui (Santa Barbara Co.), 75
	6	Alamo Pintado (Santa Barbara Co.), 76
	7	Santa Ynez Mission, 76
	8	Foxen Ranch (Santa Barbara Co.), 77
	9	Nipomo (Santa Barbara Co.), 78
	14	San Luis Obispo, 79
	24	Excursion to Los Osos (San Luis Obispo Co.), 87
	29	Santa Margarita (San Luis Obispo Co.), 92
May	4	Atascadero (San Luis Obispo Co.), 93
	8	Jolon (Monterey Co.), 93
	12	Soledad (Monterey Co.), 99
	15	Monterey, 101
	27	Pescadero Ranch (= Pebble Beach), 104; Carmel Mission, 106
June	12	Departed Monterey area, camped near Salinas, 131
	14	San Juan Bautista (San Benito Co.), 131
	20	San Francisco, 117
	26	Back to San Juan Bautista via Santa Clara and San Jose, 120
July	8	Monterey, 129
	9	Pescadero Ranch, 129
	13	Salinas, 131
	14	San Juan Bautista, 131
	15	Tres Pinos (San Benito Co.), 135
	18	Panoche Creek (San Benito Co.), 136
	20	New Idria (San Benito Co.), 138
	27	San Juan Bautista, 146
	30	Pacheco Peak (Santa Clara Co.), 149

ITINERARY

Aug.	2	Watsonville (Santa Cruz Co.), 152
	3	Santa Cruz, 152
	8	New Almaden (Santa Clara Co.), 154
	15	Excursion to "Mt. Bache" (= Loma Prieta), 161
	20	Llagas Creek (near Gilroy, Santa Clara Co.), 170
	24	San Jose, 172
	26	Excursion to Mt. Hamilton (Santa Clara Co.)
Sept.	1	Mountain View (Santa Clara Co.), 173
	4	Via Alviso and Milpitas (Santa Clara Co.) to Mission San Jose (Alameda Co.), 179
	9	Via Hayward to San Francisco, 182
	10	Return to Mission San Jose, 183
	11	Amador Valley (area of Pleasanton, Alameda Co.), 184
	13	Via San Ramon Valley (Alameda Co.) to Oakland, 185
	15	To San Francisco by ferry, 187
	16	Steamship to Sacramento, 187
	20	Steamship back to San Francisco, 190
	23	Probably by steamer to Martinez (Contra Costa Co.), and on to Clayton, on north side of Mount Diablo (Contra Costa Co.), 191, 193
	30	Martinez, 194
Oct.	4	Clayton, 195
	13	Corral Hollow, on east side of Mount Diablo (Alameda Co.), 200
	20	Via Livermore to San Ramon Valley (Alameda Co.), 210

		22	Martinez, 212
		25	By ferry to Benicia (Solano Co.), 214
	Nov.	2	Via Vallejo (Solano Co.) to Suscol (Napa Co.), 216
		3	Via Napa, 220, to "Sebastopol" (= Yountville, Napa Co.), 217, 221
		6	Knight's Valley (Sonoma Co.), 226
		7	"The Geysers" (= Geyserville), 229
		13	St. Helena (Napa Co.), 235
		14	Suscol (Napa Co.), 235
		15	Benicia (Solano Co.), 235, and by steamer to San Francisco
1862	Feb.	22	Steamer to Alviso (Santa Clara Co.), stagecoach to San Jose, 247
		24	Return to San Francisco, 247
	Mar.	6	Steamer to Sacramento, 248
		7	Return by steamer to San Francisco, 250
		23	Martinez, 255
		26	San Rafael (Marin Co.), 255
		28	Mount Tamalpais, Sausalito, 256, and by boat to San Francisco, 257
	April	4	Steamer to Petaluma (Sonoma Co.), 257, then toward San Rafael
		8	To San Francisco (by boat from San Rafael?), 258
		9	Steamer to Stockton
		11	Steamer to San Francisco
		14	Steamer to Petaluma, 259
		16	To Sonoma and Napa, 259; later returned to Petaluma and San Francisco
		22	Steamer to Martinez, 261

May	6	Clayton, 262
	28	Marsh's (Contra Costa Co.), 269
June	1	Livermore Pass (Contra Costa Co.), 267, 272
	3	Corral Hollow, 276
	6	Lone Tree Canyon (San Joaquin Co.), 277
	9	Excursion to Hospital Canyon, Mount Oso (Stanislaus Co.), 278
	10	Cañada del Puerto (Stanislaus Co.), 279
	14	Orestimba Canyon (Stanislaus Co.), 283
	19	San Luis de Gonzaga (Merced Co.), 286
	24	Pacheco Pass (Santa Clara Co.), 288
	27	San Juan Bautista, 288
July	1	San Jose, 289
	2	Excursion to Mount Hamilton, 289
	6	San Francisco and return to San Jose, 289
	?	New Almaden (Santa Clara Co.), 289
	16	Walnut Creek, 289
	18	Benicia, 289
	25	Rockville (Solano Co.), 293
	26	Up Suisun Creek to Rag Canyon (Solano Co.), 294
Aug.	8	Returned to Suisun, and by steamer to San Francisco, 294
	12	By steamer to Sacramento, 294
	13	By steamer up the Sacramento River to Tehama (Tehama Co.), 296
	16	Continuing by steamer to Red Bluff (Tehama Co.), 296
	19	Excursion to Tuscan Springs (Tehama Co.), 297

	20	To the town of Shasta [City] (Shasta Co.), 297
	23	Excursion to Middletown, Horsetown, "Piety Hill" (= Igo), and Cottonwood Creek, 297, 299
Sept.	6	Buckeye; Bass's Ranch (Shasta Co.), 303
	7	Pit River, McCloud's Fork (Shasta Co.), 304
	9	Castle Rocks (Siskiyou Co.), 306; "Strawberry Ranch" (= Sisson, Mount Shasta [City], in Siskiyou Co.), 311
	11	To timberline on Mount Shasta, 311
	12	To peak of Mount Shasta, 315
	13	Timberline on Mount Shasta, 319
	17	Sweetbriar Ranch (Siskiyou Co.), 320
	18	Dogtown (exact location unknown), 321
	19	Bass's Ranch, 322
	22	Shasta [City], 323
	23	Whiskeytown, Tower House, 325; Buckhorn Station (Shasta Co.), 326
	24	Weaverville (Trinity Co.), 326
	25	Excursion to Douglas City (Trinity Co.), 326
	26	Excursion to "Mount Balley" (Weaver Bally), 328
	27	Tower House, 331
	28	Excursion to Shasta Bally, 331
	29	Return to Shasta [City], 332
Oct.	2	Cow Creek (Shasta Co.), 332
	4	Battle Creek (Shasta/Tehama Cos.), 334 (Brewer's "north" is probably an error for "south")

	7	Red Bluff; Antelope Creek (Tehama Co.), 336
	8	Tuscan Springs (Tehama Co.), 336
	9	Deer Creek (Tehama Co.), 336
	10	Chico (Butte Co.), 337
	13	Butte Creek; "Pence's Ranch" (= Pentz), 340
	16	Oroville (Butte Co.), 344
	18	Marysville (Yuba Co.), 344
	20	Via Sacramento to San Francisco, 344
	27	By ship to Benicia (Contra Costa Co.), 348
Nov.	3	Petaluma (Sonoma Co.), 348
	4	Tomales (Marin Co.), 348
	5	Tomales Bay, Tomales Point (Marin Co.), 348
	6	San Rafael, 350
	7	By boat to San Francisco, 350
	(late)	Ferry to Oakland; Hayward, Alameda Valley (Alameda Co.); Sunol (Alameda Co.), 351; Mission San Jose (Alameda Co.), San Francisco, 352
Dec. (early)		San Jose; mountains east of San Jose; Penitencia Canyon; returning to San Jose and San Francisco, 353
	22	By boat to Alcatraz Island, 355
1863 Jan.	3	San Jose, 365
	5	San Francisco, 366
Feb.	3?	To Oakland and return, 368
Mar.	5	By boat to Alviso and by stage to San Jose, 371

	6	New Almaden, 372; later return to San Francisco
	(late)	Santa Cruz Mountains (no details), 375
April	1	By ship to Martinez, 376
	2	Clayton, 376
	3	Dutch Fred's (San Joaquin Co.), 377
	5	Camp on the San Joaquin River (probably Stanislaus Co.), 375, 378
	6	Hill's Ferry (Stanislaus Co.), 378
	7	Firebaugh (Madera Co.), 378
	8	"Fresno City" (= Tranquillity, Fresno Co.), 379, 396
	9	Kings River (Kings Co.), 379
	10	Visalia (Tulare Co.), 380
	13	Tule River (Tulare Co.), 381
	14	Deer Creek, White River (Tulare Co.); Coyote Springs (Kern Co.), 381
	15	Kern River (Kern Co.), 382
	16	Tejon Canyon, 382; "Cañada de las Uvas" (= Grapevine, Kern Co.), 383
	17	Fort Tejon (Kern Co.), 383
	22	San Emidio (Kern Co.), 385
	26	Return to Fort Tejon, 386
	28	Tejon Indian Reservation, 386
May	6	Liebre Ranch (Los Angeles Co.), 387
	8	Lake Elizabeth (Los Angeles Co.), 388
	9	Fremont Pass, Oak Creek (Kern Co.), 389
	10	Tehachapi (Kern Co.), 389
	11	Pass Creek (Kern Co.), 390
	12	Walker's Basin (Kern Co.), 390
	13	Keysville (Kern Co.), 391

	15	Walker Pass, Indian Wells (Kern Co.), 392
	18	Back to Keysville, 394
	19	Greenhorn (Kern Co.), 394
	20	"Posé Flat" (= Poso Flat, Kern Co.), "Tailholt" (= White River, Tulare Co.), 394
	21	Tule River; Visalia (Tulare Co.), 395
	25	To vicinity of Sanger (Fresno Co.), 395
	26	Millerton (Fresno Co.), 395
	27	Hornitos (Mariposa Co.), 395
	30	Knight's Ferry (Stanislaus Co.), 397
	31	Columbia (Tuolumne Co.), 397
June	2	Murphy's (Calaveras Co.), Calaveras Big Trees, 398
	4	Return to Murphy's, 400
	6	Cave City (Calaveras Co.), 400
	8	Sonora (Tuolumne Co.), 400
	10	Big Oak Flat (Tuolumne Co.), 401
	13	Return to Sonora, 401
	15	Crane Flat (Tuolumne Co.), 403
	16	Yosemite Valley, 403
	23	Porcupine Flat (Mariposa Co.), 407
	24	Excursion to climb Mount Hoffmann (Mariposa Co.), 407
	25	Lake Tenaya (Mariposa Co.), 407
	26	Soda Springs (Tuolumne Co.), 407
	27	Mono Pass (Tuolumne Co.), 408
	28	Excursion to climb Mount Dana (Mono Co.), 408
July	1	Tuolumne Meadows (Tuolumne Co.), 410
	2	Excursion to climb Mount Lyell (Madera Co.), 411

	7	Mono Pass; Bloody Canyon to Mono Lake (Mono Co.), 415
	10	Excursion to Paoha Island, 417
	11	Excursion to Aurora (Mineral Co., Nev.), 419
	15	"Walker's River" (= Walker, Mono Co.), 422
	17	Sonora Pass (Tuolumne Co.), 423
	23	Sonora (Tuolumne Co.), 425
	24	Through Columbia (Tuolumne Co.) to Murphy's (Calaveras Co.), 425
	29	Calaveras Big Trees, 430
	31	Silver Valley (Alpine Co.), 430
Aug.	1	Hermit Valley (Alpine Co.), 431
	4	Silver Mountain (town, Alpine Co.), 431
	5	Excursion to Silver Mountain (peak, Alpine Co.), 433
	8	Back to Hermit Valley, 434
	9	Calaveras Big Trees, 435
	10	Volcano (Amador Co.), 435
	11	Left Volcano en route to Lake Tahoe, 435, 436
	15	Tragedy Springs (El Dorado Co.); Silver Lake (Amador Co.), 436
	16	Carson Spur, Summit Lake (Amador Co.), 437
	18	Hope Valley (Alpine Co.), 438
	19	"Lake Valley" (= South Lake Tahoe, El Dorado Co.), Slippery Ford, 439
	20	Excursion to Pyramid Peak (El Dorado Co.), 441
	21	"Lake Valley," 441
	24	East side of Lake Tahoe (near Glenbrook, Nev.), 443

	25	Centerville, Elizabethtown, Tim-i-lick Valley (Nev.?), 444
	26	Truckee River, Knoxville (Nevada Co., Calif.?), 445
	28	Squaw Valley; Middle Fork of the American River (Placer Co.), 446
	29	Last Chance; Deadwood (Placer Co.), 447; Michigan Bluffs (Placer Co.), 448
	30	Forest Hill (Placer Co.), 448
	31	Auburn (Placer Co.), Sacramento, by steamer to San Francisco, 451
Sept.	5	Steamer to Sacramento; on horseback to Forest Hill; Iowa Hill (Placer Co.), 453
	8	Grass Valley (Nevada Co.), 453
	10	Nevada City, San Juan North (Nevada Co.), Camptonville (Yuba Co.), 455
	11	Galena Hill, North Yuba Canyon (Yuba Co.), Brandy City, Eureka, Poker Flat (Sierra Co.), 455
	12	Whiskey Diggings, Potosi, Rowland Flat, Pilot Peak, Middle Feather River, Nelson's Point, Quincy (Plumas Co.), 455, 456
	14	American Valley, Indian Valley, Genesee Valley (Plumas Co.), 456
	18	"Mormon Station" (= Portola, Plumas Co.), 457
	21	Big Meadows, Indian Valley, Feather River (Plumas Co.), 458
	22	Toward Lassen Peak, 458
	23	At end of trail on Lassen Peak, 458
	24	Boiling Lake, Steamboat Springs, Willow Lake (Lassen Peak area), 460
	25	Lassen Peak area, 460
	26	Climbed Lassen Peak, 461

Oct.	1	Hat Creek, Battle Creek, Honey Lake (Lassen Co.), 467
	2	Pit River at junction of Fall River (Fall River Mills, Shasta Co.), 468
	3	Fort Crook (Shasta Co.), 468
	6	Traveling toward Yreka, 471
	7	Elk Valley (Shasta Co?), 471; Pilgrim Camp (Siskiyou Co.), 472
	9	Shasta Valley (Siskiyou Co.), north of Mount Shasta, 473
	10	Pluto's Cave, 473; Yreka (Siskiyou Co.), 474
	14	Cottonwood, then to Klamath River (Siskiyou Co.), 475
	15	"[C]rossed the line into Oregon and climbed the Siskiyou Mtns.," 476
	20	Back to Yreka, continuing to Deadwood (Siskiyou Co.), 477
	22	Scott's Bar, Fort Jones (Siskiyou Co.), 477
	23	Via Scott's Bar to Hamburg and "Sciad" (Seiad, Siskiyou Co.), 479, 480
	25	Excursion to climb Lower Devil Peak (Siskiyou Co.), 480
	26	Down Klamath River to Happy Camp (Siskiyou Co.), 482
	27	"Sailor Diggings" (= Waldo, Josephine Co, Ore.), 482
	29	Lewis Ranch (Del Norte Co., Calif.), on north fork of Smith River, 484
Nov.	2	Low Divide (Del Norte Co.), 486
	7	Crescent City (Del Norte Co.), 490
	21	By ship to San Francisco, 496

ITINERARY

1864	May	24	Ferry to Oakland, then to Hayward (Alameda Co.), 505
		25	Warm Springs (Alameda Co.); San Jose, 506
		27	Twenty-one Mile House (Santa Clara Co.), 506
		28	Gilroy; Pacheco Pass (Santa Clara Co.), 506
		30	San Luis de Gonzaga Ranch (Merced Co.), 508
		31	Lone Willow, 509
	June	1	Firebaugh (Fresno Co.), 510
		2	"Fresno City" (= Tranquillity, Fresno Co.), 510
		3	Elkhorn Station (Fresno Co.), 511
		4	Kings River (Kings Co.), 512
		6	Visalia (Tulare Co.), 512
		8	Eastward to foothills (Tulare Co.), 513
		9	North into mountains, 514
		10	Thomas' Mill (near Grant Grove, Tulare Co.), 514, 516
		17	On divide between Kaweah and Kings Rivers (Tulare Co.), 517
		18	Big Meadows (Tulare Co.), 517, 518
		20	Excursion to climb Shell Mountain (Tulare Co.), 518
		25	J. O. Pass (Tulare Co.), 521, 532
		26	Sugarloaf Creek, Clover Creek (Tulare Co.), 522
		28	Mount Silliman (Tulare Co.), 523
		30	Continuing in Kings Canyon area (Tulare Co.), 523
	July	1	Across Roaring River near Scaffold Meadow (Tulare Co.), 524, 532
		2	Excursion to climb the peak later named Mount Brewer (Tulare Co.), 524, 532

10	Back to Big Meadows (Tulare Co.), 527
12	Lewis' Ranch (Tulare Co.), en route to Visalia, 528
13	Visalia, 528
15	Back to Lewis' Ranch, 528
16	Big Meadows, 528
17	Returning to Kings Canyon area, 529
21	Copper Creek (Tulare Co.), 531
26	Bubbs Creek, Charlotte Creek (Fresno Co.), 534, 548
27	Kearsarge Pass (Fresno Co.), Independence Creek (Inyo Co.), 534, 548
28	Owens River, 534; Bend City (Inyo Co.), 537
29	Independence (Inyo Co.), 537
30–31	Continuing up Owens River valley (Inyo/Mono Cos.), 538

Aug.
1	"Southwest Fork of Owens River" (= Rock Creek, Mono Co.), 539, 548
2	Mono Pass (Inyo Co.), not to be confused with Mono Pass to the north in Tuolumne Co.; to Middle Fork of San Joaquin River (Madera Co.), 539, 540, 548
4	Vermillion Valley (Fresno Co.), 540, 548
10?	Excursion to climb Mount Goddard (Fresno Co.), 541
15	Departed for "North Fork" (= Middle Fork) of San Joaquin River (Fresno Co.), 544
17–21	Descending San Joaquin River (Fresno Co.), 546
22	Fresno River (Madera Co.), 546
23	"Clark's Ranch" (= Wawona, Mariposa Co.), 546
26	Excursion to Yosemite Valley

Sept.	10	Mariposa, 547
	15?	Stockton, 548
	16?	San Francisco
Nov.	1	By steamer to Sacramento
	2	Placerville
	3	"Lake Valley" (= South Lake Tahoe), "Carson Valley" (= Virginia City, Nev.), 553, 554
	6	Excursion to Steamboat Springs (Washoe Co., Nev.), 561
	7	Excursion to Dayton (Lyon Co., Nev.), 562
	10	Donner Pass (Nevada Co.), 563
	11	Dutch Flat (Placer Co.), Auburn, 563
	12	Via Sacramento to San Francisco, 564
	14	Left San Francisco by ship, 564, en route via Nicaragua to New York

PREFACE TO THE THIRD EDITION

UP AND DOWN CALIFORNIA was first published in 1930 by Yale University Press, New Haven, Connecticut. It went out of print in a few years; but the demand continued, especially when it became more and more widely recognized as an important work in the field of California history. For this reason it seemed appropriate for the University of California Press to adopt the book. So, with the consent of the Brewer family and the Yale University Press, a second edition was published in 1949. This edition, too, has long been out of print. The demand has continued, however, so that a third edition seems to be required.

The text of the Brewer letters in this Third Edition, as it was in the Second Edition, is precisely the same as in the First. Only a few additions have been made in the notes, but references to "persons now living" have been left as in the 1930 edition. Appropriate changes and omissions have been made in the introductory matter.

The illustrations are substantially the same as those in the Second Edition. With one exception, they are from drawings or photographs practically contemporary with the text. In fact, the sketches by members of the Survey were made on the spot during the very journeys described by Brewer.

Professor Brewer's field notebooks are now in the Bancroft Library, University of California, Berkeley, as are the original drawings and photographs.

A good deal has been written about the Sierra Nevada since the earlier editions of this book, notably newly edited editions of Clarence King's "Mountaineering in the Sierra Nevada"; Hal Roth's "Pathway in the Sky: The Story of the John Muir Trail," Berkeley, 1965; and my own "History of the Sierra Nevada," published by the University of California

Press, 1965. There are also many new books of description and history about other regions of California, which furnish additional information and background about the places described by Professor Brewer a little over one hundred years ago.

FRANCIS P. FARQUHAR

Berkeley, California
May, 1966

INTRODUCTION

By the year 1860 California was showing signs of too much mining excitement. The days of '49 were irrevocably gone. For a decade gold mining had passed from one phase to another and disorganized individual enterprise had given way to corporate organization, capital outlay, and engineering skill. Nevertheless, the old gambling spirit persisted, stimulated by occasional rich strikes and partial successes. Moreover, the gold fever had aroused a general interest in minerals, so that there were frequent "excitements" over discoveries of silver, tin, quicksilver, and even coal. Immense resources seemed to lie all about; yet, somehow, they did not materialize with the expected abundance. Under these circumstances it became clear to certain of the more sober minds in the state that definite scientific knowledge was needed to give better direction to the development of resources.

Foremost among those who perceived this need was Stephen J. Field, at that time a justice of the Supreme Court of California, later of the Supreme Court of the United States. He realized that a geological survey of the state, in order to accomplish its purposes, must be not only competent in science, but strictly impartial and unprejudiced. He was determined, therefore, that it should be kept out of politics and that it should be free from local influences. Everything would depend upon the character and qualifications of the man to be placed in charge of the work. Accordingly, before urging the matter in public, Justice Field quietly sought advice of the leading men of science in the East and asked them to recommend a suitable director. The name that received the preponderance of endorsements was that of Josiah Dwight Whitney, a graduate of Yale, who had been engaged for a number of years in various state surveys and whose book, *The Metallic Wealth of the United States*, had attracted wide attention. Consequently, when the bill came up for consideration in the legislature, Justice Field and his associates, in spite of strong opposition from several locally supported rivals, were able to have Whitney designated in the act itself as State Geologist.

The act of April 21, 1860, in appointing the State Geologist, directed him: "With the aid of such assistants as he may appoint, to make an accurate and complete Geological Survey of the State, and to furnish, in his Report of the same, proper maps and diagrams thereof, with a full and scientific description of its rocks, fossils, soils, and minerals, and of its botanical and zoölogical productions, together with specimens of the same." Whitney accepted the appointment and set about organizing the personnel and equipment for his work.

The first man selected by Whitney for his staff was William H. Brewer. The two had never met, and did not do so until the very eve of departure for California; but so convincing was the recommendation of Professor Brush, of Yale, to whom Whitney had addressed an inquiry, and so entirely suitable were Brewer's qualifications, that the matter was arranged by correspondence. The next four years were to show how extremely fortunate Whitney was in this selection. It was of vast importance that his right-hand man should be of the strongest fiber, of unflagging energy, the soundest judgment, the utmost tact, and of unequivocal honesty and loyalty. Happily, these were the very qualifications that distinguished the character of Brewer.

Brewer's professional attainments were not those of a

geologist. He was educated primarily in the sciences centering about agriculture. But in grounding himself in these he had learned methods applicable to the study of all natural sciences. He was a very keen and careful observer, ever mindful of the importance of accuracy and of order. Moreover, he had a native shrewdness that enabled him to recognize the relative significance of things and to draw sound conclusions from his data. In these qualities he was, in fact, superior to his chief; for Whitney, in spite of his compendious knowledge and high intellectual attainments, was inclined to be dogmatic. There were other respects in which there was a contrast between the two. Whitney was forever quarreling with those with whom he disagreed; Brewer, no matter how pronounced might be his views, was always ready to let good fellowship and good humor prevail. There was a genial quality about him that proved a saving grace for the Survey on more than one occasion. Let it be said of Whitney, however, that with those whom he considered his peers and with the members of his own staff he was on the best of terms.

Notwithstanding the high rank to which he rose in the academic world, Brewer was first and last a farmer, and his life story constantly reflects his closeness to the soil. This is exemplified in his sound common sense, his farmer's handiness with everyday contrivances, his ability to keep the wheels of work going through all kinds of adversities of weather, the zest with which he engaged in hard labor, the sincerity and generosity of his relations with men, the heartiness of his humor, the wholesomeness with which he relished a salty episode, and, finally, in the sound fruition that followed his labors.

William Henry Brewer was born at Poughkeepsie, New York, September 14, 1828. The family soon afterward removed to Enfield, near Ithaca, where he grew up accustomed to the duties of a boy on a small farm. A Dutch ancestor, Adam Brouwer Berkhoven, had come to New Amsterdam in 1642, but as generations passed in the New World the name Berkhoven was dropped and *Brouwer* became *Brower*. Not until after the American Revolution did the further transition to *Brewer* occur. Ancestry on the mother's side also extended to Colonial times; the DuBois family, Huguenots, came to New York in 1662. William Henry Brewer had one brother, Edgar, three and a half years younger, who lived for most of his life on the family farm at Enfield. William Henry attended district school and then spent four winters at Ithaca Academy.

Such was his simple background when, in 1848, he secured his father's permission to study agricultural chemistry for a year at Yale under Professor Benjamin Silliman, Jr., and Professor John Pitkin Norton. When Brewer set out for New Haven in October, 1848, he traveled for the first time on a public conveyance. This journey was the beginning of an unfoldment that soon led to farther horizons than he had visioned on the farm. His year at Yale was extended to two. He applied himself eagerly to his studies and formed lasting friendships. He was one of the first members taken into Berzelius Society after its formation.

At the end of two years Brewer returned to Enfield and began his career as a teacher, first at Ithaca Academy, then at an agricultural school. In the summer of 1852 he was summoned to New Haven to be examined for the degree of Bachelor of Philosophy, which was to be conferred upon those who had studied in the "School of Applied Chemistry." On July 29, 1852, with George J. Brush, William P. Blake, and three others, he received the degree. This was the first class to be graduated from what is now the Sheffield Scientific School.

For the next three years he taught at Ovid Academy, Ovid, New York, constantly strengthening his conviction that the future development of agriculture lay in the study and application of the natural sciences. With this conviction, he resolved to go to the fountainhead of scientific teaching, and in September, 1855, he sailed for Hamburg on the bark *Ericsson*. Going directly to Heidelberg, he entered the analytical laboratory of Professor Bunsen, and a year later moved on to Munich, where he studied under Liebig. In the summer of 1856 Brewer took to the open, walking six hundred miles through Switzerland. While the study of botany was his principal motive, he did not fail to be impressed with the splendors of the mountain scenery. This journey, and a shorter one in the Tyrol the following spring, afforded experience in mountain travel that was to assist him immeasurably in California a few years later. Before returning to Ovid in the fall of 1857, he attended lectures on chemistry by Chevreul in Paris, went on a brief botanical expedition to the south of France, and saw a little of England. It is typical of his weighing of values that in order to enjoy these added travels he chose to come home "steerage" on the steamer from Liverpool.

A year after his return from Europe, Brewer was called to a professorship of chemistry at Washington College (now Washington and Jefferson College), Pennsylvania. Meanwhile, in August, 1858, he had married Angelina Jameson. His new position and his family life were, however, of brief duration, for in the summer of 1860, shortly after the birth of a son, his wife died, and a few weeks later the child followed. It was at this sad moment that the offer came from Whitney to go to California, and Brewer welcomed the opportunity of leaving the melancholy associations that a continuance at Washington College would have entailed.

The journey to California, the commencement of the field work, the day-by-day progress, the growth of a comprehensive view of the physical structure of the state of California, are described so thoroughly and so clearly in Brewer's letters that there is no need for amplification or for summary. The four years of Brewer's service with the Survey cover a distinct period, in which a very large part of its important results was accomplished. In the following years the life of the Survey became extremely precarious. At one time there was a complete shutdown because of lack of funds, and finally, in 1873, after a brief revival, it was discontinued entirely.

It can hardly be said that the original purposes of the California State Geological Survey were fulfilled. Much was indeed learned about the mining regions and the nature of the auriferous gravels; here and there a slight curb was put upon speculation; the topography of the state was fairly well mapped; and great progress was made toward an understanding of the geological history of the country. Save for the maps, however, it is doubtful whether any immediate economic advantages can be traced to Whitney's work. Certainly no new mineral fields were discovered and no direction was given to the mining industry. Whitney's excuse was that he could not produce economic results except upon a basis of scientific knowledge, and that the field was so large and so difficult that a much larger sum of money was needed than had been placed at his disposal. There is a great deal of truth in Whitney's contention; but, on the other hand, it is equally true that Whitney's own character had much to do with the diversion of the Survey from its original purposes and its consequent incompleteness. Whitney was bent upon conducting a perfect survey. He was uncompromising and unyielding in the face of practical situations that required diplomatic handling. Before trying to convince a state legislature that the study of fossils—"shells and old bones"—had a direct bearing upon the discovery of gold mines, he should have offered simpler and more comprehensible examples of the value of geological science. This he might readily have done from the multifarious material developed during the first few years of the work. He scorned such expedients, however, and refused to deviate from his nobly conceived, but extremely ambitious, plans.

Although the Whitney Survey was a disappointment to the people of California, it was, nevertheless, extremely valuable in many respects. It produced a wealth of information which was utilized by other agencies and which ultimately found its reflection in the welfare of the state. Perhaps its greatest value was in the far-reaching influence it had on the conduct of subsequent surveys throughout the United States. Out of its ranks came Clarence King, Charles F. Hoffmann, and James T. Gardiner. King proceeded to form his own Survey of the Fortieth Parallel and later developed the idea of a consolidation of all government surveys. Others were working for the same end, and presently there was a bitter struggle for control. That the United States Geological Survey, as eventually established in 1879 with Clarence King as its first Director, was a civilian rather than a military agency is directly traceable to ideas formulated in the Whitney Survey of California. Many of the methods employed by the United States Geological Survey may be traced to the same source. Hoffmann, for instance, may well be called the progenitor of modern American topography. Guided by Whitney, he taught the art to King

and Gardiner, who, in turn, developed it in the Survey of
the Fortieth Parallel. He also taught Henry Gannett, who,
with Gardiner, introduced the art to the Hayden Survey.
When the consolidation took place, the topographic work
was, therefore, almost entirely in the hands of men trained
in this school. In 1900 Brewer, in a letter to Hoffmann, reviewed this course of events and made this statement:

ALL these years I have taken pains, whenever opportunity occurred,
to keep it in mind that you introduced into America this system of
field topographical survey, which now, improved greatly, but fundamentally the same, and tho' modified and much more widely extended, is the method employed by the general Government, and
which, as I understand, has since been introduced into other countries where similar conditions occur. For this, Whitney and you
should have credit, and the fact should have a more prominent record than the mere recollections of men.

Professor Brewer's title in the Geological Survey of California was "Principal Assistant, in charge of Botanical Department." It will be observed from the contents of his journal that the botanical duties were subordinated to, and at times practically extinguished by, the responsibilities placed upon him as leader of the field parties. Nevertheless, he was able to do a considerable amount of collecting without much extra effort. With his customary precision he numbered his specimens in serial order, an aid to identification frequently neglected by collectors of his time. Classification and description was perforce left to a future occasion, so Brewer came to the close of his work in California with very little beyond his collections to show for his labors in the province of botany. For a time, after leaving the Survey, he worked on his botanical report at the Herbarium of Harvard University, where he had the benefit of the counsel of

Professor Asa Gray. In a memorandum written many years later he states:

I RECEIVED no pay whatever after the closing of my connection with the Survey in California—neither for the time nor expense in working up results. I spent an aggregate of about two years time—a little more rather than less—and over two thousand dollars in cash, besides deducting another one thousand dollars from my salary from college because of time taken out for my work—that is, absence during term time at work on my plants at the Cambridge Herbarium. After Gilman went to California as president of the State University he induced a few wealthy citizens there to subscribe money for the finishing of the botanical work and getting it printed. I got the printing started, and then employed Watson, handed over all my notes to him and the rest of my manuscript, and he finished it.

The first volume of the botanical report did not appear until 1876; the second, in which Brewer had practically no part, not until 1880.[1]

Toward the close of his fourth year with the Survey, Brewer received word of his appointment to the Chair of Agriculture in the Sheffield Scientific School at Yale. His acceptance marked the end of his roving and brought him into the full tide of his career. From the spring of 1865, when he entered upon his duties at New Haven, until his retirement in 1903 as professor emeritus, he took a prominent part in the development of the school. His influence extended far beyond its walls, however, for he was not content with academic teaching, but must needs bring the virtues of science to the farms, the villages, and the cities of his state. He promoted the establishment of agricultural experiment stations; he helped to organize the Connecticut State Board of Health and served on it for thirty-one years; he

[1] W. H. Brewer and Sereno Watson, "Polypetalae"; Asa Gray, "Gramopetalae," *Botany* (Cambridge, 1876), I, xx + 628 pp.: Sereno Watson, *Botany* (Cambridge, 1880), II, xv + 559 pp.

also served for a long time on the Board of Health of the city of New Haven. His services were also in demand in wider fields. As a special agent for the census of 1880 he reported on the production of cereals in the United States; he was a member of the United States Forestry Commission appointed in 1896 to investigate the forest resources of the country; he was chairman of the committee appointed by the National Academy of Sciences in 1903 to make recommendations for a scientific survey of the Philippine Islands; he was offered the position of Assistant Secretary of Agriculture in the Cleveland administration, but declined.

Although after leaving California Brewer never again found time for extended exploration, he by no means lost interest in such things. On three occasions he took part in shorter trips of an unusual character. The first of these was a summer trip to the Rocky Mountains of Colorado in 1869.[2] During an interim in the California Survey Whitney was teaching at Harvard and desired to bring some of his students into contact with actual field conditions. He persuaded Brewer and Hoffmann to assist him in conducting the expedition and in teaching the science of geology and the art of topography. It is noteworthy that of the four students two subsequently achieved great distinction in these fields: William Morris Davis becoming Professor of Geology at Harvard, and Henry Gannett becoming Chief Geographer of the United States Geological Survey. It was many years before Brewer made another expedition, this time to Greenland, in 1894. As a result of this trip he joined with others in forming the Arctic Club, of which he was for many years the president. In 1899 he was a member of the Harriman Alaska Expedition.

[2] Professor Brewer wrote a series of letters to his wife describing this trip. These letters have recently been published in pamphlet form by the Colorado Mountain Club, Denver, Colorado.

In 1868 he married Georgiana Robinson at Exeter, New Hampshire. To their home in New Haven four children were born: Nora (1870), now Mrs. Clifford Standish Griswold; Henry (1872); Arthur (1875); and Carl (1882).

As time went on, Professor Brewer received his share of academic distinctions. He was a member of the National Academy of Sciences and served a term as its president. In 1903 he was twice awarded the honorary degree of Doctor of Laws—by Wesleyan University and by Yale. A highly appropriate recognition came to him in 1910, when the same honorary degree was conferred upon him by the University of California. Thus, in the final year of his life, he again became associated with the state in which he had spent four of his most active years, years in which were laid down strong foundations for his vigorous and useful career.

Throughout his life Brewer was a voluminous letter writer and diarist. He recorded in his notebooks with minute punctiliousness everything he saw. His pages are filled with weather statistics, with estimates of distances, with measurements. These notebooks were for his own use, and well did he use them, again and again. But when he came to write out his impressions for the benefit of others, he clothed the bare bones of his statistics and created something pulsing with life. Yet he never altered his facts to make an impression. The statistics in his letters agree with those in his notebooks; and, if one were to go back to the scene today and remeasure with the same instruments and the same resources, one would in all probability find the facts to be much the same as Brewer said they were. If the altitudes that he gives for mountains are not quite the same as those shown on our latest maps, it is only because his means were inadequate, not because he failed to observe accurately. It is this accuracy of observation, coupled with his devotion to truth, that gives to his letters unusual historical value. Moreover, in all his writings he rarely goes beyond the limits of his own experience—there is very little "hearsay" in Brewer's journals.

During his four years in California he exercised his recording faculties to the fullest extent. In the midst of a most prodigious activity he found time to keep several distinct sets of notebooks, to prepare elaborate scientific reports, to engage in a miscellaneous correspondence, and to write the vigorous and comprehensive letters that constitute his personal journal. These letters are the more remarkable in that they were sometimes written late at night by firelight or candlelight, sometimes in the blistering heat of a summer noon, sometimes in a leaky tent with cold rain and wind outside. Numbered serially, they were sent to his brother, Edgar, with urgent instructions that after they had been passed around among family and friends they should be held for him until his return. Happily, only two or three numbers, all of lesser importance, failed of delivery.

Brewer probably never intended these letters for publication. At least, he never edited them or took any steps in that direction. Nor would they, perhaps, have attracted much attention if they had been published in the years immediately succeeding the events described. They were not "literature"; they were not written in the style of certain superficial travelers of the day whose animated accounts of what they saw and what they didn't see in California still cumber our shelves. Clarence King, Brewer's young *protégé*, could write "literature," however, and did, with a brilliancy that marked his course in many fields. His *Mountaineering in the Sierra Nevada*, published in 1872, after appearing in part in the *Atlantic Monthly*, was the only publication resulting from the California State Geological Survey outside of the official reports and Whitney's scientific by-products. In King's delightful book there is glamor and entertainment; in Whitney's reports, voluminous information ably presented. But in the ripeness of time Brewer's letters will come to fill a place quite as important as either Whitney's reports or King's essays. They are an unabridged, undecorated record of the times, as replete with significant facts as the reports, often as vivid in descriptions as the essays, yet devoid of the obsolete deductions of the former and the occasional exaggerations of the latter.

In preparing these letters for the press, the editor has taken certain liberties with the text which he believes Professor Brewer would have cordially sanctioned were he alive. It would be unfair to a scholar of high standing to perpetuate errors of spelling, hastily contrived sentence structure, unwitting repetition, and other trivialities, resulting from the trying conditions under which the writing was done. Moreover, there are portions of the letters in which considerable condensation has been possible without the sacrifice of anything of permanent value. Better balance and facility in reading has been brought about by abandoning the original letter lengths and substituting chapters. There seems to be not the slightest advantage in reproducing here a precise facsimile. Should any question arise upon which the exact text is desired for comparison, reference can be readily made to the original manuscript which has been deposited in the Yale University Library. There is also a carefully compared typed copy in the files of the California Historical Society in San Francisco. The editor confidently believes, however, that in no instance has he altered Brewer's meaning or impaired his accuracy and that no matter of importance has been omitted.

For readers who may desire to pursue farther the subject

matter of these letters, references to other publications will be found here and there in the footnotes. Foremost among these is the *Geology* volume of the Whitney Survey.[3] A reading of Brewer's letters makes it clear that a considerable portion of this work was written by him, or at least composed substantially from his reports. Part of the material contained in the *Geology* is also to be found in the several editions of *The Yosemite Guide-Book*.[4] These are by no means all of the publications of the Survey, but they are the ones most likely to interest the non-scientific reader. Of Brewer's associates on the Survey, Whitney and King have been the subjects of biographical volumes.[5]

In the many years since the field party of the California State Geological Survey set out with its mules and wagons over dusty roads and incredibly steep grades, enormous changes have come upon some portions of the scene. Where these changes have obliterated all traces of earlier conditions, Brewer's vivid descriptions will serve to summon a vision of the past with all its picturesqueness and romance. But there are some spots, a little off the main highways,

---

[3] J. D. Whitney, State Geologist, Geological Survey of California, *Geology*, Vol. I. "Report of Progress and Synopsis of the Field Work from 1860 to 1864" (1865, xxvii + 498 pp., woodcuts). [Vol. II, dealing with later work, was published in 1882.]

[4] (a) *The Yosemite Book* (1868, 116 pp., 28 photographs, maps). [A handsome gift book, limited to 250 copies, containing photographic prints of Yosemite and the Tuolumne Meadows.]

(b) *The Yosemite Guide-Book* (1869, 155 pp., woodcuts, maps). [Many copies bear the date 1870.]

(c) *The Yosemite Guide-Book* (pocket edition, 1871, 133 pp., maps).

(d) *The Yosemite Guide-Book* (new pocket edition, revised and corrected, 1874, 186 pp., maps).

[5] (a) Edwin Tenney Brewster, *Life and Letters of Josiah Dwight Whitney* (Boston and New York: Houghton Mifflin Company, 1909, xiii + 411 pp., illustrations).

(b) *Clarence King Memoirs—The Helmet of Mambrino*, published for the King Memorial Committee, Century Association (New York and London: G. P. Putnam's Sons, 1904, vii + 429 pp., portraits).

(c) Thurman Wilkins, *Clarence King. A Biography* (New York: The Macmillan Company, 1958, ix + 441).

where, even today, the reader of these letters will have little difficulty in identifying the landmarks and where he may, if he chooses, tread in the very footsteps of Brewer, Whitney, Hoffmann, Gardiner, King, Averill, and the other bearded and sunburned men whose story is told in these pages. Historian, traveler, and general reader alike, will, I am sure, thank Professor Brewer for his pains in writing so faithfully of what he saw as he traveled up and down California during those four years, 1860 to 1864.

F. P. F.

*San Francisco,
March, 1930.*

BOOK I
1860–1861

1860–1861

CHAPTER I

TO CALIFORNIA VIA PANAMA

Panama—Acapulco—San Francisco.

New York.
Sunday Evening, October 21, 1860.

I SHALL sail at noon tomorrow, and drop a line before starting. I went to Chickering's[1] at Springfield, Vermont, Saturday night, October 13, after I left home and stayed until Monday noon. Monday night I spent in Boston, and went to Greenland Tuesday afternoon. I left there again for Boston early Wednesday morning. Wednesday night I went to Northampton, and Thursday, October 18, I came on here with Professor Whitney and his family, stopping a few hours at New Haven where I met old friends and had a short but pleasant visit. At Springfield I met the publisher of the valuable work on the trees of America, in five volumes, worth seventy dollars.[2] He wants to learn more about the trees of California and offers to send me a copy of his work—quite a gift certainly. I also met Gray, the botanist, at Cambridge, and got much valuable information. He also gave me a useful book.

Whitney has been back to Boston and I have had to get the baggage on board ship including our apparatus, some $2,500 to $3,000 worth, belonging to him and to the Survey.

I took dinner with Mr. Brush[3] today and tea at Hunter's this evening. We sail on the *North Star*, the steamer Vanderbilt made his famous pleasure trip with. It has a full load of passengers. I have met quite a number of scientific men this evening who will see us off in the morning. I have had many letters to write and it is now past two o'clock.

My baggage from Washington has not arrived. If it is not here in the morning, I shall have it shipped by Freeman's Express. It will be as cheap as to take it with me, but not so convenient.

At Sea.
October 26.

WE sailed Monday noon, punctually as we expected. There was the usual crowd, and partings, and tears, and blessings—I was glad I had parted with my friends at home, not here in this crowd. It was a dark, nasty day, but it did not rain much. We passed out of the harbor, ran down the Jersey coast and before night we had left the land out of sight. We have since then kept due south. The weather has been growing steadily warmer. Today we crossed the Tropic of Cancer and we are now in the tropics in earnest. Our ship is a good one, but terribly crowded. I think there must be 1,000 or 1,200 on board—certainly the former number of passengers.

Caribbean Sea, off the south coast of Cuba.
Saturday, October 27.

CUBA has been in sight nearly all day, and it is now, at 5 P.M., vanishing from view beneath the sea. We now take our way across the Caribbean Sea and the next land we see will be the Isthmus. We have had a most lovely day with a fine breeze from the east, the trade wind, yet it has been intensely hot, like the hottest August day. We ran east of Cuba, about three miles from the shore. San Domingo (or Haiti) was in the dim distance, but Cuba was most beautiful. The east end of the island is very rough, high mountains rise from the interior, nearly as high as the White Mountains of New Hampshire.

Monday, October 29.

CUBA is a most picturesque island. I regret that I could not see more of it. We are now nearly across the Caribbean Sea; will reach the Isthmus in the morning, when I must send this. This is a poor place to write because of the intense heat and the crowds on deck, which is now piled with trunks that have been brought out in the night to be weighed. Around is the clear deep blue Caribbean Sea. My baggage did not arrive at New York, so it will be sent by an express company.

Ship *Golden Age*.
Pacific Ocean, near Acapulco.
November 6.

WE arrived at Aspinwall on Tuesday noon, a week ago today, and crossed to Panama the same day. The weather, although intensely hot, was cooler than usual. I wish I could adequately describe the gorgeousness of the vegetation of the Isthmus—it even exceeded the pictures imagination had painted—cocoanut trees, palms, bananas, plantains, oranges, lemons, limes, and other tropical fruits. Tall palms, mahogany, and such woods, vines hanging like ropes from the branches, parasitic plants growing over everything, made a scene I wish you all might see. We were detained more than half the night before we got on the ship at Panama, then found we could not sail all the next day, Wednesday; so I went ashore again and visited the town, for the ship lay a mile or two off, anchored. Panama is the most picturesque place I ever saw. It was once a place of much importance, walled about like the cities of the old world in the Middle Ages, but its glory has departed. Its walls are in ruins, and also many of its fine old churches are entirely in ruins, and over all is a most vigorous growth of tropical vegetation. Trees and shrubs flourish wherever they can get a foothold. This exuberance of vegetation is *the* feature that strikes me, reared in a colder clime. You have no idea of the dampness, especially at this, the rainy, season of the year. An old resident told me that clothing must be aired every day—even books on the shelves wiped every day, or they mold and rot.

The town seems quaint enough. Its tile roofs—no chimneys, for fires are only needed to cook by—its open houses—all make it a queer old place. There are a few Americans connected with the railroad. The better class are Spanish. The masses are black but speak Spanish. They are scantily clothed, the children entirely naked. All along the railroad we passed villages where the houses were only mere roofs upon poles or posts, no sides, the roofs steep and covered with palm leaves, the women with a skirt only and the men with only pants, naked from the waist up, the children naked as cupids. The blacks have become the dominant race, and it occurred to me that perhaps in some of our warmer states we were following in the same way.

Wednesday at midnight we were off. We have had hot weather and a smooth sea ever since. Tonight we stop at Acapulco for coal, but it will be after dark and I do not know whether we can go ashore. We have a much more comfortable time on the *Golden Age* than we had on the other side. In eight days more we will be in San Francisco.

A word more about the Panama railroad. It is well built, its bridges of iron—indeed, iron is used wherever possible, for the wood rots in a year or so. The length is forty-eight and one-half miles, the fare twenty-five dollars, and freight accordingly; so you can well believe that it pays well. Most of the hands are blacks, but all the conductors are Americans. It does an immense business and is a great enterprise, but it cost four thousand lives to build it amid the swamps and miasma of that climate. It follows up the Chagres River about twenty-five miles, then crosses the ridges to the Pacific. The Bay of Panama is fine and large.

San Francisco, California.
Thursday Evening, November 15.

TUESDAY, November 6, the day of the great election, was a quiet day on shipboard, and at about half-past eight or nine in the evening we ran into the harbor of Acapulco, in Mexico, where the ships coal. Much to the regret of us all we ran in by night and thus missed the beauties of that beautiful harbor. It is so completely locked in by hills that when in, one cannot see the ocean at all; it is completely secure from the storms of the ocean outside. In fact, it is the best harbor between Cape Horn and San Francisco.

The ship anchored in the bay, for there are no wharves; the coal and washing is taken on from lighters. Notwithstanding that it was night, a number of us went ashore. It is a picturesque place of about four thousand inhabitants, hemmed in by hills, and hot as Brazil. All the fruits of the tropics grow there, and I bought a basket of oranges, limes, plantains, and bananas; the last two fruits are very similar. The houses are all of the most open description—lattice doors and often lattice sides. The public square or *plaza* was filled with natives, some with fruits, shells, and liquors to sell to the passengers, others gambling with each other. I spent an hour looking at them gamble. They are nearly all Indians, but there is some negro and some Spanish blood, and Spanish is the language. Some of the girls were quite pretty.

The men who put the coal on the ship were in strong contrast with those we saw at Panama, in just so much as the Indian form is finer than the negro. The coal was in bags, on an old hulk moored in the bay, and was brought on board by these natives, and the effect of the light of the bright torches on these half-naked men in the gloomy night was picturesque beyond description. The coal is brought from the Atlantic states and costs the company from eighteen to twenty-five, and even at times thirty, dollars per ton in these Pacific ports. We coaled, took on cattle, etc., for provisions, and before morning had passed outside and were again on our way on the broad Pacific.

Wednesday, November 7, we kept nearly all day in sight of land, the high Cordilleras often in sight; in fact, they were always in sight when we could see the land at all, and some of the views of them were grand. I had never heard that the views of the mountains of Mexico and Central America were so fine from the ship, so I was entirely unprepared for them.

Thursday they were even finer, and Colima, the volcano, was in sight all day, although it was ninety miles inland and we not less than twenty miles off the coast. We saw it plainly when we were 150 to 180 miles distant, its massive cone-like peak rising against the clear blue sky. Sometimes it stood out clear, at other times clouds curled about its summit. The weather continued magnificent, indeed it has remained so ever since.

Friday, November 9, we crossed the mouth of the Gulf of California and had more wind and a rougher sea, although still comparatively smooth. The next two days we skirted along the coast of Lower California; its barren shores, parched and burned, looked desolate beyond description. On Sunday afternoon two passengers died, one a Mexican who got on at Acapulco, the other a child. Both were sick when they came on shipboard, and both died quite suddenly. They were buried that night, both at the same time. A hard rain squall happened just at that moment, making the scene doubly sad and impressive.

Monday and Tuesday we were along the coast of California proper; the shore often, but not always, in sight, and the weather much cooler. Thick coats were comfortable, and I could no longer sleep on deck. All along in the tropics I slept on the deck, wrapped in my blanket, with the sky overhead—it was not only delicious but glorious, considering the contrast between that and my stateroom, where four persons were crowded into a place scarce six feet square.

Wednesday morning (yesterday) we sailed into the Golden

Gate, as the entrance to the harbor is called.⁴ It is a narrow strait between steep rocks, with a crooked channel. Inside, the broad bay spreads out, a calm placid lake, so large that all the ships of the world might ride there at anchor with room to spare, and yet so placid that it seemed like a mere pond. It is by far the most beautiful harbor I have ever seen, not even excepting New York.

As we had instruments to carry we did not hurry on shore, so it was nearly noon before we were fairly at our hotel, and our journey of nearly 6,000 miles brought to an end. It is over 2,000 from New York to Panama, and about 3,300 from Panama to San Francisco. And on the whole of this journey we had fine weather, indeed lovely weather; twenty days out of the twenty-three were as fine as our finest Indian summer.

This place I will not speak of now—it has astounded me. First, after our long trip by sea and by land we are still in the United States; second, a city only about ten years old, it seems at least half a century. Large streets, magnificent buildings of brick, and many even of granite, built in the most substantial manner, give a look of much greater age than the city has.

The weather is perfectly heavenly. They say this is a fair specimen of winter here, yet the weather is very like the very finest of our Indian summer, only not so smoky—warm, balmy, not hot, clear, bracing—in fact I have not words to describe the weather of these two days I have been here. In the yards are many flowers we see only in house cultivation: various kinds of geraniums growing of immense size, dew plant growing like a weed, acacia, fuchsia, etc., growing in the open air in the gardens.

Our arrival was anticipated by the Pony Express. All the papers had announced that the members of the Geological Survey were on their way, and yesterday and today all the city papers have noticed our arrival. Whitney and I get most of the puffs, some of which are quite complimentary.⁵ I have been introduced to many prominent citizens, tendered the use of libraries, etc. The Coast Survey offers many facilities, free passage on its vessels along the coast, etc. This afternoon Judge Field and the Governor of the State leave Sacramento to come down here to see us; we shall see them tomorrow.

We found the news of Lincoln's election when we landed, an unprecedented quick trip of news. I have been out to see fireworks, processions, etc., in the early part of the evening, so now it is late. Good night.

NOTES

1. J. W. Chickering taught at Ovid Academy with Brewer, 1857–58.
2. F. Andrew Michaux and Thomas Nuttall, *The North American Sylva* (Philadelphia, 1857).
3. George Jarvis Brush, a classmate of Brewer at Yale (Ph.B. 1852); was Professor of Metallurgy at Yale, 1855–71; Professor of Mineralogy, 1864–98; emeritus, 1898 until his death in 1912.
4. This name was given by Frémont before the discovery of gold. In his *Geographical Memoir upon Upper California, in Illustration of His Map of Oregon and California,* 1848, he says: "Called *Chrysopylae* (Golden gate) on the map, on the same principle that the harbor of *Byzantium* (Constantinople afterwards) was called *Chrysoceras* (Golden horn). The form of the harbor and its advantages for commerce (and that before it became an entrepot of eastern commerce), suggested the name to the Greek founders of Byzantium. The form of the entrance into the bay of San Francisco, and its advantages for commerce (Asiatic inclusive), suggest the name which is given to this entrance" (p. 32).
5. An editorial in the *Daily Alta California* speaks of Brewer as "a gentleman of splendid scientific attainments."

CHAPTER II

LOS ANGELES AND ENVIRONS

By Steamer to San Pedro—Los Angeles—Benito Wilson's—Santa Monica Mountains—Personnel—San Gabriel.

Los Angeles, California.
Sunday, December 2, 1860.

Professor Whitney returned from Sacramento Wednesday, November 21, with the "sinews of war," and with orders to get off immediately. The next two days were spent in the greatest activity, buying blankets, getting tents made, getting harness, saddles, some groceries, tea and coffee, etc., and Saturday, November 24, we were on board a steamer for San Pedro, 380 miles southwest of San Francisco. Four of us started. Professor Whitney went to Mariposa with Colonel Frémont, intending to come down by the next steamer, about three weeks later. As first assistant, the company was placed in my charge, a heavy responsibility I would like to have had placed on someone else. We were to come down here, buy mules, provision and equip fully, and go into camp and await Professor Whitney.

Well, we sailed on Saturday, a most lovely morning, and took our course down the coast to the southeast. There are two small rocky islands just at the Golden Gate and they were completely covered with sea lions, a kind of large seal, apparently nearly as large as a walrus. They barked at us as we passed and many tumbled into the sea, but hundreds were basking in the sun or moving about with awkward motions.

The next morning we arrived at San Luis Obispo. The port is but a single house, and the village is about four miles distant. As we lay there all day, I went ashore with a friend, a doctor from the United States Army, and spent several hours, saw the country and collected some plants, all strange to me. The steamer was anchored a mile from the shore, and the freight, some seventy or eighty tons, had to be landed in yawls or rowboats. There is no dock, so it is landed on a rock.

The land was very dry, the rains had hardly begun, so the vegetation looked very scanty and the land desolate. Scrub oaks, crabbed sycamores, and scrubby undershrubs composed the scanty vegetation. We wandered along the beach and picked up a few shells, some of great beauty. The sea has worn the rocks in fantastic shapes; there are several natural arches, one of great size.

On our return to the ship, we found the passengers playing cards, singing songs, drinking whiskey, etc.—a Californian sabbath. The Boundary Commission,[1] to run the line between California and countries east, were aboard, a hard set, who were making much noise and drinking much whiskey. They are now encamped near here. One of their men died this morning, killed most probably with bad whiskey. He was out yesterday, walked to camp last evening, and died this morning.

Well, we started that evening. The next morning, after stopping a few hours at Santa Barbara, we arrived at San Pedro, the port of Los Angeles, about twenty-five miles from here. We got in about sundown, rode six miles up the river on a small steamer, then disembarked for this place by stage. It was a most lovely night, but there were more than three times as many passengers as there was stage room, so two of us came up and left two other men with the baggage. They came up the next day. We have been here since, looking at mules, harness, bacon, stores, etc. We hope to be in camp in two days more. I have been to church once today, we had a congregation of about thirty or forty, I should think.

In Camp at Los Angeles.
December 7.

WELL, we are in camp. It is a cold rainy night, but I can hardly realize the fact that you at home are blowing your fingers in the cold, and possibly sleighing, while I am sitting here in a tent, without fire, and sleeping on the ground in blankets, in this month. We are camped on a hill near the town, perhaps a mile distant, a pretty place.

Los Angeles is a city of some 3,500 or 4,000 inhabitants, nearly a century old, a regular old Spanish-Mexican town, built by the old *padres*, Catholic Spanish missionaries, before the American independence. The houses are but one story, mostly built of *adobe* or sun-burnt brick, with very thick walls and flat roofs. They are so low because of earthquakes, and the style is Mexican. The inhabitants are a mixture of old Spanish, Indian, American, and German Jews; the last two have come in lately.[2] The language of the natives is Spanish, and I have commenced learning it. The only thing they appear to excel in is riding, and certainly I have never seen such riders.

Here is a great plain, or rather a gentle slope, from the Pacific to the mountains. We are on this plain about twenty miles from the sea and fifteen from the mountains, a most lovely locality; all that is wanted naturally to make it a paradise is *water,* more *water.* Apples, pears, plums, figs, olives, lemons, oranges, and "the finest grapes in the world," so the books say, pears of two and a half pounds each, and such things in proportion. The weather is soft and balmy—no winter, but a perpetual spring and summer. Such is Los Angeles, a place where "every prospect pleases and only man is vile."

As we stand on a hill over the town, which lies at our feet, one of the loveliest views I ever saw is spread out. Over the level plain to the southwest lies the Pacific, blue in the distance; to the north are the mountains of the Sierra Santa Monica; to the south, beneath us, lies the picturesque town with its flat roofs, the fertile plain and vineyards stretching away to a great distance; to the east, in the distance, are some mountains without name, their sides abrupt and broken, while still above them stand the snow covered peaks of San Bernardino. The effect of the pepper, fig, olive, and palm trees in the foreground, with the snow in the distance, is very unusual.

This is a most peculiar climate, a mingling of the temperate with the tropical. The date palm and another palm grow here, but do not fruit, while the olive, fig, orange, and lemon flourish well. The grapes are famous, and the wine of Los Angeles begins to be known even in Europe.

We got in camp on Tuesday, December 4. We had been invited to a ranch and vineyard about nine miles east, and went with a friend on Tuesday evening. It lies near San Gabriel Mission, on a most beautiful spot, I think even finer than this. Mr. Wilson,[3] our host, uneducated, but a man of great force of character, is now worth a hundred or more thousand dollars and lives like a prince, only with less luxury. His wife is finely educated and refined, and his home to the visitor a little paradise. We were received with the greatest cordiality and were entertained with the greatest hospitality. A touch of the country and times was indicated by our rig—I was dressed in colored woolen shirt, with heavy navy revolver (loaded) and huge eight-inch bowie knife at my belt; my friend the same; and the clergyman who took us out in his carriage carried along his rifle, he said for game, yet owned that it was "best to have arms after dark."

Here let me digress. This southern California is still unsettled. We all continually wear arms—each wears both bowie knife and pistol (navy revolver), while we have always for game or otherwise, a Sharp's rifle, Sharp's carbine, and two double-barrel shotguns. Fifty to sixty murders per year have been common here in Los Angeles, and some think it odd that there has been no violent death during the two weeks that we have been here. Yet with our care there is no considerable danger, for as I write this there are at least six heavy loaded revolvers in the tent, besides bowie knives and other arms, so we anticipate no danger. I have been practicing with my revolver and am becoming expert.

Well, to return to my story, and to Mr. Wilson's. We found a fine family, with two lovely young ladies. The next day, Wednesday, December 5, we went up into the mountain, followed up a canyon (gorges are called *cañons* or canyons), and then separated. I climbed a hill 2,500 or more feet, very steep and rocky, gathered some plants, and had one of the most magnificent views of my life—the plain, and the ocean beyond. The girls went with us into the canyon, but did not climb higher. After our climb and a lunch, a ride of eight miles over the fields (for no fences obstruct the land) brought us back; then dinner and return here. We had a delightful time—I ought to say "we" were the field assistant Mr. Ashburner and I. We will try to visit them again when Professor Whitney comes.

It is cold, wet, and cheerless, so good night! Rain patters on the tent and dribbles within.

Sunday Evening, December 9.

YESTERDAY was rainy and cheerless enough in our tents—cold, damp, wet—but it cleared up by noon, and today is most lovely, yet cool, thirty-nine degrees in our tent this morning.

I like camp life thus far. I had expected to take cold and all that, but not so. I have slept well and have eaten well, and am well, only have more responsibility than I wish. By tomorrow night I shall have paid out over nine hundred dollars in the last two weeks—but Professor Whitney will be here Tuesday.

Camp No. 2, Sierra Santa Monica.
Monday Evening, December 17.

MONDAY, December 10, I sent a wagon and two men to San Pedro, on the coast, for instruments, and that night Professor Whitney arrived. He stayed in town; we kept in camp. The next day (Tuesday) it rained very heavily all day. It began before daylight and drowned us out at dawn. Soon the water was ankle deep in the tent. Oh, the comforts of camp in these tropical rains, when it doesn't rain but pours! Ditches were dug, but were insufficient, stakes were freshly driven to keep the tents from blowing down in the wind, then blankets, instruments, books, maps, etc., were transported to the driest tent—lucky that we had but four in camp! We breakfasted on raw bacon and dry bread. An unoccupied hut was found, where we built a fire and spent a part of the day, and two spent the night there. I stuck to the tent, along with the cook.

At ten at night the men arrived with the wagon, but I sent them into town to sleep. They brought me some papers and two letters from San Francisco, the first and only ones I have yet received. I sat up in bed—that is, in my blankets—and read them. But how the rain came! It poured, it battered through the canvas until I was wet; yet I slept well that night, although between the letters and the novel situation, my dreams carried me back to other scenes with other friends around me.

On Wednesday and Thursday, December 12 and 13, we explored the region round about and completed our equipment. We have nine fine mules, saddles, harness, spurs, and all. The morning of the fourteenth we raised our camp and came here to explore this range—a small range north of Los Angeles.

We had some most amusing incidents on this trip. A four-mule team drew our wagon, in which two rode; the remaining five were mounted on similar brave animals, some of them scarcely half broken, just half wild from the ranches, with these queer Mexican saddles, still queerer Mexican bridles, and most queer of all Mexican spurs. By a grand streak of good luck, no one was thrown. But there was kicking and jumping, and mules persisting in going the wrong way, and whipping and spurring, then fresh kicking and some swearing. We were in camp by dark. A gentleman from Los Angeles had come out with us, but his horses ran away that night and he is after them still.

Saturday morning four of us started, muleback, for the center of the Sierra. The gentlest mule carried the Professor, with a barometer to measure heights; I went to botanize; another to guide; another (Spaniard) to hunt. With no little trepidation, which I was ashamed to show, I mounted my mule. He is one of the most spirited in the crowd—our driver says the best in the lot—and quite wild. I expected a scene when I took my botanical box (tin) on him, but managed him better than I anticipated or even hoped for; but, cunning brute, he knows I am both awkward and green, and takes advantage accordingly.

Well, we rode up the canyon a few miles, rugged (but not high) mountains on either side, with here and there a crabbed tree and a stunted, shrubby vegetation. We at last tied our mules, ascended a ridge, took some observations (but did not reach the highest peak), found some fossils, or rather, found where to get them, then returned.

Tuesday Evening, December 18.

ON our return in the trip mentioned, we came near having an accident. It was necessary to jump our mules over a log. The first two mules required much urging, but when mine came he not only jumped the log but sailed over a steep bank, the steepness and depth of which were concealed by bushes. Visions of his rolling over me popped into my head as I caught a glimpse of where I was and of the distance below me, but thanks to his wisdom and strength we got out, I don't know how even yet. No horse could have saved himself as that mule did. He was hurt some, and I have not ridden him since, nor will I for a week to come, except just to move our camp.

Sunday, December 16, it rained all day and a part of the night. Monday, it was clear, and after getting nine o'clock observations for longitude, I started again with Ashburner to the peaks. Professor Whitney remained to get more observations for latitude, and to watch the station barometer. We carried the other along to measure the heights. This time we went on foot—went up the canyon a few miles, then climbed again the peak visited on Saturday. Ashburner almost gave out on the steep ascent, but he rested while watching the barometer. Another peak a few hundred feet higher was near. We climbed to that, but only to see one still higher an hour's climb farther on. I was for going to it. Ashburner objected, so we set up the barometer and watched its height for an hour. A most grand and magnificent view was beneath us, unlike anything I have seen before. We got a load of fossils, and, tired enough, returned after dark.

This morning we sent four men, with wagon and one tent, to find another camping place near the end of the ridge at the sea, while Professor Whitney and I remained here, with our cook. After the morning observations, a cloudy sky coming up, we followed. He walks much better than Ashburner, so we did much more, but a fog came on and so enveloped the peaks that nothing was to be seen. However, we found some new fossils, traced up the granite core of the mountains, the backbone as it were, then returned this evening. We leave here tomorrow.

And now, of our company—I believe I have not yet told you of them.

First, Professor Whitney, a capital fellow—I think the best man in the United States for this gigantic work. I like him better each day.

Second, the botanist, etc., of the Survey, your humble servant will not describe.

Third, Mr. Ashburner,[4] of Stockbridge, Mass., a good fellow, graduate of the School of Mines, in France, about my age, or younger. He is field assistant.

Fourth, Averill,[5] a young man, a graduate of Union College; then spent a year and a half on a voyage around the world, visited South America, East India, China, etc., and is now here seeking his fortune. He is a capital fellow. He keeps accounts and assists in general at whatever he can do.

Fifth, Guirado, a Spanish-Mexican-Californian, about twenty, a brother-in-law of the Governor,[6] a regular Spaniard, a good fellow, just the one to ride a wild mule and to shoot our game, yet by far the least valuable of our crew.

Sixth, Mike, a jolly young Irishman, our cook, just getting broken into the harness, and I think with practice will do well.

Seventh, last, but by no means least, Pete,[7] our jolly mule driver—a capital fellow in his line—young, game, posted as to mules, can tell a story, sing a song, shoot rabbits (and dress, cook, and eat them)—a most valuable man. Has been over the plains, was with Colonel Lander on his wagon-road expedition, etc. I pride myself on choosing him out of the host of applicants.

Oh, how still it is! No sound but the hooting of owls, or the sound of other night birds. No house near, and but few signs of civilization. Good night!

Camp 3.
Sunday Evening, December 23.

ANOTHER rainy Sunday, or at least it rained all the morning, but has cleared up tonight, most lovely. We came to this camp on Thursday. It is to the west of our last, on the seacoast, or rather, within a mile and a half of it. We hear the surf continually; it is the last sound at night, breaking the stillness of the night, but not the solitude, for it seems more solitary than ever. This effect is increased by the doleful hooting of numerous owls all night long, and the occasional bark of the coyote, or California wolf, a small, sneaking but not dangerous animal.

Friday two of us went several miles along the seashore to observe the outcropping of rocks there. On Saturday we took mules, rode up the canyon about eight miles, rising a thousand feet in that distance, then leaving our animals and climbing to a peak about eight hundred feet higher. It was the hardest climbing I have done yet. It was very steep, many precipices obstructed us, and when there were no rocks there was an almost impenetrable thicket, or chaparral, as it is here called. We carried up a barometer, which increased the labor, and got back at dark, tired enough.

Today we have loafed around tent, cleaned pistols, etc., but it bids fair for a good day tomorrow.

Los Angeles.
December 26.

WE came in here the twenty-fourth and stopped over night at a hotel. It commenced raining and has rained until this afternoon. It has now stopped and we shall probably go on in a day or two, certainly as soon as the weather and streams will allow.

As I had lost my mule (he is since found), I walked in from Camp 3, about twenty miles, over the plain—a most lovely rolling plain, only wanting water to make it of the greatest fertility. Now, during the rainy season it is most green and lovely, thousands (probably thirty thousand to fifty thousand) of horses and cattle are grazing there. One of our mules is among them still, or else stolen. One of our men has been hunting him a week. We hope to get him, however.

Rain interfered with Christmas festivities, but it was still quite lively. I stepped in a *fandango* a little while in the evening and looked on to see the dancing, which did not come up to my expectations. In the next room they were playing *monte* for large piles of silver—the stakes not large, but the silver accumulated.

The rain has been the severest for eleven years. Probably as much as six or seven inches fell in about forty hours. You can imagine the effect. It is very hard on these *adobe* houses. Several have fallen, one row of stores, among the rest, involving a loss of many thousand dollars. It has been the rainiest season since '49—lucky we were not in camp during this siege, it is decidedly better at the hotel. As a sample of how damp the air is, when I am writing, fine as my writing is, the first line of the page is not yet dry when the last is written.

Camp No. 6, mouth of San Gabriel Canyon.
January 3, 1861.

I SENT my last letter from Los Angeles a week ago and have

been too busy to write any since. The rain ceased, and we sent an advance camp out on Friday, December 28. Professor Whitney and I followed on Saturday. We camped on the ranch of Mr. Wilson, and only left there this morning. His ranch is in a beautiful plain, hemmed in on all sides by hills or mountains, except for one narrow opening to the sea, like a bay filled up. The hills on the west, toward Los Angeles, are not high. This valley plain is perhaps fifty or sixty thousand acres; his ranch is four leagues, about fourteen thousand acres. He also owns very extensive vineyards, the "Lake Vineyards," which have as fine a reputation as any in the state.

When this part of the state belonged to Mexico, it was settled by the old Spanish missionaries, or *padres* as they are here called, who converted the Indians and formed great missions, wealthy and powerful. One was near Mr. Wilson's, the Mission of San Gabriel. Thousands of natives were the voluntary slaves of these priests, vineyards of great extent were planted, and at the time of the confiscation of their property by the Mexican Government they had twenty thousand horses, eighty thousand cattle, etc. All now is in ruins.[8]

Sunday morning I rode over to the Mission, about three miles from camp, with Guirado, who went to Mass. The old church and a few houses still stand—the church bells are by far the sweetest I have heard in California—six are left in the old tower, two are gone. Extensive ruins of *adobe* buildings, now the abode of myriads of ground squirrels, told how large the town once had been. Long lines of *tuna*, or prickly-pear hedges, now all ruined, told of ancient enclosures and vineyards, but now a waste. Immense labor had once wrought this lovely valley into a veritable paradise, but now it is desolate again. A few tall date palm trees are there, but the fruit does not ripen. We went into a garden owned by the priests, still enclosed. It was still kept up, the finest orange trees were laden with golden fruit, so that the trees were propped up to keep them from breaking under the load. They were most beautiful

—the graceful form of the tree, the intense dark green of the foliage contrasting with the rich golden fruit, produced a beautiful effect. We bought some oranges, also lemons and limes. Olives abound—many of the trees are large—and English walnuts grow as fine as in Europe. Water was brought for irrigating from a neighboring stream in a long ditch from the San Gabriel River, over twenty miles distant from the remotest part. Such *was* San Gabriel Mission—now it is a ruin and cut up into ranches.

To the east of us lay a very precipitous chain of mountains.[9] I had before been in a canyon in them and had climbed a few hundred feet. We now determined to ascend. We four, chief men of the Survey, drew cuts to see who should stay in camp to observe the station barometer, as one must stay, and all were anxious to go. The lot fell on me, but Averill most generously yielded his right, and I went in his place. No one had been on the highest peak, but a native agreed to pilot us to the second highest point.

We arose on Monday and breakfasted before dawn, then waited over an hour for our guide. We rode to the base, and into a canyon about five or six miles from camp, tied our mules, and after a barometrical observation, commenced to climb. It was the steepest and hardest climb I have ever had by far. We carried up barometer, compass, and botanical box. The chaparral became almost impenetrable. It was terribly hard to climb. As we crossed from one peak to another on a very narrow ridge, the third hour, Ashburner gave out. I took his load and we left him behind. In four and a half hours' climbing we found it impossible to make the highest peak, so we planted our tripod and put up compass and barometer. We had risen 4,200 feet above camp and 5,000 feet above the sea. Another peak rose 1,500 feet, at least, above us yet. The view was magnificent. I will attempt no lengthy description. All the lower hills to the west sank into the plain that was spread out beneath us to the very sea, and we could see a great distance, probably fifty or sixty miles, out to sea. Los Angeles, with its vineyards and all, was a mere speck on the landscape.

A little snow lay around us, and the summits above were very white. We built a fire, melted snow in my botanical box for drinking, ate our lunch, took the bearings of the most important points, and descended. All the region to the north and east was very mountainous, yet it was hard to realize that I was nearly as high as Mount Washington, and higher than the celebrated Rigi. We carried down my box full of snow for the ladies at the ranch. It was a great curiosity—the younger ones had never seen it before. One man went off in ecstasies on tasting it; he had not had snow in his hand since leaving Europe fifteen years ago.

It was long after dark when we got back. A hearty supper so much refreshed me that I spent the evening, New Year's Eve, at Mr. Wilson's, and spent it very pleasantly. He has a large family; there are several ladies there.

Tuesday (New Year's), we sent the wagon to town (Los Angeles), nine miles distant, for supplies. Professor Whitney and I, along with Mr. Wilson, rode a few miles to visit some old quarries and see some other things of interest. It was a lovely day. We dined at five at Mr. Wilson's—a most sumptuous dinner. A small party was there and we spent a pleasant and lively evening, notwithstanding that our "rig," just from camp, was hardly fashionable. The evening was lovely, as was the last. The midnight bells at the old Mission the night before, tolling out the old year and in the new, were sweet, but no sweeter than the nine o'clock bells of that New Year's night.

Yesterday, January 2, we sent our men to another camp, while Professor Whitney and I stopped for observations on latitude and longitude. It so happens that I can observe time on the chronometers much closer than the assistant who was brought along for that purpose, so I stay in camp some days to do it. The day was lovely—thermometer 67° F.—spring weather. Think of such weather for the holidays! Today we raised our camp, and rode on here, about twelve miles; the other camp will be here tomorrow.

Saturday, January 5.

YESTERDAY Professor Whitney and I went a few miles up the San Gabriel Canyon. Silver and gold are worked. A silver mine is being opened, and the stockholders desired to go up with us. Some came along and we went up to their mine, about six or seven miles. It is a wild canyon, granite rocks from two to four thousand feet high on each side, very steep but nowhere perpendicular. Many side ravines come in, and many small streams swell the San Gabriel to a river. We had to cross it twelve times each way, twenty-four in all, very easy to those who had horses, but not quite so easy with our short-legged mules as the water often came up to their sides. The mine has been commenced by running a tunnel into the mountain, but we found it caved in by the recent heavy rains, and as we could not get in we returned.

On our way we met four more of the stockholders—they urged us to go back, as they had the "materials" for a jolly night coming. We, however, kept on our way. Soon we met some men with pack-mules. One carried blankets; another a basket of champagne and other wine; another, a Spaniard, a guitar. So I imagine that they are having a jolly time in their cabin this rainy morning.

The path was up a mere mule path, over rocks, logs, among bowlders, where you would think no horse or mule could get. All provisions, etc., must be packed, that is, carried on the backs of mules. It was a wild scene all the way. Yet one I had never heard of before, nor is the canyon laid down on any map, although it is forty miles long.

This want of maps, as well as incorrect maps, is a very serious evil which we feel much. We have to make observations all the way. Professor Whitney does work splendidly. Two sets of observations at the last camp, where he used the sextant and I the chronometer, agreed to within the *one-tenth of a second*, while our last barometrical observations, for altitude, two sets, agreed to within an inch and a half, although the camps were eight or nine miles apart. We have very fine barometers, reading to the thousandth of an inch, and we carry them with care, hence the precision. When we measure any height we use two barometers, one at the camp, the other carried with us.

Sunday, January 6.

It rained most of the day yesterday, all last night, and thus far today without any prospect of a cessation. Yesterday I fixed plants, wrote up descriptions, mended shirts, drawers, etc., made oilcloth cases for compasses, etc. My cheap woolen shirts don't prove well made—have had to sew on buttons and work all the buttonholes over again; the work is now securely done even if it is not ornamental.

Monday, January 7.

It has now rained about seventy hours without cessation—for forty hours of that time, over twenty consecutive, it has rained like the hardest thundershower at home. No signs of clearing up yet—fire out by the rains, provisions getting rather scarce —one meal per day now. But our tent is dry—we have it well pitched, and in a dry place.

I have been studying Spanish, writing up letters, notes, etc. I have written thirteen letters, or about eighty pages, during this rain, to be mailed when we can get to town, but it will be a number of days, for the streams will be impassable. Lucky we did not stay up in the canyon Friday night as they wanted us to, we could not have got down yet. I never saw such rains before, and it has not rained so much before of a winter since 1848, so the people say.

Tuesday, January 8.

Rain has stopped—the San Gabriel River is impassable, so we can neither get to town for supplies, nor visit up the canyon, nor move camp toward Temescal. The stream is high and swift. An empty stage was carried off night before last, and a man was drowned about a mile from our camp yesterday while attempting to swim the river. We shall get letters when we can get to Los Angeles again. It is a fine evening.

Wednesday Evening, January 9.

WE have had a lovely day. Three of us climbed a ridge about two thousand feet above our camp to measure height and get the bearings of the various points around. We got a most magnificent view. The rain here had been snow on the peaks behind; they lay in their silent grandeur, so white and massive, while on the opposite side were the lovely plains of San Gabriel, El Monte, and Los Angeles. But I am too tired to write. Possibly a team can get to town tomorrow to mail this as the river has fallen some.

NOTES

1. The records of this commission are obscure. It appears to have had some connection with work done by Lieut. Joseph C. Ives on the Colorado River in 1861, but the group mentioned here by Brewer was probably not under Ives's command. Contemporary newspapers mention a party of fourteen men, with transport, including three camels, engaged in the eastern boundary survey, traveling from Los Angeles, via Mohave Desert, to Owens Valley.

2. Harris Newark, in *Sixty Years in Southern California, 1853–1913* (2d ed., revised, 1926), gives a picture of Los Angeles life at this period.

3. Benjamin Davis Wilson, familiarly known as "Benito" (1811–78), a native of Tennessee, came to Los Angeles in 1841 after a career as trapper and trader in New Mexico. In 1852 he purchased the Lake Vineyard property and made his home there. In 1864 he built a burro path to the top of the mountain which bears his name.

4. William Ashburner (1831–87) became a prominent citizen of California. After leaving the Geological Survey he entered the practice of mining engineering in San Francisco and was for a time Honorary Professor of Mining Engineering in the University of California. In 1880 he was appointed a Regent of the University. He was also a trustee of the Leland Stanford, Junior, University. During his later years he was active in banking. His wife, Emilia Field, whom he had married in Stockbridge in 1856, was a niece of Justice Stephen J. Field of California, who had been most influential in causing the State Geological Survey to be established. Mrs. Ashburner is now (1930) living in San Francisco.

5. Chester Averill, also from Stockbridge.

6. John G. Downey married, in 1852, María Jesús Guirado, daughter of Don Rafael Guirado, who had come from Sonora to Los Angeles in 1853.

7. His full name was John Peter Gabriel.

8. Misión San Gabriel Arcangel was founded in 1771 by the Franciscan missionaries, Fray Angel Somera and Fray Benito Cambón. Secularization took place from 1832 to 1840.

9. The San Gabriel Range, which includes Mount Lowe (5,650 feet), Mount Wilson (5,700 feet), San Gabriel Peak (6,152 feet); and farther east, Mount San Antonio (10,080 feet), and others. The climb appears to have been upon one of the lesser peaks near Mount Lowe or Mount Wilson.

CHAPTER III

MORE OF SOUTHERN CALIFORNIA

Chino Ranch—Santa Ana Mountains—Temescal.

Camp No. 6, San Gabriel Canyon.
Sunday Evening, January 13, 1861.

SINCE the heavy rain ceased we have had fine warm weather—lovely days, some slightly cool (55°), and one night some frost. The peaks just above us are covered with snow, and at times cold raw winds come down from them. Such is the case tonight. The wind howls most pitilessly, shakes the tent, whistles under its edge, and flaps its sides in a decidedly lively manner.

We have the most lovely sunsets I have ever seen. I have watched sunsets from my own native hills, the finer sunsets from Ovid over Seneca Lake, from shipboard on the Atlantic and on the Pacific, finer than either on the Caribbean Sea; I have watched the sun set behind the Rhine Valley, over the plains of Bavaria; I have climbed the matchless Alps and Tyrol, and have even seen it setting over the Mediterranean, that land and sea of sunsets—but have never seen these surpassed. Professor Whitney, who has been three times to Europe and has traveled from Norway to Italy and from England to Moscow, says he has never seen them equaled. In the Alps of Switzerland, just at sunset and sunrise, a peculiar rosy pink light illuminates the snowy peaks and glaciers—the *alpenglühen* of the Swiss mountaineers. That peculiar and lovely tinge is even more marked here than I ever saw it there.

During the recent rains, much snow fell on the mountains above us, and the contrast of their cold tops, not ten miles distant (they seem but one mile), with our green plain, is most lovely.

Our wagon brought out a "pile" of newspapers—a real lot—two dollars worth, for New York papers cost twenty-five cents each, and San Francisco papers ten cents. Two dollars bought near a dozen. How we have perused them—the exciting political times, etc.

We heard possibly of our lost mule—may, and may not yet get him, probably the latter—but bought a horse to take his place. Professor Whitney takes him, so I get his well-trained mule, a little white fellow, for all the world like an ass, I should judge about ten hands high.

Four men were drowned near here in the recent rains, and much damage done. In town vast damage was done—*adobe* houses hurt—the *adobe* cathedral which has stood over half a century is nearly ruined, some of its walls fallen.[1]

When we came from San Francisco near two months ago, the United States and California Boundary Commission (to run the east line of California), came down with us—a decidedly hard set. They camped near Los Angeles before we did, and have loafed there ever since—on pay of course. Their mules and camels came three weeks ago to take them across the desert, but they started not. Well, they were just ready as the rains came. They were camped near the river, which rose, swept away their tents, saddles, equipage, all—one camel was killed.[2] Most of them were green hands, many of them the personal appointees of President Buchanan. They started at last yesterday.

Tuesday Evening, January 15.

It is a cold evening—some frost last night. It is only seven o'clock, but the rest are in bed—gone to bed to keep warm. I will write until my fingers get cold, then go too.

Yesterday we started on the advance camp, all but Professor Whitney, myself, and the cook. They took one tent and a load of effects. The Professor and I rode a few miles to visit a curious hill of volcanic rock that rises like an island about six or seven hundred feet in this plain, the smooth plain like a sea about it.

This morning, after taking longitude observations, Professor Whitney stayed in camp to calculate them and at the same time take observations on the barometer to compare with those being taken at the next camp, while Mike (the cook) and I rode a few miles to climb a peculiar looking sharp peak that rises like a spire on the edge of the regular mountain chain. It is a little over 2,000 feet high above the camp, quite steep, and to our surprise we found it a great mass of lava, very perfect and very marked. It had a few large granite veins in it. We climbed to the top, but the good weather was deserting us. We had that same magnificent view I have described before, but it began to rain some and we hurried down and hastened our mules back to camp. But little rain fell. It has again cleared off cold, and the wind blows again tonight very hard. Our driver returned, and we shall move on in the morning—weather permitting.

Camp No. 8, Santa Ana River.
Thursday, January 17.

WELL, so much farther on our "winding way." All are in bed but me. I will write half an hour—it is only eight o'clock—ten hours in bed is more than I can stand.

It was blowing hard when I went to bed Tuesday night. The wind increased. We got up about eleven and packed the instruments. While getting ready for the worst, a hard gale took the fly of the tent and tore a rent fifteen feet long. We ran out and unfastened it, and over it went. A few minutes later the tent followed. We held on, drew it over our goods, piled carpetbags, saddles, etc., on the edges to keep it down so the wind could not get under it, crawled under again, rolled in our blankets, and whiled away the night as best we could. It was exceedingly laughable, but by no means comfortable. The wind howled and shrieked above us, the heavy tent flapped and whacked and slapped, shaking dust into our faces all the time, and last, but by no means least, multitudes of flies which had been in our tent were now caged close to us, buzzing and humming and crawling as we warmed them into life. The night was cool, but we slept some. Morning came, but the wind still howled fiercely, and, sweeping down from the snowy mountains, was raw and cold enough. However, we packed up and left and rode across the plain about twenty-two miles to the next camp at Chino Ranch. The wind was so fierce at times as to almost blow us off from our mules.

Chino is much colder than where we left, with mountains on both sides. We could see dense clouds of dust on the desert twenty-five miles east of us. A regular sand-storm was raging there. We stopped at Chino all night. Ice froze over half an inch thick, but I slept warm and well. It is a most beautiful ranch, much of it level, with mountains for the background on either side, and with two streams running through it. It is some twenty or thirty thousand acres and supports many cattle. We left there this morning for Temescal. But my fingers are too cold to write legibly any longer. I must turn in.

Sunday, January 20.

WE left Chino Ranch Thursday morning, but had not got over two or three miles when, in crossing a stream, we were brought to a dead stop by the breaking of the tongue of our wagon. A stream which runs across the plain had worn a channel about eight or ten feet deep—in one place wagons get down and up, but it is very steep. Our wagon went into the slough *ker-chug*, and off went the tongue, the wheels up to the axles in the mud. We unloaded, got out the wagon, and sent back to a ranch about two miles, where we got a piece of board and some rawhide. With two or three hours' work, all was repaired and we were ready to proceed. I had carried a barometer on my mule for about forty miles, but had handed it to the man in the wagon just a mile or two back. We took it out, found to our dismay some mercury in the case, but on taking it apart found the tube still whole. Professor Whitney and I spent the time in mending that—the kid-leather packing had given way. We sent Guirado to a ranch two miles off; the lady gave us a kid glove, and in time all was in order again, but we could only get on about seven miles that day.

We are in a most interesting place for a geological camp. High bluffs rise above us, while across the river rise the Santa Ana Mountains. Friday the Professor and I geologized along the foothills, and yesterday three of us climbed the highest point near, about three thousand feet. The climb was not so steep as some we have made before, but the chaparral was so exceedingly dense that it seemed as if no progress could be made. We carried up compass, tripod, and barometer. I never before appreciated the difficulty of carrying a barometer in such places. I carried it first on my mule as far as we could go, then on my back. On our return Pete and I ruined our pants in the bushes—they hung in ribbons. Averill had severe rents in his, and Pete's shirt was literally torn from him—there was scarcely anything left. But we got back all safe just at sunset.

The river here is quite a stream—several rods wide and up to the sides of our mules as we ford it. I have not seen a bridge in southern California, nor is there timber enough here to make them. Talk about a railroad! Timber for bridges and ties would cost at least a hundred dollars per thousand here, and much more farther east. The Overland passes here—that miracle of undertakings—over plains and deserts, over mountains, through gorges, into rivers, yet always inside of schedule time.

Today is a lovely day, but the nights are cold enough. It freezes thick every night here, temperature 27° or 28° in the tent—decidedly cool—and it makes my rheumatism complain some, yet I feel otherwise in most excellent health.

Camp No. 12, San Gabriel.
Sunday, February 3.

I HAVE not written for some time—the nights have been too cold for me to write without bringing on my rheumatism, and my days have been entirely occupied, much of the time with fatiguing travel or field work. Today is lovely again. I am writing this sitting under a tree, minus coat, vest, and hat, thermometer at 80° in shade, and sky as clear as August. We are back in good weather again, and I will take up my journal where I left it.

Monday morning, January 21, it looked like rain. We measured the cliffs above camp (about four hundred feet high). Then it cleared up, and I went on with advance camp to find a good place to camp at Temescal, where we expected to spend several days. We took our way southeast, over a sort of plain which sloped each side from the mountains. Most of it was very green, but alkali covered a part, and to the northeast many hills of granite rose like islands in it. The flying clouds shading this great plain in places, the snowy mountains on the horizon, the grassy carpet beneath, the fine hard road we were on, all conspired to make it a lovely view and pleasant ride. We got in camp before night near the Temescal Overland station, at the foot of the Temescal hills, a splendid place to camp, wood and water plenty, and protected from the winds.

Tuesday, January 22, I stayed in camp all day and observed barometer, to compare with observations at Santa Ana River. Pete went back and brought up the rest of the party. Wednesday, January 23, while the Professor went to the tin mines, Averill and I went across to the hills on the opposite side of the valley to observe rocks. The Temescal hills are a range some two thousand feet high, lying east of the Santa Ana Mountains, and are celebrated now as being the locality of fabulous mines and quantities of tin. People are "crazy" about tin ore, every man has from one to fifty claims, while poor devils with ragged clothes and short pipes talk as they smoke of being the wealthy owners of one hundred or two hundred tin claims, each in time to rival Cornwall or Banca. It was to see these mines and the formation around that we came here.

The Santa Ana Mountains rise between us and the coast. It was desirable that we should ascend them, so Thursday, January 24, Averill, Pete, and I started on our mules to ride a few miles and examine the base and find some practical way to get up them. Hot springs issue from the base, where a bathhouse has been erected. Here Averill and Pete stayed, not wishing to climb, while I went alone to a height of 2,500 or 3,000 feet above the sea, found a practical way of ascent, and made some interesting geological observations. The stillness was profound, the solitude almost oppressive. I found no grizzlies, of which I had heard so much, but on my way back, trying to explore a canyon, got into chaparral so thick that I tore my pants off almost. They were completely ruined—my last pair, but I bought a pair of Averill to get back to Los Angeles with.

Camp at San Gabriel Canyon.
February 3.

FRIDAY, January 25, we rode to the principal tin mine, four miles distant—found it a splendid humbug. These hills are desolate beyond description, rough and dry, no trees, scarcely a bush, very little water and that quite strongly alkaline and nauseous. Many black streaks are found in the rocks; some of which contain *some* tin. Many claims are made and entered. One man has invested $14,500, and has commenced mining operations, that is, has sunk a shaft in the granite to look for richer ore. All thus far is mere speculation, and will end in that, I think. We carried a barometer and measured the height of the hill at the mine—found it about a thousand feet above the sea.

We were back soon after noon, when Professor Whitney, Pete, Guirado, and I took our small tent and went about seven miles to Camp 10, at the base of the mountains across the valley from the ridge of tin, and camped in the mouth of a wild canyon, *Cañon Agua Fria*, or "Cold Water Canyon." A fine stream of pure cold water here issues from the mountain. We were in bed early for an early rise. The Professor and I were to climb the mountain the next day, so all was got in readiness.

Saturday, January 26, we were up and breakfasted at dawn, and as soon as it was light enough we started. Professor Whitney carried barometer, hammer and bag, and canteen of water; I, a compass and tripod, level, spyglass, provisions, and another canteen. We had heard such big stories of animals that, by the advice of all, each carried a heavy revolver loaded with slugs, and a heavy bowie, for emergencies. We carried some bread, a little meat, and some *panoli*. This last is made by the Indians here, and is pulverized roasted corn—when mixed with some water and with some sugar is very refreshing as a drink, sustaining one as hardly anything else will, as we found that day. These, with a few other small items made up our burdens.

We struck up the canyon a short distance. The granite sides were generally inaccessible, but at last we took up a slope exceedingly steep, some thirty-three degrees (the roof of an ordinary house is but twenty-three degrees). Vigorous climbing in due time brought us on a ridge, where the rising sun greeted us, first gilding the snowy peaks in the distance, and then flooding the valley below us in light and casting dark shadows in the canyons of the Temescal Range opposite. Several very steep slopes were surmounted until we gained a ridge at 2,200 or 2,500 feet, which runs laterally from the central chain. Here we found a few pine and fir trees. The chaparral became more dense, but by following the ridge we found in places a trail worn by deer and other animals. Tracks of deer, wildcats, and coyotes (small wolves), with their other traces, were numerous. Here we reached the highest point I had climbed to three days before, and we made it in a little over two hours.

The sun came out hotter, thawing the frozen ground, making it slippery in places, and increasing our thirst—our canteens were often used. After another hour's climbing I lost the plug to my canteen, so we stopped and mixed the water left in it with *panoli*, and after dispatching it and resting, felt decidedly refreshed and in fine climbing trim.

Now commenced the real hard work of our ascent. We had

risen much over half the height to be gained, but in places we had to climb over cragged granite rocks, and then walk over a steep slippery slope of decomposing feldspar. The stones which were loosened by our feet went bounding away into the canyons sometimes hundreds of feet beneath. But the real difficulty was the chaparral, which in places seemed absolutely impenetrable—a tangled mass of stiff, interlaced, thorny shrubs. Sometimes we broke them down (our hatchet had been lost), sometimes tore through, sometimes crawled on our hands and knees a long distance. At one time nearly an hour was consumed in making probably sixty or eighty rods. I had rigged up an old pair of pants for the occasion, but they were "nowhere"—they were torn to shreds. My drawers "followed suit" and left my legs to the mercy of the thorns. I had to go ahead, as the Professor carried the barometer. His shirt fared as badly as my pants, but my shirt stood it with only a few tears.

At last, after over six hours of the most vigorous climbing, we reached the summit.[3] We had found snow for the last two hours of our climb, and a cold, piercing, raw wind fairly shrieked over the summit. We went about thirty feet below to hang the barometer in the bushes. After half an hour's observations, eating our lunch, and drinking the rest of our *panoli* we put up the barometer, and planted our compass on the summit to get the bearings of the conspicuous objects around. It was so cold that I could scarcely write the bearings as read off.

But the view more than repaid us for all we had endured. It was one of the grandest I ever saw. Not less than ten or twelve thousand square miles were spread out in the field of vision; or, if we take the territory embraced within the extreme points —land and sea—more than twice that amount. We were on the highest point of the Santa Ana Range. To the west and south lay the sea, 150 miles of the coast in full view, from Point Duma to the islands off Lower California, Los Coronados. Table Mountain in Lower California was in full view on the horizon. The whole plain along the sea lay to the northwest:

the plain of Los Angeles and beyond, some eighty or ninety miles to the Sierra Santa Monica, with the Santa Clara Mountains much farther still, the tops covered with snow, 125 or 150 miles distant (much more by road). We could see out on the Pacific a hundred miles from the coast, with the islands of Santa Barbara, San Miguel, etc., visible.

To the north lay the chain we were on, gradually growing lower, and at last sinking into the plain; beyond that the snow-covered Sierra Madre, the highest peaks nine thousand or more feet. A desert at the base, although eighteen or twenty miles across, seemed but a brown level field. In the northeast the great mass of San Bernardino with its many ridges shut out the farther view; it was probably fifty miles distant. A sort of plain stretches from its base to us, like the sea, with numerous rocky hills rising from it, like islands, some twenty or thirty miles in extent, but far beneath us. The scene in this direction, as well as to the southeast, was desolate in the extreme—dry, almost desert, broken into rough, rugged, rocky ridges, or dry valleys—no forests, no water or rivers to amount to anything—a country nature had not favored.

Time was short, we must hurry. We had to pass the same chaparral. Trying an easier way in one direction for a short distance, we found trails, but the traces of grizzlies grew so very numerous, that we took to the ridge again.

The sun set while we had a thousand feet still to descend. We saw it gild the snowy peaks on the horizon. Tired as we were, we were not too tired to admire the beauties of that sunset. It was just dark as we got back. Pete had shot some quail and rabbits, and had them served in their best style, and how that meal tasted! We cut a sorry figure as we came back—clothes torn, parts out, boots ruined, scratched, bleeding, bruised, dirty, and tired, I was nearer used up than I had been before; the Professor stood it better. But a hearty supper, early to bed, and late up in the morning, worked wonders. More quail for breakfast, then a most luxurious bath at the warm sulphur springs, and, save bruises and scratches, I was myself again. Our barometrical observations made the height of the peak 4,900 feet above camp and 5,675 above the sea. The next morning after breakfast we raised our tent, loaded up soon, and being short of provisions, went back to the other camp, about six miles distant.

On the way we stopped at the Temescal hot sulphur springs and bathed. The warm sulphur-water issues from the rock in large quantities at a temperature of 93°, very soft water, and slightly mineral. A rude bathhouse has been erected over it, and a bath in that warm water was refreshing in the highest degree. An Indian village, the old village of Temescal, lies at the spring, although the Overland station five miles distant is now the place called Temescal. The Indians speak Spanish and Indian. They are a miserable, thieving set. I saw a half-breed squaw, the prettiest I have seen in California thus far. We spent the rest of the day resting in camp, quietly enough, and had fine roasted wild ducks for dinner.

Los Angeles.
February 10.

Monday, January 28, I rode with Averill to the north end of the Temescal Range, a series of granite ridges, covered with bowlders, some of immense size. One was seen forty feet high and many twenty-five or thirty feet.

Tuesday, January 29, the advance camp started to return to Los Angeles, on our old route. Professor Whitney and I rode with the proprietor of several tin leads to see them, several miles in the mountains, up steep and narrow trails, where we at home would think no horse or mule could climb, much less carry a rider. I rode my mule, the Professor our new horse, which had shown many signs of tricks and had thrown him two or three times. When about seven or eight miles from camp his horse threw him and got away. Our companion lassoed him once, but he got away again. He lost his saddle, which was stolen by the Indians, and we never recovered it. I tied my mule to some sagebushes while we climbed a hill, and when I returned it had pulled up the bushes and left. We footed it back. We found my mule at camp, but it was two days before we got the horse.

During the next two days I climbed over those barren hills with Professor Whitney looking at tin leads and studying the interesting geology of the place. I found some exquisitely beautiful flowers of very small size, several species being less than an inch in height, as small as any alpine vegetation.

The loss of our horse detained us a day longer than we expected, but we were off Saturday morning, February 2, for San Gabriel Canyon, about thirty-eight miles distant, where the rest were encamped. We were detained, however, by a "slough," and failed to make it. Although we had started very early, night found us five or six miles short. So we pitched camp, and early the next morning, Sunday, were up, raised our tent, and got into camp before the rest had got their breakfast.

Monday, February 4, the Professor and I rode a few miles, and climbed a ridge about 2,500 feet high, where there were extensive outcroppings of rock. We saw four fine deer, but of course had no guns to shoot them with.

Tuesday, February 5, we came on to Los Angeles and camped again on the site of our first camp. We passed over the lovely plains of San Gabriel, El Monte, and Los Angeles, with their thousands of cattle, horses, and sheep feeding; tens of thousands were seen, in pleasing contrast with the barren hills of Temescal. We saw on our return quantities of wild ducks and geese in the ponds along the road. The plains and hills were green, men were putting their vineyards in order, and fruit trees were coming in bloom here, fruit plenty in the streets, but very dear—apples four to fifteen cents each, oranges five to ten cents.

Thursday, February 7, Professor Whitney left for San

Francisco, leaving the party in my charge, a responsibility I by no means desire, but I will make the best of it. That night three of our men came in very drunk, for the first time, a bad beginning of my rule. I had been in town in the evening, and on returning put things in order, as it looked like rain. At two o'clock they came in, and disarranged some of the things. Soon after it began to rain hard, and rained until near daylight, the wind cold and increasing. Just before daylight the wind changed, and a fierce squall carried over both tents in the rain. *Such a pickle!* Instruments and blankets and books were hurried in the wagon, clothes lost in the intense darkness. I worked barefooted, in shirt tail and drawers—*ugh!*—how cold it was! It completely sobered the men, I assure you. It was fortunately near daylight, and the wind and rain lulled and finally ceased. When day dawned we found our clothes (wet, to be sure), put up our tents, dried ourselves by the fire, and, after a hearty breakfast, laughed at our mishaps. No serious damage was done.

The next two days were spent in making preparations to leave for the north.

NOTES

1. This building, dedicated to *Nuestra Señora la Reina de Los Angeles,* was erected in 1822, succeeding an earlier structure located nearby. It was remodeled and rebuilt in 1861 and again restored in 1912.

2. Camels were introduced into the southwest in 1857 by the United States War Department (Jefferson Davis, Secretary of War) under supervision of Lieut. Edward F. Beale. The camels used by the Boundary Commission were doubtless loaned by the War Department. The camel experiment was abandoned by the Government shortly afterward and the animals were sold at auction or turned loose in the desert. (See *Uncle Sam's Camels*, by Lewis Burt Lesley, Harvard University Press, 1929.)

3. To this mountain they gave the name Mount Downey, in honor of the then governor of the state (Whitney Survey, *Geology*, I, 177). The name has not persisted, however.

CHAPTER IV

STARTING NORTHWARD

The San Fernando Valley—San Buenaventura—Carpinteria.

> Camp No. 13, Sycamore Canyon,
> San Fernando Valley.
> Sunday, February 17, 1861.

Our course before leaving Los Angeles the last time had been from San Pedro on the coast to the Temescal Range about eighty miles east. Either a plain or a valley runs this whole distance, with high, steep, rugged, barren mountains on one or both sides, nowhere covered with timber or of any value for agricultural purposes. These mountains so far as yet examined are mostly of porous granite, or other porous rock which absorbs most of the water that falls on them. The streams that run off in places follow narrow canyons or gorges down their sides, and as soon as they strike the plain or valley spread out wide and generally sink. These "washes" or dry beds are often two or three miles wide, covered with bowlders and sand, supporting only a vegetation of stunted shrubs, from five to ten feet high.

It is surprising how large a stream will soon sink. The Santa Ana River, 150 or 200 yards wide and nearly up to the bellies of the mules, sinks in a few miles after leaving the mountains, leaving only dry drifting sand in its bed. In this way a man may travel a great distance and see no water, yet cross the beds of streams every little while, beds sometimes a mile wide, over which the streams in high water shift their courses, sometimes following one channel, sometimes another. Already, only the middle of February, we have to follow up the canyons into the base of the mountains to find water for camp. We are now about a mile or a mile and a half up a stream; here is water, but below only sand.

This is one of the proposed routes for a Pacific railroad, yet from San Pedro to Temescal, eighty miles, to even *fence* the road, would take all the available timber from a strip *ten miles wide*, five miles each side of the road, unless a brush fence was made. Of course there are places where there is some timber, but I have asked each of our men if they thought a strip ten miles wide would do it, and they think not, and the thing grows worse as we go farther, for here we can get timber from the north but it costs enormously to get it inland. Fence posts of redwood, split four by six inches and about seven feet long, cost there fifty cents each; what must they cost a hundred miles farther on!

The Temescal region is so barren as to be practically useless, and will ever support only a very sparse population, and this whole country will only be used for stock raising on large ranches. In a few places, of limited extent compared with the whole, will be lovely fertile spots where there is water, but the agricultural capabilities of the region are small.[1]

Even here the San Fernando Valley looks fertile, yet you could take a patch in the middle of a hundred or a hundred and fifty thousand acres, where it does not touch the hills, where there would be no water for over half of the year. Hence the land is owned in large ranches, and those only in the more favored places. On these ranches, as there are no fences, the cattle are half wild, and require many horses to keep them and tend them. A ranch with a thousand head of cattle will have a hundred horses. The natives here are lazy enough, but are slowly giving way before the Americans, with whom they do not assimilate.

There is a knotty political question here which causes no fuss now; but make southern California a slave state and people it with southerners, and it may become complicated. Our treaty on obtaining this region guaranteed to the Mexican citizens all the privileges of American citizens on entering this republic. Mexico recognized Indians, negroes, etc., as citizens under certain circumstances, so there are actually negroes *citizens* of the United States.

Now to my journal. On Monday, February 11, we left Los Angeles and came on about twelve miles and camped in the Cahuenga Pass, where the Overland road passes through the Sierra Santa Monica, there a range of hills about 1,600 feet high. It is not much of a pass. We stopped there until Wednesday morning, then entered the San Fernando Valley. We went along the north side of the Sierra Santa Monica, at its base, and camped here, some twenty-five miles from Los Angeles. We intended to leave here Friday morning, but on starting, had not gone three rods when a wheel of the wagon broke down. We had to unload, camp again and send the wagon to Los Angeles for repairs. Here we are yet, but the wagon got back late last night and we will move on in the morning. We have reduced our load, and having two extra mules, we "pack" them; that is, place loads on their backs—blankets, bags of barley, carpetbags, etc.

Camp 15, near Triunfo Ranch.
Sunday, February 24.

ON Monday, February 18, we crossed the valley to the north side, a stretch of about ten miles across a plain, a part of it almost desert for want of water, the rest covered with grass. It all belongs to the San Fernando Mission, under old Spanish grants. On the north side of the valley there is a great chain of hills—mountains, I should say.[2] The rocks are all broken up, and rise in ridges, some of them two thousand feet or more above the plain, the broken edges of the strata forming lines of rock or high precipices in places, visible for many miles. We camped in a quiet canyon at the base.

Tuesday, February 19, we sent the wagon again to Los Angeles to take in Ashburner, get letters, etc. Ashburner has given out, and the Professor called him from the field to work in the laboratory. He cannot stand camp life. He got the scurvy a year and a half ago while on geological excursions in Newfoundland, and it broke out again in camp, and we feared it would use him up. I advised Professor Whitney to send him to the city some weeks ago. I think there is no danger now, as he was much better before he left, but he has not at all been able to stem the hard work of the field.

This increases my labors some, as he has made all the mathematical calculations; some of them I must make now. I have practiced Guirado in making observations with the barometer, and shall teach Averill soon to compute them; this will lessen the work for me again.

Wednesday, February 20, I started back in the hills alone, got into a wild rocky canyon and followed up it for three or four hours—I don't know how many miles. The stream made its way through the red sandstone rock, which often rose in high precipices—a lovely walk. At last I climbed a high ridge, some two thousand feet above camp. Here a stratum of rock comes out filled with large shells in fine preservation. It rises in a ridge, ending in a precipice to the north. In places these fossil shells had been weathered out in immense numbers. The ridge was strewn with them, as thick as any seabeach I have ever seen, and in as good preservation—oyster shells by the cartload, clam shells, in fact many species. They have not lost their character as shells yet, that is, they have not turned to stone. The shell of lime was as when fresh, and the scar where the muscle was attached was as plain as if it had stood the weather but a few years. Some were worn by the waves. Oyster shells had grown together in that old ocean as now, and the pebbles of the beach were bored into by shells as I see them here on the coast now.

I cannot describe my feelings as I stood on that ridge, that shore of an ancient ocean. How lonely and desolate! Who shall tell how many centuries, how many decades of centuries, have elapsed since these rocks resounded to the roar of breakers, and these animals sported in their foam? I picked up a bone, cemented in the rock with the shells. A feeling of awe came over me. Around me rose rugged mountains; no human being was within miles of me to break the silence. And then I felt overwhelmed with the magnitude of the work ahead of me. I was at work alone in the field work of this great state, a territory larger than all New England and New York, complicated in its geology.

But the real soon roused me from reveries—I must get back. I was alone, far from camp—grizzlies might come out as the moon came up, for the weather was warm. I made my way back into the canyon, and at dark arrived at camp, tired enough. Peter brought back from Los Angeles a pile of letters, and after supper how I devoured them!

Friday, February 22, I started to examine a peak a few miles to the north. Averill and Mike (our cook) went with me. I carried a barometer. Averill shot an eagle with his revolver. It measured fifty inches from tip to tip. He was a savage fellow, and as he was not killed entirely, he fought most vigorously. We had a difficult climb. My companions both showed signs of giving out and finally stopped at the foot of the last slope we had to rise. I went on alone, but they finally followed and succeeded. The peak was 2,700 feet above camp, or some 3,800 feet above the sea. We had a glorious view from this point and collected some more fossils on our way back. We celebrated Washington's birthday, in the evening, with a glass of toddy.

Saturday, February 23, we raised camp and started on our course. We had an accident or two in crossing the plain, but came on. On getting into the hills, in coming down a very steep one, that wheel began to crack again. We went on cautiously a mile or so and found a good place to camp, a small running stream and some trees for wood. Here we are now, about three miles from Triunfo Ranch.[3] Good grass around.

Our animals are now gaining. We have had to feed barley up to this time, but I hope that the grass will soon be big enough. It is good here. We had in Sycamore Canyon a week ago thermometer from 70° to 80° (once 86°) for several days. It is every day nearly up to 60°. Think of that for February! What must June, July, and August be! Whew!

Tuesday, February 26.

YESTERDAY I visited the country southeast of camp and had a hard climb over rocky and precipitous hills. We rode to the foot, about four miles, and there left our mules. We came upon four fine, large deer, almost within pistol shot—graceful and beautiful animals. This makes eleven deer I have seen this month.

We intended to move on this morning, but before day it began to rain hard, and rained at intervals all the morning. It cleared up at noon, however, and we will move in the morning. Peter has repaired the weak wagon wheel with that universal plaster for ailing implements, rawhide, and says it will now go. We will try it. I had no idea of the many uses to which rawhides are put here. I was in a house on a ranch, where a rawhide was spread before the beds as a carpet or mat. Bridle-reins and ropes or lassos (*riatas*) are made, fences are tied—everything is done with rawhide.

It is a clear, cold evening, all the men are smoking around the fire except me; my fingers are cold and I must go out too. I wish I could be in "The States" this and next week—exciting times! Next week must tell what the South will do on Lincoln's inauguration. We get the political news pretty well; we have got New York papers nearly every steamer and will get them again on reaching San Luis Obispo, if not before. I hope to be at Santa Barbara in ten days.

Camp 18, Carpinteria.
March 5.

A WEEK has passed since I wrote anything to anybody. Wednesday, February 27, we raised camp and went about eighteen miles, first passing the lovely Triunfo Ranch, a large grassy valley surrounded by high hills. Then we crossed a high rocky ridge and descended a hill about five or six hundred feet. It was terribly steep, but Peter managed the wagon with a skill to be praised—all down safely. We then struck west a few miles, in a valley, and by a stream near Cayeguas Ranch. Here we stopped over one day and gathered some fine fossils. A hill was as full of large clam shells, barnacles, conch shells, oyster shells, etc., as any modern beach—much more so than the beach of the Pacific here now.

Friday, March 1, we came on to San Buenaventura, on the seacoast. Soon after leaving Cayeguas we entered the plain, which there lies along the sea, and crossed it to the sea about twenty miles. It is a fine grassy plain, with here and there a gentle green knoll, with a few dry creeks or alkaline ponds, and one fine stream, the Santa Clara River, running through it. We stopped for an hour on its bank and rested our mules, lunched and refreshed ourselves in a grove of cottonwoods which came nearer to a forest than anything I have yet seen here. We forded the river and came on. At San Buenaventura the hills come up to the sea, the plain ceases, but a fine stream comes down from a pretty valley, green, grassy, and rich.

Here is the old Mission San Buenaventura, once rich, now poor.[4] A little dirty village of a few inhabitants, mostly Indian, but with some Spanish-Mexican and American. The houses are of *adobe*, the roofs of red tiles, and all dirty enough. A fine old church stands, the extensive garden now in ruins, but with a few palm trees and many figs and olives—the old *padres'* garden. Ruined buildings, two or three old fountains with lions and horses sculptured on them, now dry and ruined, told of former luxury. An old threshing floor stood, a circular wall of stones laid up in mortar, about forty or fifty feet in diameter, the wall about four or five feet high, where they used to put in wheat and drive in wild horses to thresh it.

Saturday we roamed over the hills, went down to the beach, took a bath in the surf—decidedly refreshing, but cold. The roar of the breakers hushed us to sleep at night and was the first sound heard in the morning.

Sunday we strolled down to the beach in the morning and made the discovery of multitudes of cockles (small clams) in the sand. Soon we were all digging with our fingers; we got so many in a few minutes that they are not eaten yet. After a feast on raw clams, we went into town to Mass.

The church was precisely like the others seen here, only in better condition. A description of it will do almost as well for Los Angeles or San Gabriel. The walls are very thick, built of *adobes;* the ceiling is of timber and boards laid across, painted; the walls are painted rudely in pilasters, festoons, etc.; the floor is of large square bricks. The room was a parallelogram in shape, three times as long as broad, probably 120 or 130 feet long, ceiling high, with two entrances, one at the end, one at the middle of one side. Over the end and main door is a small wooden gallery for the singers; opposite the door is the altar decked with tinsel and silver, a few old images standing in various places about the altar and about side shrines— once brilliant with paint, tinsel, and gilding, now faded, dingy, and dilapidated. A few pictures hang on the walls, some really quite good, but dingy, their gilded frames worm-eaten and tarnished. All speaks of decline, decay.

At the altar was an Italian priest, saying Mass. Kneeling, sitting, or standing on the floor, was the congregation of about fifty women and half as many men, reverent, devout, and attentive. The effect was very picturesque. The women, sitting or kneeling, had shawls over their heads, hanging down behind and held or pinned beneath the chin, as is the custom here—some black, but most of very gay colors. A few children were among them; here a babe in arms—black hair, blacker sparkling eyes, and dark skin—peeping over some mother's shoulder; there a little girl, just learning to cross herself and read her prayers, beside her mother or larger sister. Most of them were Indians, but there were a few Spanish or other whites. Some of the half-breeds were really pretty. Some of the men wore moccasins, leggins, and Indian costume, others the Spanish or common. The women wore frocks, some with many flounces, while a few hoops and flat hats told of inroads of modern fashion in this place. The women were better looking as a set than those at the Indian villages or missions before visited. Several Indians visited our camp, quite intelligent looking fellows. All the Indians here are much blacker than those in the East, and with flatter noses and less intelligent faces.

After Mass we went back to camp, a mile from town, but Guirado stayed with some friends during the afternoon and came back telling of spirited horse races—six horses changed owners by betting—billiards, dancing, and a fight—common accompaniments of a Spanish Sunday. What things we purchased there we paid a most exorbitant price for, except *cigarritos*, the only cheap thing found in the place.

Yesterday, Monday, March 4, we came on here, about eighteen miles. The road followed the seacoast; high bluffs or cliffs rose from the shore all the way. Sometimes we rode on the sand close to the water, sometimes over sand-hills, half knee-deep, sometimes over a little flat beck with deep, steep gulches, and, worse than all the rest, over big bowlders for a mile or two—bowlders piled thick together as only the sea can pile them. This was too much for our invalid wagon wheel; it showed signs of giving in. We stopped in this fine valley—good grass and wood, but poor water.

Tuesday Evening, March 5.

LAST night as we were sitting around our cheerful camp fire, the sound of a cannon came booming on the still night air, above the roar of the surf. How it startled us, for it told of the arrival of the steamer at Santa Barbara, eight miles distant, probably with the Professor, or if not with him, with letters and *funds*.

This last was an important item. He left four weeks ago, leaving with me but three hundred dollars. Several bills which had been left unpaid and our unexpected break-downs reduced this. I sent for more by Ashburner when he went up two weeks ago, for when the wagon returned from Los Angeles I was horrified to find that our treasurer had paid out all but about twelve or fifteen dollars.

This afternoon Guirado returned, brought letters, but no Professor, and what was worse, *no funds*. I counted up and found $3.25 in the treasury and $3.00 in private hands—total, $6.25. Five men in camp, two weeks before another steamer, flour all gone, jerked beef ditto, onions ditto, potatoes ditto—long ago—have forgotten how some of them looked—bacon, small chunk, and even beans only a meager, lonely few left in the last corner of the sack. We have lived poorly the last two weeks, looking forward for better fare on reaching Santa Barbara; decidedly a poor prospect ahead! I shall stay here one day more, then go to Santa Barbara and try to make a raise of fifty or a hundred dollars by an order on Professor Whitney to sell to someone, by borrowing, or otherwise. Can't tell exactly how, but I will make it go in some way or another.

Rode along the beach a few miles today through the fog, visited some rocks of interest, frightened a large seal off from some rocks, saw thousands of gulls, geese, cranes, and other sea birds. After dinner at four o'clock, took a fine bath in the surf.

NOTES

1. Brewer overlooked the possibilities of irrigation and of the importation of water by long aqueducts, developments which have completely transformed this region.
2. The Santa Susana Mountains.
3. This spot was visited January 13, 1770, by Portolá's party on the return from the expedition to San Francisco Bay, and was named on that

occasion *El Triunfo del Dulcísimo Nombre de Jesús* (Fray Francisco Palóu, *Historical Memoirs of New California,* ed. Herbert Eugene Bolton [1926], II, 254–255). It is interesting to follow the course of the Portolá expedition of 1769–70 over much of this same ground through the medium of Doctor Bolton's translation of Palóu's *Noticias.*

4. This mission named in honor of the Doctor Seráfico San Buenaventura (Giovanni Fidanga, 1221–74), was founded by Junípero Serra in 1782. The name of the locality has been shortened to Ventura.

CHAPTER V

SANTA BARBARA

A Decadent Town—The Old Mission—A Remarkable Grapevine—Rough Trails—Inspecting a Coal Mine—Holy Week.

In Camp at Santa Barbara.
Sunday, March 10, 1861.

WE came on here Thursday, March 7, arriving in the afternoon. The steamer was to leave that night for San Francisco, the only public communication with the outer world. I tried to make a raise and get some money from express agents, merchants, etc.—no go—so wrote on to Professor Whitney that we would wait here until either funds or he arrived. Friday we visited the Mission, examined the foothills, etc. More of the Mission anon.

Saturday, with Averill, I visited a hot spring about five miles from here. First a good road, past some pretty ranches, then up a wild ravine by such a path as you would all put down as entirely impassable to horses, but it was mere fun for our mules. They climbed the stones and logs, now between these bowlders and now over this rock, as if it were their home. We found several copious springs, making together a fine brook, issuing from the rocks at the base of a very steep rocky mountain. This is just near the base of a rugged peak, at perhaps five hundred feet above the sea. The water was sulphury and had temperatures varying from 115° to 118° F. In the States, or near a large city, it would be a fortune to some enterprising man. There is more timber here, as at Carpinteria, than we have seen south, along the streams and in the valleys.

Santa Barbara lies on the seashore, and until lately it was isolated from the rest of the world by high mountains. No wagon road or stage route ran into it from without, only mere trails or paths for horses over the mountains. For a few years they had had a mail once in two weeks by steamer from San Francisco—two mails per month was the only news of the world outside. But the Overland has been working the road—or the county has—and will run this way after the first of April. Here is a village of about 1,200 inhabitants. A wealthy Mission formerly existed here, but like all the rest, is now poor after the robbery by the Mexican Government. I have not seen before in America, except at Panama, such extensive ruins.

The Mission was founded about the time of the American Revolution—the locality was beautiful, water good and abundant. A fine church and ecclesiastical buildings were built and a town sprang up around. The slope beneath was all irrigated and under high cultivation—vineyards, gardens, fields, fountains, once embellished that lovely slope. Now all is changed. The church is in good preservation, with the monastery alongside—all else is ruined.[1]

It was with a feeling of much sadness that I rode through the old town. Here were whole streets of buildings, built of *adobes*, their roofs gone, their walls tumbling, squirrels burrowing in them—all now desolate, ruined, deserted. Grass grows in the old streets and cattle feed in the gardens. Extensive yards (*corrals*) built with stone walls, high and solid, stand without cattle. The old threshing floor is ruined, the weeds growing over its old pavement. The palm trees are dead, and the olive and fig trees are dilapidated and broken.

We went into the church—a fine old building, about 150 feet long (inside), 30 wide, and 40 high, with two towers, and a monastery, sacristy, etc., 250 feet long at one side, with long corridors and stone pillars and small windows and tile roofs. The interior of the church was striking and picturesque. Its walls were painted by the Indians who built it. The cornice and ornaments on the ceiling were picturesque indeed—the colors bright and the designs a sort of cross between arabesques,

Greek cornice, and Indian designs, yet the effect was pretty. The light streamed in through the small windows in the thick walls, lighting up the room. The floor was of cement. The sides and ceiling were plastered with the usual accompaniment of old pictures, shrines, images, altar, etc. The pictures were dingy with age, the tinsel and gilt of the images dull and tarnished by time and neglect. Some of the pictures were of considerable merit; such were two, one of the Crucifixion and another of the Conception.

On either side of the door, beneath the choir, were two old Mexican paintings: one of martyrs calm and resigned in fire; the other, the damned in hell. The latter showed a lurid furnace of fire, the victims, held in by iron bars, tormented by devils of every kind. In front was the drunkard with empty glass in his hand, a devil with the head of a hog pouring liquid fire upon him from a bottle. The gambler, ready to clutch the money and the cards, was held back by a demon no less ugly. An old bald-headed man stood with a fighting-cock in his hand, but tormented now. A woman had a serpent twined about her and feeding upon her breast, another was stung by scorpions.

Although the picture attracted the attention and imagination, it had none of the merits of Rubens' "Descent of the Damned." The victims had not that expression of remorse and anguish which he could paint so well, nor the demons that fiendish diabolical expression he conceived and expressed.

The same was true of another picture of Judgment Day, the separation of the just from the unjust—an elaborate work of the imagination, but not good as a work of art. Much better was a picture of the Virgin with broken scales of justice in her hand, an angel on each side pointing and directing the penitents at her feet to her look and mercy.

There were old tombs beneath the church, and a churchyard by the side. A few monks still occupy the place and preserve the church and monastery from utter ruin. They were kind to us. I got much information from the old *padre*, nearly seventy years old, a fine old benevolent-looking man, who had known the Mission in the days of its prosperity and who could tell of wildernesses reclaimed and works of art erected, of savages converted and taught the arts of civilized life, and of heathen embracing the gospel. One of the monks, an Irishman, with the strongest Celtic features, showed us through the building, took us up into the towers, where we had a good view of the Mission and its ruins, the scene of its former greatness and present desolation.[2]

Up the canyon two or three miles a strong cement dam had been built, whence the water was brought down to the Mission in an aqueduct made of stone and cement, still in good repair. Near the Mission it flows into two large tanks or cisterns, reservoirs I ought to call them, built of masonry and cement, substantial and fine. These fed a mill where grain was ground, and ran in pipes to supply the fountains in front of the church and in the gardens, and thence to irrigate the cultivated slope beneath. But all now is in ruin—the fountains dry, the pipes broken, weeds growing in the cisterns and basins. The bears, from whose mouths the water flowed, are broken, and weeds and squirrels are again striving to obtain mastery as in years long before.

I find it hard to realize that I am in America—in the *United States*, the young and vigorous republic as we call her—when I see these ruins. They carry me back again to the Old World with its decline and decay, with its histories of war and blood and strife and desolation, with its conflict of religions and races.

Tuesday, March 12.

STILL foggy and wet. This weather is abominable—now for nearly two weeks we have had foggy, damp weather, tramping through wet bushes, riding in damp, foggy air, burning wet wood to dry ourselves, no sun to dry our damp blankets. I find that it makes some of my joints squeak with rheumatic twinges. Went out this morning, found it so wet that we had to return to camp. I have been writing labels and packing specimens, and now will write letters, hoping that it will dry off some after dinner.

Yesterday, with two citizens of the place, a lawyer and a surveyor, who were going to survey a ranch, I rode about six miles west along the coast. We rode over grassy hills, with some timber, where many cattle and sheep were grazing. We struck the coast about six miles from here, where asphaltum, a kind of coal-tar, comes out of the rocks and hardens in the sun. It is used for making roofs, by mixing with sand, boiling, and spreading on hot. It occurs in immense quantities and will eventually be the source of some considerable wealth. We found some fossils, stayed there several hours, and then rode back along the beach, it being low tide. It was an interesting ride to us. The strata which come out to the sea have been twisted and torn by volcanic forces, and then worn into fantastic shapes by the waves.

Sunday Evening, March 17.

WE have had a clear hot day, after a two weeks' fog, and have improved the opportunity to dry our blankets and clothes, botanical papers, etc.

Yesterday three of us rode again to the hot springs five miles east, and took a refreshing bath in the hot waters. On the way we passed the most remarkable grapevine I have ever seen. Although not quite so large at the main stalk as a wild one at Ovid, and none of the branches so long, yet it was much more remarkable, as it was pruned and under good cultivation. It was at Montecito, about four miles east, in the garden of José Dominguez. It was planted by his mother about thirty years ago. It stands in the center of a sort of garden, and its branches occupy the whole of it. It is trained up in a single stalk, like a tree, about six feet, then branches off into about twenty branches from six to twenty inches in circumference, running in every direction. The main stalk is from thirty-one to thirty-five inches in circumference in its various parts—the branches extend over a horizontal framework about seventy feet in diameter each way. In summer the foliage is very dense over the whole of this surface, some 3,600 to 4,000 square feet, or about one-tenth of an acre. The vine was well pruned, and the yield of grapes is as extraordinary as its size, being from three to four tons per year—good years the latter quantity is estimated. One year 6,300 bunches were counted and that was hardly more than a third—*sixteen thousand bunches* was considered a low estimate for that year. Single bunches have weighed as high as *seven pounds*, as can be attested by many witnesses! I question if the world can produce its equal, especially if we consider its youth. None of the old vines of the Old World are as great, so far as I can remember. The woman who planted it was old at that time—she is now about a hundred years old. She sat watching it like a child, with a stick to keep the fowls away. It is not yet in leaf for this year. A little *sancha* (artificial stream) runs near it, from which it is irrigated by hand. It is about three miles from the sea, high, steep mountains rise to the north of it to shelter it from the north winds. Men have visited it from all parts of the world, all pronounce it the king of vines.[3]

In Camp at Santa Barbara.
Monday, March 25.

THE foggy weather that had lasted for over two weeks ceased, the sky cleared up on Sunday night, and on Monday morning, March 18, I started to climb and measure the ridge lying north of us. Averill was somewhat under the weather, so I took Peter and Guirado with me. We rode to the hot springs, about five miles, left our mules in charge of Guirado, while Peter and I made the ascent. To the first peak, about 1,500 or 2,000 feet above the hot spring, was very steep, rocky, and hot. The sultry sun poured down floods of heat on the hot, dry rocks. The sun falling on the thermometer for scarcely a single minute ran it up to 120° F., and as it was graduated no higher I could not measure the temperature; it must have been 140°, or more, in the direct rays of the sun.

Reaching the first peak, we struck back over a transverse ridge, down and up, through dense chaparral, in which we toiled for seven hours. This is vastly more fatiguing than merely climbing steep slopes; it tries every muscle in the body. We reached the summit of the ridge at an altitude of 3,800 feet above the sea—over 3,700 above camp. Our lunch was useless, for in our intense thirst we could eat nothing except a little juicy meat. Our only canteen of water gave out long before we reached the top, although we had husbanded it by taking merely sips at a time.

I never before suffered with thirst as I did that day. What must it be on the deserts! I have heard tales of suffering here, on the deserts of California, Utah, Arizona, etc., as touching as those of Africa or Arabia. Peter found relief by chewing a quarter of a dollar for several hours, the means they use on the plains, but I could find no relief that way.

About sundown we reached the hot spring. A small pool of bad water was there. How I wanted cool water; hot sulphur water (118°) for a thirsty man is hardly the thing, yet we found it good. We ate our lunch, sat by the spring for half an hour, drinking small quantities often, then bathed in the hot waters and were more refreshed than one could have believed. But night closed in on us then. Guirado had brought the mules up into the canyon. The moon was bright as we struck down the wild dangerous trail. The wild dark canyon, rugged rocks, the dark shadows under the bushes and behind the rocks, the wild scene on every side, conspired with the hour to produce a most picturesque effect. Refreshed, we were lighthearted. Peter rode ahead, I followed on my sturdy mule with the barometer, Guirado bringing up the rear. Occasionally a snatch of song would awaken the echoes above the clattering of the hoofs of the mules over the rocks.

As we approached the most dangerous place, where the path went down a steep slope, over and among large bowlders, as high as the horses on each side, and piled in the path, we were stiller. Suddenly a crash—Peter's mule caught his foot between two rocks and fell, Peter pitching headlong over his head on the rocks. How he escaped unhurt, I cannot imagine, yet he was but slightly bruised. The poor mule fared not so well. His forefoot was held between two rocks as in a vise. He had fallen over below, and was hanging much of his weight on that foot. We could budge neither the rocks nor his foot. We thought his leg broken, and saw no way of releasing him. He was a valuable mule, worth $150 or more. We tugged, toiled, pried with levers, dug, all to no purpose. He made a tremendous effort, but only made matters worse, twisting his leg nearly around. After lying so for some time, while we worked frantically, he made another effort, tore off his shoe, and got out—strange to say, *uninjured*. A horse would have been ruined. We washed his foot and leg in the brook, led him a mile or so, and soon he scarcely limped. Peter then mounted him and rode him home to camp.

It is in such places that the superior sagacity of mules over horses is seen. Much as is said and written about the sagacity of horses—poets sing of it and romance writers harp on it—it is far inferior to the much abused mule. This fellow, as he lay so helpless, instead of struggling frantically, would get all ready and then coolly exert his greatest strength to get his foot loose, but not when we were working with it. Although he groaned pitifully and gnawed the ground and rocks in his intense pain, he did not bite us, but would put his head against us and look up most wistfully.

And while on the subject, a word more about our mules. I have an old white mule, I think the oldest in the lot, but can't tell her age. She is only thirteen hands high, but is very stout. It would take two whole letters to give the instances of her sagacity. How sure-footed she is on a mountain trail—she never treads on a loose stone or on a smooth one, never treads in a hole where her feet may get caught, never puts her foot in a mud-hole until she tries if it is miry or not. I carry a barometer on her; she is just the mule for my use, gentle, sure-footed, true, sagacious, but *awful* homely. Some of our mules are fine ones; it is considered a valuable lot.

We got back at nine o'clock in the evening, and found that the steamer had arrived, and with it Professor Whitney. He stayed until Saturday morning early, then left again. We expected that he would remain with the party, but I think he was decidedly pleased with our work, and concluded to leave the party in my command in the Coast Range, while he looked up the coal regions of Mount Diablo and the gold regions farther north. We were very busy the four days he was here, packed up eleven boxes of fossils we had collected, and did much work, explored some, he going over the ground we had before visited.

On Wednesday, March 20, we walked along the beach to the asphaltum beds, and over the hills, a long walk of eighteen or twenty miles. Some interesting things turned up during the day. We found a whale stranded on the beach. I had no idea how huge they look when fresh. He was forty-five feet long, and about thirteen to fifteen through from back to belly. Such a pile of flesh I never saw in one mass, it was equal to at least half a dozen large elephants. We also found a crab that was just shedding its shell. We secured it in its soft, velvety, new shell, and the old one alongside. Not the least—a half-naked Indian fishing on the shore had caught two of the remarkable *vivaparoa* fishes, which instead of laying eggs bring forth their young alive, a thing nowhere known except on the coast of California. We saw the mother fish with a number of little ones.

Monday Night, March 25.

THIS is so notorious a place for horse stealing and robbery that we have kept guard since we have been here. We divide the night into three watches, the first until twelve, the second until three, the next until morning. I have the midnight watch, and it is so clear and light by the bright moon that I will add a word. These clear nights it is rather pleasant than otherwise. The clear sky above, the twinkling stars—to watch them rise over the mountains in the northeast and sink out of sight in the west, to watch the moon rise from the waves as it does in its wane—all this is pleasant. And the roar of the surf, coming up like that of some mighty waterfall, is continual music. But not so the foggy nights. For near two weeks the air was very foggy and wet—unpleasant wet days followed by wetter nights. Then the watches were anything but pleasant—sky black, nothing visible at any distance, beard and clothes dripping with wet, the camp fire light scarcely penetrating the gloom for a few feet, the roar of the ocean coming up dull and sullen on the thick air. Watching such nights is by no means poetical, and it awoke musings and memories of a very different fireside but a short year ago.

Sunday Evening, March 31 (Easter).

WELL, I have another week's experience to detail—rich enough in events to make two letters, but I cannot write all. A trip in the mountains of three days; Holy Week among the Catholics; men getting drunk, fighting, and in jail, etc. I have spent half of today getting one of my men out of jail where he got put yesterday for fighting. But to take things in order.

A coal mine is reputed to exist a short distance from here in the mountains; a company has been formed, some hundreds of dollars expended, and many look ahead to speedy wealth. It was desirable that we visit it. "Rich indications," "Not more than fifteen miles distant," "Possibly farther by the trail," "Good trail over the mountains for a horse," "Railroad practicable," etc., were a few of the statements of anxious stockholders. We decided to visit it. Rains and bad weather prevented for some time—the good weather we had the week that

Professor Whitney was with us left the day that he did—but it looked better, so Tuesday, March 26, Averill and I started. We had an experienced mountaineer guide, the original discoverer and, as consequence, owner of several "shares." He rode a good horse for such a tramp; I rode my trusty mule; Averill took "Old Sleepy," quite a noted mule in our flock, one supposed to be peculiarly adapted to such a trip. We were up at dawn and off with the sun—saddlebags on the saddle, our blankets, three days' provisions, coffeepot, etc., strapped behind, knives, pistols, and hammers swung to our belts—all equipped in good style.

We rode directly east about six miles, past some fine ranches and across two or three small streams that issued from the mountains, with some timber—almost a forest in places in the wetter soil—then struck up a canyon into the heart of the ridge. Such a trail as we found that day! The worst I had traveled before was a turnpike compared with that. Now following along a narrow ledge, now in the brook over bowlders, now dismounting and jumping our mules over logs, or urging them to mount rocks I would have believed inaccessible—yet this was "pretty good yet," our guide told us. Arrived near the head of the canyon, high, steep slopes hemmed us in. I saw no means of getting farther, but the "trail" ran up that slope. I saw the rocks rising near or quite a thousand feet above us, at an angle of forty-five to sixty degrees in many places, and up this the trail wound. I dismounted, but our guide said, "Oh no, ride up, man, it's not bad!" Averill drew the girth tighter on Old Sleepy and started. I preferred leading my mule. Up a few hundred feet, going up over a steep rock, down went Old Sleepy. For a moment I expected to see him and Averill roll into the canyon two or three hundred feet beneath, but he caught against a bush. We helped him, but after that Averill took it afoot. The trail ran up by zigzags, at an actual angle of thirty degrees average, and in places over forty degrees! We measured one slope of several hundred feet where the trail was at an angle of thirty-seven degrees, the slope itself much steeper. You will appreciate this better when you remember that the roof of a house with "quarter pitch," the usual slant, has an angle of but about twenty-four degrees.

We crossed the summit at an elevation of 3,500 to 3,700 feet, but clouds had enveloped us for the last thousand feet—damp, drizzly, and thick, decidedly unpleasant—and shut out the fine view we had when we returned. The north slope of the ridge was less steep. It was covered with a very dense chaparral, about twelve feet high, so dense that no animal could get through it in many places, but a good trail had been cut through. We descended about two thousand feet into a deep canyon, then struck into another, finally crossed the Santa Inez River and struck up another canyon in this wild labyrinth of mountains. We found deer and wolf tracks in abundance, and a few grizzly tracks. We rested an hour and lunched on a little grassy spot, then pushed on. Near night we came up to a deserted cabin, an old Indian "ranch" called Nahalawaya (*Nah-hah-lah-way-yah*). As it looked like rain and no other shelter could be expected, we concluded to stop until the next morning. We soon had a fire in its old fireplace, a good lunch, and a sound sleep that night under its hospitable shelter; our animals found some poor grass.

We were up early in the morning, and pushed for the "mines." We were getting used to a "hard road to travel," but this beat our yesterday's experience. We passed up a canyon, in which we surmounted obstacles I would have thought entirely impassable. It was perfectly astonishing how the mules would go. We would get off, tie up the reins, lead them up to a rock; they would eye it well, and coolly, with a spring or two, mount rocks nearly perpendicular, six to ten feet high, if they could only get a foothold. I wondered how we were to get back, but on returning they would slide down coolly and safely.

We followed this canyon a few miles, then crossed another ridge near or quite four thousand feet high, possibly more.

From this summit we had a grand view of the desolate, forbidding wilderness of mountains that surrounded us. We then sank into another canyon, 1,500 or 2,000 feet, followed it up, and at last arrived at the mines near noon—thirteen or fourteen hours in the saddle to overcome a distance of about twenty-four or twenty-five miles from camp, half of that time on not over six miles of trail.

We found the mines positively nothing. A few seams of coal from one-eighth to three-quarters of an inch thick, and those short, standing in perpendicular strata of rock, were the "indications." A sort of "pocket" had furnished about a peck of coal or less, on which the company had been formed, a shaft commenced, four hundred dollars expended, and great prospective wealth built up—to such a feverish state is the whole community worked up here about mines. I did not tell the stockholders how *very* slim the indications were, on my return, but slicked it over by merely telling them that they would not find the coal in profitable quantities, that the difficulty of access, position of the strata, and necessary thinness of the beds would prevent the mines being profitable.

We found tools, drills, picks, shovels, hammers, crowbar, tent, provisions, etc., which had been left by the men when the work was deserted some months ago. We saw bear and wolf tracks along the stream—one bear must have been truly huge —and deer tracks without number. We once came on a flock of ten beautiful deer. Averill tried to get a shot with his pistol but could not succeed. We returned to the cabin again that night, as we found no grass for our hungry mules beyond that point; and lucky we were, for it rained nearly all of that night —decidedly damp to be out.

The next day (Thursday) we returned. We got a fine view from the ridge—the plain and ocean on the south, the mountains on the north. We were up to the height of snow on other peaks, but we found none. This trail was cut two or three years ago to carry the mail on horseback to Fort Tejon, but never used, and rains and neglect had reduced it to its present condition. The trip was a tiresome one, but most interesting. Could it all be put on one ascent and descent, I doubt not that it would make a twenty thousand or twenty-five thousand foot climb up, and the same back.

Only six things were lacking to make it a very "thrilling tale." First, to have had it rain hard, and we, with no shelter, to lie out in it—it rained enough but we had the shelter—and to have had said rains so swell the water in the canyons that we could not get out, but have to subsist on mule meat, roots, and "yarbs." Second, to have met and vanquished several grizzlies, and to have returned triumphant with their skins, and lots of wounds and bites, as trophies—we saw only their tracks. Third, to have had a mule and its rider go tumbling down some precipice, both to be food for the buzzards and a warning to the venturesome—but Old Sleepy only slipped. Fourth, to have killed sundry deer with our pistols and returned fat and portly on eating so much venison—we only saw the deer, and got back hungry. Fifth, to have had our mules get away, and we have to foot it home, packing our saddles, blankets, and specimens on our own backs—alas, they were well picketed. And sixth, to have lost our teapot on the first slope and be obliged to drink cold water—we lost only the lid. Owing to these failures I have no thrilling tale to write. Good night!

April 4.

I PROMISED to tell something about the festivities of Holy Week in Santa Barbara. The whole week was a week of festival, but I was in town only the last three days. Friday I was in camp most of the day, but there was the ceremony of "lying in the sepulcher," "washing of feet," etc. The town seemed like a true sabbath day. Among the true Catholics, men are not allowed to ride on horseback—formerly policemen prevented any from so riding on Good Friday—and but few horsemen are seen now.

I went into town in the afternoon. In the church the altar was trimmed off with a profusion of flowers around the sepulcher, tapers were burning, the windows were partially darkened, and a few of the devout were praying to their favorite saints. We rode to the Mission church—its windows were darkened by thick curtains and the many candles at the altar did not light up the obscurity. Many Indians were about. Within, a number of Indian women were kneeling before a shrine; one would lead off with the prayers and all join in the responses. Their pensive voices, the darkened vast interior, the pictures and images obscurely seen in the dim light, the tapers of the altar, the echoes of their voices, the only sounds heard breaking the stillness, produced an effect I can easily conceive most touching to the imagination of the worshipers. Some of the Indian girls and half-breeds were quite pretty, but the majority were decidedly ugly.

Saturday I attended Mass in the morning. The curtains were removed from the altar, and more ceremonies were gone through with than I can detail, but they differed very materially from the ceremonies at Munich on a similar occasion. The music was the best I have heard in California. It began with an instrumental *gallopade* (I think from *Norma*), decidedly lively and *un*devotional in its effect and associations. But other parts were more appropriate. As the priests chanted the long list of saints in order, the response, "Ora pro nobis," by the audience (I can hardly say congregation) and choir was very pretty indeed. At the unveiling of the altar, two lovely little girls dressed as angels, with large white swan wings upon their shoulders, one on each side of the altar, looked most lovely. They stood there as watching angels during the ceremonies.

Not the least interesting to me were the costumes. Standing, kneeling, sitting over the floor were the people of many races. Here is a genuine American; in that aisle kneels a genuine Irishman, his wife by his side; near him some Germans; in the short pew by the wall I recognize some acquaintances, French Catholics, also an Italian. But the majority of the congregation are Spanish Californians. Black eyes twinkle beneath the shawls drawn over the heads of the females, and glossy hair peeps out also, and the responses show sets of pearly teeth that would make an American belle die with envy should she see them. A few bonnets and "flats" tell of American or foreign women mingled with the crowd. Here is a group of Indians, the women nearly conforming to the Spanish dress, only their calico dresses are of even brighter colors—all are dressed in holiday clothes. Here is a man with Parisian rig; there one with the regular Mexican costume, buttons down the sides of his pants; beside him is an Indian with fancy moccasins and gay leggins; behind me, in the vestibule, looking on with curiosity, are two Chinamen. No place but California can produce such groups.

In the afternoon there were horse races, etc., but I did not attend them. Thereby hangs a tale, but of that more anon.

Saturday evening I was in town again. A gay, jolly crowd were in the streets, the *buccaros* on their horses, and such horsemen as only Mexico or similar countries can show! Such feats of horsemanship one cannot see in a circus—trying to throw each other from their horses, or throw their horses—it looked as if somebody must be killed, but of course nobody was seriously hurt in their rough sport. I will tell another time of the horsemanship here. Now for less poetry.

Saturday afternoon I was busy at camp, but the men were in town. Peter returned—said one was "jolly," and the two others getting decidedly "mellow." Soon they returned, minus Mike, one of them decidedly "over the bay"; Mike had got in the jail. He had not been drunk before with us, but perfectly sober and steady. He is a zealous Catholic, and today celebrated too hard—got pugnacious like all drunken Irishmen, pitched into everybody, whipped and rolled ignominiously in the dust the fat Dutch justice of the peace who came to arrest him, much to the amusement of the crowd, but was finally overpowered, bound, and taken off to jail. Now this excited one of the other men. He was cool at first, but seeing Mike tied, he took out his knife, cut him loose, and was about to take him to camp when the sheriff came and carried Mike to jail. The others then came to camp.

I was decidedly annoyed by this turn of affairs, and while ruminating on this new episode in my company, I was waited on by the sheriff and a deputy who came to arrest the other man on charge of "assault and battery, aiding in escape, prevention of arrest, etc." I kept him out of the lock-up after much palaver by going his bonds for appearance. I tried to get Michael, but did not get him out on bail until the next morning (Sunday). Finally, on Monday, with the aid of an ingenious American lawyer, I got them off with the payment of costs, about twenty or twenty-five dollars. A decidedly unpleasant affair all around, but a severe lesson for both of them.

During our stay in Santa Barbara we bought milk at a house on the edge of town. Sunday afternoon the woman and baby, little girl, and daughter (young woman), came into camp and paid us a visit of several hours. They sat down in the tent and chatted away very lively—Spanish, of course. The mother was middle-aged, very dark, as dark as a dark mulatto, but with Spanish features, with a shade, perhaps a quarter, Indian—hair black as jet, eyes even blacker, teeth like pearl. She had a lively babe of eight months, which she fed in the "natural way," decidedly unreservedly. The senorita was quite pretty, hardly as dark as her mother, with clear, olive skin, hair like a raven (not *crow*), and the finest sparkling eyes, and pearly teeth. She chatted, smoked *cigarritos*, and apparently enjoyed the visit as much as we did. Some of these Spanish girls are pretty, especially the hair, eyes, and teeth.

NOTES

1. The Presidio of Santa Barbara was founded in 1782, but the founding of the Mission was delayed until 1786. The first buildings were destroyed by earthquake; the main building standing in 1861 dated from 1820. Missionary activities declined until they ceased in 1856, but the building continued to be used by a missionary college.

2. "The old Spanish Franciscan mentioned in the narrative of Brewer was the Very Rev. Fr. José María González Rúbio, O.F.M., who died here November 2, 1875. The young Irish Franciscan was the Rev. Fr. Joseph Jeremiah O'Keefe. He was a native of San Francisco, Calif. He departed from this life at St. Joseph's Hospital, San Francisco, on August 13th, 1915" (letter to the editor from Rev. Fr. Zephyrin, O.F.M.; see also Fr. Zephyrin Engelhardt, O.F.M., *Santa Barbara Mission* [San Francisco, 1923]).

3. This vine appears to have been older than Brewer supposed. It is said to have grown from a slip cut from a vineyard at San Antonio Mission, Monterey County, and planted before 1800, perhaps about 1796. The planter of this vine was Doña Marcellina Feliz de Dominguez, wife of an old soldier, José María Dominguez, who came up to Alta California with one of the earliest expeditions from Sonora, before 1780. He died in 1845 at the age of nearly 100 years. Doña Dominguez died in 1865 at the age of 102, or, according to some, 105 years. The couple had fourteen children, and at the time of her death there were three hundred descendants. (San Francisco, *Daily Evening Bulletin,* May 26, 1865.)

CHAPTER VI

THE COAST ROAD

California Caravans—Gaviota Pass—Santa Inez Mission—Foxen's Ranch—A Wagon Wreck—San Luis Obispo—The Santa Lucia Mountains—On "The State of the Union."

Camp No. 22, near Santa Inez.
April 7, 1861.

WE are camped about four miles north of Santa Inez Mission at the ranch Alamo Pintado. It is a quiet, hot Sunday afternoon, 98° in the tent, so I go out and write in the shade of a tree, where it is cooler, only 90°; yet last night was a cold chilly night, almost cold enough for frost.

Tuesday, April 2, early in the morning, we started north. Santa Barbara County until now has been nearly isolated from the country around by rugged mountains. During the last few months thirty or forty thousand dollars have been expended by the county on getting a good wagon road through from San Luis Obispo on the north to Los Angeles on the south. The southern part of the road is not yet finished but the north end is, and a fine road connects it with San Luis Obispo. This road we are following—sometimes it is a mere obscure trail across the grassy plain, scarcely visible yet for want of travel, at others well engineered, built over and along high hills and through deep canyons at great expense and labor. Fine bridges of wood span the streams and gulches, the first bridges we have seen in the southern country. Our mules are shy of these, to them, strange structures.

We came on about twenty-five miles and camped on the seashore, where a fine stream emerges from a canyon, on the ranch of Dos Pueblos.[1] During the day's ride, the high, rugged mountains ran nearly parallel with the coast, from one to six miles distant from the sea. The space between was made of gentle slopes, and very green grassy hills, on which were a profusion of wild flowers with brilliant colors. Immense herds and flocks of cattle, horses, and sheep were feeding. We passed one herd of over 2,000 head, kept in a close body by a large body of *buccaros* (herdsmen on horseback), while the owners were separating out cattle for some drover to take north. This fertile, lovely strip is well watered by frequent streams that come down from the mountain at intervals of every two or three miles, and is all occupied, either by *rancheros* under old Spanish grants or by the recent, wandering, worthless American "squatters." I found a fine mastodon (or mammoth) tooth during the day's ride.

We camped in a lovely spot, where the sea was unusually rough, just at a point. The surf was heavy, and its thunder lulled us to sleep.

We spent the next day there looking up the adjacent hills. The road for the first three days from Santa Barbara was more traveled than any we had seen before. The first *Overland* through Santa Barbara, on Monday evening, April 1, was celebrated with the firing of cannon, etc. Many emigrants were passing over the road. One long train was bound for Texas, sick of California. One meets many such uneasy families who have lived in Ohio or Michigan, then Kansas or Iowa, then California or Oregon, and now for Texas or somewhere else. Several small companies of five to ten passed us on horseback, natives (Spanish Californian), traveling for pleasure or business, on horseback with one or two pack-mules along with baggage. The women wear black hats with feathers, much like a Kossuth hat, ordinary (not long "riding") dresses, often of gay colors. They ride with the feet on the *right* side of the horse, sitting nearly squarely crosswise, both feet hanging down as if they were sitting on a bench. Often a strap ornamented with silver and tassels, or a mere red sash, is tied over the lap, holding them firmly to the saddle. No horse can throw them; they would go sweeping past us at a California gallop. We came on two or three parties at their noon lunch. They will ride sixty or seventy miles in a day and not complain.

Thursday, April 4, we came on about twenty miles, through the Gaviota Pass, crossed it, and camped in a most lovely valley on its north side, at the Nojoqui Ranch. After leaving the previous camp the mountains began to approach the sea; the green hills were scarcely half a mile wide, the barren, rugged, sandstone hills rising immediately back of these slaty green hills. This sandstone ridge is a continuous one, and has but one break, the Gaviota Pass, for a hundred miles or more. At the Gaviota a rent or fissure divides the ridge, but a few feet wide at the narrowest part and several hundred feet high. The road passes this "gate" and then winds up a wild rocky canyon, the wildest pass I have yet seen here. The mountains rise very rugged about 2,000 feet on each side. The narrowest part is not the highest; the road continues to ascend for about six miles where we cross the summit. A horrible trail ran through this formerly, but now the road is good.

As I could not agree with Dr. Antisell, of the Pacific Railroad Survey,[2] in his notions of the geology of the pass, we camped, and the next day Averill and I rode through it, and climbed two high hills. We had fine views of the ocean over the pass, and the labyrinth of hills to the north of the Santa Inez River. We got some fossils and I killed a rattlesnake that we came upon—he was inclined to get away at first, but fought bravely when attacked. He was not very long—two and a half feet—but was very thick—half as large as a large man's wrist—and had eight rattles.

Saturday, April 6 (yesterday), I visited the hills two or three miles north of camp in the morning, alone, for Averill was under the weather. In the afternoon we raised camp and came about ten or twelve miles to our present camp, near Santa Inez Mission.

Camp No. 24, Nipomo Ranch.
Wednesday, April 10.

We camped on Saturday, April 6, about four miles from the Mission at a little ranch owned by an American—the Ranch Alamo Pintado. It was a lovely spot. Large oaks scattered here and there, the green grass beneath, and the great profusion of flowers, made it look like a fine park. There are two species of oaks here.[3] One is an evergreen, with great spreading branches, gnarled and knotted trunks, worthless for timber because it is never straight and it has so many branches, but beautiful, as a tree, with its dark green foliage. The other is a deciduous tree. Like the first it branches low down, so it, too, is useless for timber. It is a most beautiful tree, however, the large limbs branching in great curves—not Gothic arches like the elm, but great round curves, great Roman arches of thirty to fifty feet span, coming down again near the ground. Sometimes such a limb will be thirty feet high twenty or thirty feet from the tree, and again near the end almost touch the ground. A tree close by camp, under which I wrote on Sunday, had a head of over a hundred feet in diameter, and the trunk was about fifteen feet in circumference in the smallest place below the branches. A trailing lichen hangs from every branch, delicate as lace, of a greenish gray color, swaying with every breeze—the effect is beautiful.

On Sunday morning Guirado and I rode to the Mission.[4] Here was quite a town in former times, but, like the rest of the missions, it is in ruins now. A large, old church stands, but there were scarcely more than a dozen persons—two or three Californians, and a few groups of Indians—kneeling in the vast church. It looked desolate and lonely. The church was highly painted, pictures hung on the walls, but all was dilapidated. The bells were of sweet tone—we could hear them at our camp.

Alongside of the church is a college, which once had a hundred or more students.[5] It now has but eleven, three of whom are Guirado's brothers. The place is in complete ruins. Not over half a dozen houses are inhabited, the rest going to ruin. Some are roofless, and the *adobe* walls are crumbling with every rain; some, mere banks of dirt or clay, the abode of great numbers of ground squirrels that burrow in the ruins. The old corral is torn down in places, the old threshing floor broken in —all in decay. Long lines of water courses, *sanchas* or small aqueducts, some of them miles in length, laid in stone and cement, to supply the town and irrigate the fields, are now dry and broken. The vineyards are all gone, now dry pastures, and the olive and pear trees are dead. No town is growing up in its stead. A fine cement reservoir and a mill alongside are in ruins. It is the same story that I have written before of other missions.

Here, in this county, is a great field for missionary labor— not a single Protestant church or congregation in the county, not even a mission station, the prestige of the Roman church failing, the *padres*' power lost, a race growing up more wicked, desperate, immoral than any that has gone before. The religious destitution and moral state of the county (Santa Barbara) is not easy to describe. It is the most Spanish, or Mexican, in its character and inhabitants of all counties in the United States.

Monday, we went on to Camp No. 23, at Foxen's Ranch, about twelve miles. Foxen is an old Englishman who came to America a mere boy—came as a sailor to the western coast, was hunter and trapper, then married a Spanish wife and settled on a ranch. He has been in California over forty years. He was decidedly an original character.[6] We camped near his house, for there is only water at the ranches, at intervals of six to ten miles on an average.

The hills we passed among during the day's ride were covered with pasture, or grass, with a great profusion of flowers. Sometimes we went along a valley with fine scattered trees. But the road was worse and our erring wagon wheel once more began to show signs of weakness and Pete mended it again with thongs of rawhide. I examined the region around and found many fossils, among them a portion of a fossil whale, dug up at the ranch, the bones very stony.

Tuesday, April 9, we came on here, to Nipomo Ranch, about twenty-two miles. Our road first wound through some valleys, then struck into the valley of Santa Maria River. This river is now entirely dry, not a drop of water, its valley a perfectly level plain, with the exception of an occasional terrace or old riverbank, about six or eight miles wide. We struck down and across this valley about ten or twelve miles, a most tedious ride. We were dry, but no water was met with for the twenty-two miles traveled except a sink-hole with stagnant, alkaline, dirty, stinking water. Our lunch of dry bread and drier cheese, which we ate as we rode along, was hardly "sumptuous."

The ride was very tedious as we wound our slow way over the plains, here a drifting sand, there a partial pasture. Nothing relieved the eye; the senses tired with the level scene. The profusion of flowers, beautiful elsewhere, now tired us with their abundance and their sameness; wind filled the air with gray dust, sometimes shutting out the sight of the hills like drifting snow. Lovely green hills lay on each side at the distance of a few miles. Many cattle and horses were feeding on the hills or on the plain. Water every four to six miles in the side canyons was sufficient for them. They seemed mere specks on the plain—a herd of a thousand like a few flies on the floor. This valley runs to the sea, and in that direction a mirage kept ahead of us in the hot air—a very good appearance of water, but not nearly so perfect as I saw on the plains in Bavaria.

How we hailed the first tree of shade we came to, a fine sycamore on the dry riverbank, with fine shade—the first we had seen for fourteen miles. We stopped a few minutes, then pushed on, crossed the dry bed of sand half a mile or more wide, and struck up a side canyon about two miles, to water, at this ranch. To be sure, the water is alkaline and stinks from the droppings of the many animals, but made into tea it is drinkable, and we can stand it if those who live here can. They, however, have a "spring," so called—a hole dug in the bank half a mile or more from here, where the water is cleaner. Bad water has affected the bowels of most of the party except me—I escape any material bad effects.

Today, Averill and I have been over the hills near here, exploring the geology and botany, quite a ride and walk. We came once on a large coyote, or wolf, and got a pistol shot at him but did not hit him. He was a big fellow, and two more were seen near camp by the other men. A snake five or more feet long, but harmless, was killed near our tent just at dark.

I forgot to mention that I killed a rattlesnake at Camp 22. He was within a rod or two of the tent, a small one, of another species from the first. There are several species of rattlesnakes found in this state, but all are dangerous. This fellow had fangs sharp as needles. We examined them. When not irritated they are covered with skin, like the claw of a cat, but are erected when required for use. This fellow, like the last, did not show fight until after he was attacked.

Camp No. 26, near San Luis Obispo.
Sunday Morning, April 14.

WE were at Nipomo Ranch when I last wrote. Thursday, April 11, we came on. After leaving Santa Barbara County the roads were again horrible—no road in fact, but a mere trail, like a cow path, hardly marked by the track of wheels, and often very obscure. We crossed gulches down almost straight on one side, then "ker-chug" in the bottom, then up as steep on the other.

Our wagon is like the Overland stages, square covered body, hung on straps or "thorough-braces," as they are called. It is too light for our purpose, although it stood the road, but that weak wheel groaned and complained at times, notwithstanding its rawhide supports.

We wound among hills, and at last at the Arroyo Grande,

had a bad hill to descend. We had come a longer road because the "hill was easier" this way. Well, we got to the "easy" hill. It was about five or six hundred feet high, the sides at an angle of about thirty degrees, down which the road ran in "crooks" —now one side up, now the other. No work had been expended on it, so it was always very sidling, and very steep at the same time. We chained both hind wheels, and for a time all went well. We had descended about one-third of the way, sliding, slipping, dragging, when, quick as a flash, over went the whole concern. Pete and Mike escaped from under the pile by a miracle of agility that would astonish a circus performer. Such a pile! The wagon caught when completely upside down, the wheels high in the air. The mules were tangled in the harness, one on his back, his mate standing over and astride him. One of the wild leaders got loose, and was lassoed by Guirado a mile distant.

We got up the mule, then attended to the wagon. I never before unloaded a load from the bottom—carpetbags, instruments, tools, provisions, tent-ropes, botanical papers, etc. Two or three large boxes had been filled with rocks and fossils, each specimen carefully wrapped in paper and packed, now in one promiscuous pile. Frying pans, pails, basins, soap, etc., completed the picture. Michael had, at last camp, providently boiled a huge dish of applesauce for our supper that night. It, too, played its part in the confusion, and sundry very suggestive looking spots as a consequence adorned our carpetbags and furniture generally. (Themes for more papers on "The Distribution of Species" than even the famous antiquarian stone of Mr. Pickwick.)

We unloaded, turned the wagon up again, found the top a total wreck with no insurance, but no other serious damage, loaded up a half, and camped at the foot of the hill on a pretty, grassy bottom by the finest stream of water we had seen for some time. After dark we sat by our cheerful fire and talked over the adventures of the day and laughed at our mishaps, troublesome though they were. I had the curiosity to go back to the hill the next day, when we packed down on our backs a part of the baggage, the wagon top, etc., and measured the angle. In one place for some distance the road descended at *an angle of twenty-nine degrees!* Yet this is the "better" road to San Luis Obispo.

Friday, April 12, I sent the wagon on here with a part of the load, about twelve miles. Mike and I remained. The wagon returned and we came on yesterday afternoon.

The camp was in a pretty spot, on Mr. Branch's ranch. He is an American and has a ranch of *eighty thousand acres*, well stocked with many thousand choice cattle and horses, comparatively well watered, and fertile.[7] I explored the region around and called on him at his house. He lives quite stylishly for this county—that is, about half as well as a man would at home who owned a hundred-acre farm paid for.

The advance camp carried the tent, so Mike and I had to take the open air. Rolled in our blankets on the green sod, the stars above in the clearest sky, we slept better than if beds of down supported us and a canopy of silk covered us. I love to watch the stars in the open air as I go to sleep, and see them greet me if I awaken in the night. But the nights are cold here under this clear sky. The thermometer sinks generally forty or fifty degrees lower than it was by day—90° in the shade in the afternoon, and 38° or 40° at night. As a consequence, dew falls, very heavy, almost like rain, which is the most serious drawback in sleeping out. We put an India rubber or oilcloth over us, and the water flows from this like rain, yet it is not so bad as you would think.

We are camped about two miles from San Luis Obispo, and will remain here two or three weeks. We must meet Professor Whitney about 250 or 300 miles north, near Monterey, about the middle of June. He is north now, where severe rains have deluged the country.

My health is excellent. The chaparral was so bad for pants

that I bought three buckskins. Peter "smoked" them as the Indians do, and from them I have made a splendid pair of pants, which defy chaparral, are healthy for rattlesnakes and tarantulas, and please me much every way, except that they are *not* particularly ornamental—in fact, I would hardly attend a party East in them. The hot sun has given the color of well-smoked ham to my hands and face; my hair nearly came out, so I have it cut short, the longest scarcely half an inch long. How I would like to happen in on you. See if you would recognize the *captain* of our geological party.

Camp at San Luis Obispo.
April 23.

SAN LUIS OBISPO town lies in a beautiful, green, grassy valley, about nine miles from the sea. A ridge of the Coast Range lies to the north, a continuous ridge, about three thousand feet high, with a single pass through it near town. The pass is about 1,500 or 1,800 feet high. This valley is more like a plain, from four to six miles wide and fifteen or twenty long, running northwest to the ocean. A range of hills lies to the south, separating it from the sea in that direction.

Through this plain rise many sharp peaks or "buttes"—rocky, conical, very steep hills, from a few feet to two thousand feet, mostly of volcanic origin, directly or indirectly. These buttes are a peculiar feature, their sharp, rugged outlines standing so clear against the sky, their sides sloping from thirty to fifty degrees, often with an *average* slope of forty to forty-five degrees! One near camp is beautifully rounded, about eight or nine hundred feet high, and perfectly green—scarcely a rock mars its beauty, yet the rock comes to the surface in many places. A string of these buttes, more than twenty in number, some almost as sharp as a steeple, extend in a line northwest to the sea, about twenty miles distant, one standing in the sea, the Morro Rock, rising like a pyramid from the waters.[8]

We arrived on Saturday, April 13, in the afternoon. Sunday I remained in camp until the afternoon, when I went into town, about two miles. The old church is much like the other missions, except that the ceiling is made of short but wide split boards, and these are alternately painted in different bright colors, probably an Indian fancy, but by no means pleasing to the taste of Americans. The town looks more South American or Spanish than even the others we have seen. It is a small, miserable place.

Monday, April 15, we climbed a butte east of town, 1,200 to 1,500 feet high. A most lovely view we had from the top. The mountains to the north were covered with clouds at their summits, but their green sides, the great green plain to the south and west at our feet, the curious old town, the rugged buttes rising from this plain, the winding streams in it, all aided in making a lovely picture. A range of hills along the coast terminated the valley, but we were higher than they and could see the ocean beyond, covered with a fog near its surface, white, and tossed by the wind into huge billows.

That night fog again settled over the plain, as indeed it did every night during the week, but the fog cleared up sometime during the forenoon. The nights, however, were cold, wet, and disagreeable.

Tuesday we rode to the sea, and examined the coast hills. Wednesday we examined some of the buttes on the plain. Thursday we rode on to the summit of the pass, nine or ten miles, and visited the adjacent hills. Friday we visited a ranch ten miles distant, but as we expect to go there again I will defer description. We got somewhat wet by a rain that day, and rode the ten miles in wet clothes. Saturday was another wet day, but in the afternoon we examined and climbed a very rocky butte about four miles northwest of camp. A fog came on and shut out the view just as we reached the top. Sunday was a better day, but I spent it quietly in camp until the afternoon, when I rode into town and mailed letters; then rode to a ranch near camp to see about an orphan child, at the request of a lady in San Francisco.

We had been waiting for better weather for climbing and measuring the Santa Lucia Mountains. As Monday, April 22, was a fine day, I got an early start, taking Guirado with me, and leaving Averill to observe barometer at camp—of course, carrying another barometer along with me. We rode about five miles to the base, left our mules, and climbed to the summit in four hours. For the first two thousand feet the way was up a very steep but perfectly grassy slope, covered with wild oats about a foot or foot and a half high, green as the greenest meadow. Then we struck a low chaparral. We gained the summit of the first ridge, but as usual a higher one rose farther toward the center of the chain, so we descended about five hundred feet, got on a transverse ridge, and in due time reached the highest peak. It was 2,605 feet above camp, or about 2,900 feet above the sea. The day was lovely, cool, and the air clear— not so clear as it often is here, but it would be called very clear at home. Objects twenty or twenty-five miles distant seemed as plain as they would through four or five miles of our air at home. For example, the breakers on the shore were *perfectly distinct twenty miles distant!*

The view was very fine, finer than we shall have again soon. To the south we could see plain beyond plain, and hill beyond hill, although beyond the Cuyama Plain, thirty-five miles distant, things were indistinct through the dust from that plain. To the southwest and west lay all the lovely plain of San Luis Obispo, the buttes rising through it—over twenty were visible —brown pyramids on the emerald plain. Beyond were the coast hills, while beyond all was the blue Pacific, stretching away to the horizon. To the northwest was our chain of mountains; north, the valley of Santa Margarita and Salinas Valley, bordered with myriad hills, stretching away for sixty or seventy miles. We sat and contemplated the scene for over an hour before leaving.

Each mountain ascent has something peculiarly its own to distinguish it from the others. The feature of that day's trip was the unpoetic one of *rolling rocks down the slope.* Nature seemed to have made it for that—a smooth, grassy slope, with few obstructions on it, and plenty of rocks at the right place near the top. We could start them, they would go about six hundred to nine hundred feet at an angle of forty-five or fifty degrees, then roll down a slope of twenty-five to thirty degrees, going a mile from their starting place and falling probably nearly two thousand feet. Their velocity was incredible. As they would roll, large, angular fragments bounding in immense leaps through the air, they would whistle like cannon balls. *We could hear them whistle half a mile!* Their leaps would surpass belief. After rolling many, I went down to the foot of the first slope to see them come by—Guirado starting them. Some came within thirty feet of me; their whistling exceeded my belief. They would leap through the air on meeting slight obstructions—pieces flying off would fly a hundred feet in the air, whistling like bullets. One stone of over a hundred pounds leaped close to me. I measured the leap; it was *sixty feet!* Another, much larger, perhaps four hundred pounds, came thundering down, struck a flat stone bedded flat in the soil, which it crushed into a thousand pieces, then bounded *one hundred feet*, and then took its straight course down the slope.

Sunday, April 28.

I WILL answer some inquiries made in letters from home.

First—as to whether we have "camp bedsteads?" No, by no means—*State* officers can't afford such luxuries, only Uncle Sam's men can indulge in them. Each man has two pairs of heavy blankets, and an India-rubber sheet or oilcloth. The latter is spread on the ground, to keep us from the wet, and we sleep on that, rolled in our blankets. The colder the night the more we use above and the less below. When we sleep out in the open air we generally put an oilcloth or old coat over us to keep our blankets as dry as possible, for the dews are like rain these clear nights. One soon gets used to the ground, but it is often hard, and oftener rough with stones or cattle-tracks. This last is the most serious inconvenience. Often a great hummock or hollow is found just under one, and one must adapt himself to the ground. For pillows we use coats, saddlebags, or something of the kind—one learns to sleep on a hard pillow, only it makes the ears sore and bruised.

Second—tent. We first used two; the larger is discarded now. We use a Sibley tent, of government model, built after the style of an Indian "lodge," round, with one pole only, in the middle; and after our experience of blowing down in the rain we strengthened this with three guy ropes or stays. These latter are also handy to hang shirts on to dry, towels, etc. The canvas closes into a ring at the top, about two feet in diameter, which is suspended to the top of the pole by short ropes. This leaves a hole in the top for ventilation on hot days. It is closed by a hood or fly.

Third—the barometers. These are mountain barometers. The glass tube is enclosed in a tube of brass; the cistern is so arranged as to be closed with a screw, the air expelled, and the mercury made to fill the whole tube and cistern. This is then inverted, put in a wooden case, and this again in a leather case. This last is round, about three inches in diameter and three feet long, and is carried by a strap over the shoulder. They are admirably packed, but it requires much care to carry an instrument with so long a glass tube filled with mercury. We have, however, not broken one yet, except one of the thermometers attached, which burst with the heat. It was graduated to only 120°, which is entirely insufficient for open-air use in this climate, where reliable men have told me that they have seen it 167° F. in the sun. We have lost two thermometers by leaving them where the sun would come on them; in a few minutes they would burst at 120° to 125° F.

Another week's labors have closed, and we have finished all that we have time to do here. Professor Whitney has gone to Washoe. He will be back in San Francisco about June 15, and soon after rejoin us. We will be up to Monterey by that time.

I brought matters up to Monday night. Tuesday we intended to go about twelve miles for fossils, but our mules got away and too much of the day was spent in getting them to go then.

Wednesday, April 24, we went. The fossils occur on a ranch of Mr. Wilson, an Englishman. Our road lay down the valley of Osos, toward the sea, west of San Luis Obispo. Mr. Wilson has several ranches together, about 80,000 to 100,000 acres, keeps 20,000 head of cattle, 1,000 or 1,500 horses, etc., living in patriarchal style, monarch of all he surveys. His "farm" is about thirteen or fourteen miles long and nearly as wide. He seemed like a close-fisted old fellow, but treated us well. The fossils lay on a high hill. We could not get within two and a half miles of them with our wagon, so we camped by a brook. We packed the specimens down on mules.

In rough, broken hills, at about 1,800 feet elevation are these immense beds of fossil oysters. The shells are as numerous as in a modern oyster bed, all grown together, and of gigantic size, a foot to fifteen inches long, half as wide, and the thickest shell, sometimes *five inches thick*. (See also *Pacific Railroad Reports*, Vol. VII, Pt. 11, p. 45, for a description of similar ones.) They would weigh from ten to thirty pounds each. We packed down several mule loads of these and other fossil shells.

As we stayed over night, we camped. We declined an invitation to stop with Mr. Wilson a mile distant, as a child died the day we arrived and was to be buried the next. We sat by our bright camp fire until the bright moon rose, then went to bed on the green grass, in our blankets. The wind blew up fresh from the sea a few miles distant. We could hear the breakers, although they were five or six miles off. It is glorious to watch the stars and moon before going to sleep, but unpoetical to turn in the night and bring yourself in contact with a portion of the blanket soaked with dew, and *ugh*, how cold! But I have always slept gloriously in the open air, whenever I have tried it.

We returned Thursday evening. Friday we packed up our specimens, and Saturday (yesterday) took them to the landing for shipment. We had sixteen boxes, enough for quite a cabinet.

Last night a mail arrived bringing the first and scanty news of the attack on Fort Sumter. The eastern troubles have worried me much of late, although I have not written. We get papers often, a package nearly every steamer. I fear the prestige of the American name is passed away, not soon to return. We are doing and reaping as monarchists have often told us we would do—put designing, immoral, wicked, and reckless men in office until they robbed us of our glory, corrupted the masses, and broke us in pieces for their gain. But four and five short years ago I often argued this could never be—at the very time that we were pampering the knaves that could do it. I hope and trust that we may yet be united, but the *American Union* can never exist in the hearts of the entire people again as we have fondly dreamed that it did. I have long been prepared for anything that southern politicians would try, demoralized as they have become, but I expected a much more conservative force there than has shown itself.

This state is eminently for Union. The people almost unanimously feel that all that California is she owes to her nationality. I don't know a single Secession paper here. Of course, there are many desperadoes who would do anything, hoping to gain personally in any row that might arise, but the masses feel that their only safety is in the Union. Without protection, without mails, what would California be? A "Republic of the Pacific" is the sheerest nonsense. A republic of only about 900,000 inhabitants, less than a million, spread over a territory much larger than the original thirteen states, scattered, hostile Indians and worse Mormons on their borders—what would either sustain or protect such a country? And the people feel it.

But bad men are in power here as well as elsewhere in the United States. I have heard good citizens say that there was but one honest officer in this county. Court adjourned one day last week because both judge and district attorney were too drunk to carry it on. It is a *common thing* to see the highest officials of this county drunk on the streets here in town, but this is a notoriously hard place. I assure you, we never go to sleep without having our revolvers handy.

But the masses of the state are farther north. The whole south is sparsely populated, and will so remain so long as it is mostly divided into ranches so large that they are never spoken of by the *acre*, but always by *the square league*. A has four leagues, B ten leagues, C twenty leagues, etc.

NOTES

1. The Dos Pueblos grant was made to Dr. Nicholas A. Den in 1842. The site is now called Naples. The name *Dos Pueblos* dates back to the landing of Cabrillo, 1542.
2. The remarks of Thomas Antisell, M.D., geologist of Lieutenant Parke's expedition of the Pacific Railroad Survey, 1854–55, are found in *Reports,* Vol. VII, Pt. II.
3. The first is the coast live oak (*Quercus agrifolia*); the second is the valley oak (*Quercus lobata*).
4. La Misión Santa Inéz (or *Inés*, i.e., Agnes) was founded in 1804.
5. Founded in 1844 (Bancroft, *History of California,* IV, 425–426).
6. Benjamin Foxen came to California as a sailor on the British ship *Courier* in 1826, settled in Santa Barbara, was naturalized (baptized William Domingo), and married into a Spanish family. He died in 1874. Bancroft says: "He was a rough and violent man, often in trouble with other rough men and with the authorities, being sentenced to four years in prison in '48 for killing Augustin Davila—yet accredited with good qualities, such as bravery and honesty" (*ibid.,* III, 746).
7. Francis Zida Branch came to California in 1831 with the Wolfskill party of trappers from New Mexico. He settled in Santa Barbara, married Manuela Carlou in 1835, and obtained the grant of the Santa Manuela Rancho in 1839. There he spent the rest of his life. He died in 1874 at the age of seventy-two (*ibid.,* II, 727).
8. These buttes are concisely described in a bulletin of the United States Geological Survey ("The Shasta Route and Coast Line," *Guidebook of the Western United States,* No. 614, Pt. D, p. 115) as follows: "The most

prominent topographic feature in the vicinity of San Luis Obispo (Spanish for St. Louis the bishop) is the row of conical hills that begins with Islay Hill, on the right (east), a little over 2 miles southeast of San Luis Obispo, and extending to Cerro Romualdo, about 4 miles northwest of the town. There are eight of these hills (Spanish *cerros*), the four larger northwest and the four smaller southeast of the city. These hills, of which The Bishop (1502 feet) is the highest, are composed of igneous rock and are the cones of small volcanoes which broke through the Franciscan sedimentary rocks. The eastern part of Islay Hill consists of a surface flow of basaltic lava."

CHAPTER VII

SALINAS VALLEY AND MONTEREY

"Chores"—Paso Del Robles—San Antonio River—Animal Life—Soledad—Guadalupe Ranch—Monterey—Pescadero Ranch—Mission of Carmelo—A Trip in the Monterey Hills—Finch's Ranch.

Camp No. 28, Nacimiento River.
Saturday Afternoon, May 4, 1861.

IT is a lovely afternoon, intensely hot in the sun, but a wind cools the air. A belt of trees skirts the river. I have retreated to a shady nook by the water, alike out of the sun and wind; a fine, clear, swift stream passes within a few rods of camp, a belt of timber a fourth of a mile wide skirts it—huge cottonwoods and sycamores, with an undergrowth of willows and other shrubs. We have been here three days.

I returned from a long walk at noon and concluded to devote the afternoon to writing and "chores." First, dinner; next, put on clean clothes and wash my dirty ones. A few buttons sewn on, and rents repaired; then the garments lay in the water to soak while I wrote a letter of three sheets to headquarters, during which time a flock of sheep trod my shirts into the mud. Then the wash, that I so much abominate. But clothes must be cleansed, and there is no woman to do it. Were I to describe the abominable operation it would take a whole letter. I can't do it—just some items only. First, I get a place on the bank and begin. A huge gust scatters sand over the wet clothes, which are in a pile on the bank. Stockings are washed—I congratulate myself on how well I have done it. An undershirt is begun—goes on swimmingly. Suddenly the sand close to the water where I squat gives way. I go in, half boot deep, and in the strife to get out, tread on the clean stockings and shove them three inches into the mud and sand. A stick is got and laid close to the water. On that I kneel, as do the Mexican and Indian washerwomen. This goes better, and the work goes bravely on. Next, the slippery soap glides out of my hands and into the deep water—here a long delay in poking it out with a long stick, during which performance it goes every way except toward shore. At last the final garment is washed. With a long breath I rise to leave, when I find the lowest of the clean pile is all dirty from the log I laid them on—the cleanest place I could find. But soon all difficulties are surmounted, and the clothes are now fluttering in the wind, suspended from one of the guy ropes of our tent. The picture is underdrawn rather than exaggerated—just try it by taking your clothes to the creek to wash the next time.

Monday morning, April 29, we left San Luis and took our way on—we had been there two weeks. We crossed the San Luis Pass of the Santa Lucia Mountains, a pass about 1,500 or 1,800 feet high, and entered the Santa Margarita Valley. North of the Santa Lucia chain, which trends off to the northwest and ends at Monterey, lies the valley of Salinas, a valley running northwest, widening toward its mouth, and at least a hundred and fifty miles long. This valley branches above. One branch, the west, is the Santa Margarita, into which we descended from the San Luis Pass. We followed down this valley to near its junction with the Salinas River and camped at the Atascadero Ranch, about twenty-two miles from San Luis Obispo and six from the Mission of Santa Margarita.[1]

On passing the Santa Lucia the entire aspect of the country changed. It was as if we had passed into another land and another clime. The Salinas Valley thus far is much less verdant than we anticipated. There are more trees but less grass. Imagine a plain ten to twenty miles wide, cut up by valleys into innumerable hills from two to four hundred feet high, their summits of nearly the same level, their sides rounded into gentle slopes. The soil is already dry and parched, the grass already as dry as hay, except along streams, the hills brown as a stubble field. But scattered over these hills and in these valleys are trees every few rods—great oaks, often of immense size, ten, twelve, eighteen, and more feet in circumference, but not high; their wide-spreading branches making heads often over a hundred feet in diameter—of the deepest green foliage—while from every branch hangs a trailing lichen, often several feet long and delicate as lace. In passing over this country, every hill and valley presents a new view of these trees—here a park, there a vista with the blue mountains ahead. I could never tire of watching some of these beautiful places of natural scenery. A few pines were seen for several miles, with a very open, airy habit, entirely unlike any pine I have ever seen before, even lighter and airier than the Italian pines common in southern France by the Mediterranean. They cast but little shade.

The Mission of Santa Margarita was in ruins. It is the seat of a fine ranch which was sold a few days ago for $45,000. The owner, Don Joaquin de Estrada, lives now at Atascadero Ranch, where we camped. This last ranch is all he now has left of all his estates. Five years ago he had sixteen leagues of land (each league over 4,400 acres, or over 70,000 acres of land), 12,000 head of cattle, 4,000 horses, etc. Dissipation is scattering it at the rate of thousands of dollars for a single spree. Thus the ranches are fast passing out of the hands of the native population.

Camp No. 29, Jolon Ranch, on San Antonio River.
May 8.

I DID not write last Sunday as there was an American ranch near our camp and we borrowed some magazines, rare luxuries for camp, and I read them all day. The American who has this ranch, keeps fifteen or sixteen thousand sheep. He is a very gentlemanly Virginian and was very kind to us. He says that the loss of sheep by wolves, bears, and rattlesnakes is quite an item. We are in a bear region. Three men have been killed within a year near our last camp by grizzlies.

Monday we came on here, about twenty-five miles. The day was intensely hot, and as we rode over the dry roads the sun was scorching. We crossed a ridge by a horrible road and came into the valley of the San Antonio, a small branch of the Salinas, and followed up it to this point, where we are camped on its bank. We passed but one ranch and house in the twenty-five miles. In one place, two bears had followed the road some distance the night before—their tracks were very plain in the dust.

We are now over sixty miles from San Luis Obispo. Here is a postoffice, the only one between the latter place and San Juan, a distance of about two hundred miles. We found the "office" at the ranch; not one person could speak a word of English, so we searched out our letters from a handful that lay on the mantel, the whole stock on hand.

The last two days we have been exploring the hills. Yesterday, with Averill, I climbed some hills. Today he had to go to a store a few miles distant for flour, so I took a long tramp of eighteen to twenty miles alone. We got an early breakfast, and I started in the cool of the morning, with a bag of lunch, compass, canteen of water, and knife, pistol, and hammer in belt. As one is so liable to find bears and lions here, it is not well to be without arms. I pushed back over the hills and through canyons about ten miles from camp to the chain of rugged mountains west of us. I was indeed alone in the solitudes. The way led up a canyon about four miles, with high steep hills on each side, then a ridge to be crossed, from which I had a fine view, then down again and among gentle hills about three miles farther to the base of the mountains. Here a stream was crossed by pulling off boots and wading, and then up a canyon into the mountains. This last I followed as far as I considered safe, for it was just the place for grizzlies, and I kept a sharp lookout.

Here I climbed a ridge to get a view behind. The slope was very steep, the soil hot, no wind, and the sun like a furnace. I got the view and information I desired. A very rugged landscape of mountains behind, steep, rocky, black with chaparral, 3,500 to 4,000 feet high. In front was the series of ridges I had crossed; beyond, the Salinas Valley, with blue mountains on the distant eastern horizon. Some very peculiar rocky pinnacles of brown rock rose like spires near me, several hundred feet high—naked rocks.

I started to return, and had reached the stream, when a crash in the brush near by startled me, and in a moment two fine fat deer, small but very beautiful, sprang out. I shot at one with my pistol, but only wounded him—so he got away. I have had the sight recently repaired, and I find to my disgust it is all wrong; had it been correct I certainly would have killed him, for it was a very fair shot, not over twenty-five or thirty yards. I shot twice more, but the deer were too far off, and my balls went still wider from the mark. I lunched in the cool shade of a fine oak.

The cattle here over the hills are very wild; they will run if they see a man on foot at the distance of forty or fifty rods off. Sometimes an old bull will boldly make an attack, so it is unsafe to go through a herd alone and on foot. The *rancheros* consider it desirable that their cattle be thus wild—they are less liable to be stolen or to be caught by wild animals. I passed near a herd. They first ran, but an old bull took for the offensive-defensive, and made for me. I did not dare trust to my pistol, so took "leg bail" and made for a tree, reached it, and then stood my ground, resolved to shoot him when he got near; but a club at last brought him to a stop, and finally he fled. On my way back I met Averill about two miles from camp, coming up the canyon to meet me with a mule.

While speaking of animals—the grizzly bear is much more dreaded than I had any idea of. A wounded grizzly is much more to be feared than even a lion; a tiger is not more ferocious. They will kill and eat sheep, oxen, and horses, are as swift as a horse, of immense strength, quick though clumsy, and very tenacious of life. A man stands a slight chance if he wounds a bear, but not mortally, and a shot must be well directed to kill. The universal advice by everybody is to let them alone if we see them, unless we are well prepared for battle and have experienced hunters along. They will generally let men alone, unless attacked, so I have no serious fears of them.

Less common than bear are the California lions, a sort of panther, about the color of a lion, and size of a small tiger, but with longer body. They are very savage, and I have heard of a number of cases of their killing men. But don't be alarmed on my account—I don't court adventures with any such strangers. Deer are quite common. Formerly there were many antelope, but they are very rapidly disappearing. We have seen none yet. Rabbits and hares abound; a dozen to fifty we often see in a single day, and during winter ate many of them.

There are many birds of great beauty. One finds the representatives of various lands and climes. Not only the crow, but also the raven is found, precisely like the European bird; there are turkey-buzzards, also a large vulture something like the condor—an immense bird.[2] Owls are very plenty, and the cries of several kinds are often heard the same night. Hawks, of various sizes and kinds and very tame, live on the numerous squirrels and gophers. I see a great variety of birds with beautiful plumage, from humming birds up.

But it is in reptile and insect life that this country stands preëminent. There are snakes of many species and some of large size, generally harmless, but a few venomous. Several species of large lizards are very abundant. Salamanders and chameleons are dodging around every log and basking on every stone. Hundreds or thousands may be seen in a day, from three inches to a foot long. Some strange species are covered with horns like the horned frogs.

But insects are the most numerous. They swarm everywhere. House flies were as abundant in our tent in winter as at home in summer. Ticks and bugs get on us whenever we go in the woods. Just where we are now camped there are myriads of bugs in the ground, not poisonous, but annoying by their running over one. Last night I could scarcely sleep, and shook perhaps a hundred or two hundred out of my blankets this morning.

I shall sleep outdoors tonight—in fact all the rest are asleep but me, and only one is in the tent. We are under some cottonwood trees which so swarm with ladybugs that Mike yesterday counted how many he brushed off of him in an hour. They amounted to 250—but he sat still under the tree. Scorpions occur farther south and are much dreaded. The equally dreaded tarantula abounds here. It is an enormous spider, larger than a large bumblebee, and has teeth as large as a rattlesnake's. I killed one by our tent at Camp 27, and saved his teeth as a curiosity. Their holes in the ground are most ingenious.

Camp 31, Guadalupe Ranch.
May 12.

WE left San Antonio Thursday morning, May 9, and followed up the valley a few miles, then crossed a high steep ridge over one thousand feet high, which separates the San Antonio from the Salinas, and then descended and struck down the great Salinas plain. Dry as had been the region for the last sixty or seventy miles, it was nothing to this plain.

The Salinas Valley for a hundred or more miles from the sea, up to the San Antonio hills, is a great plain ten to thirty miles wide. Great stretches are almost perfectly level, or have a very slight slope from the mountains to the river which winds through it. The ground was dry and parched and the very scanty grass was entirely dry. One saw no signs of vegetation at the first glance—that is, no green thing on the plain—so a belt of timber by the stream, from twenty to a hundred rods wide, stood out as a band of the liveliest green in this waste. The mouth of this valley opens into Monterey Bay, like a funnel, and the northwest wind from the Pacific draws up through this heated flue with terrible force. Wherever we have found a valley opening to the northwest, we have found these winds, fierce in the afternoon. For over fifty miles we must face it on this plain. Sometimes it would nearly sweep us from our mules —it seemed as if nothing could stand its force. The air was filled with dry dust and sand, so that we could not see the hills at the sides, the fine sand stinging our faces like shot, the air as dry as if it had come from a furnace, but not so very hot— it is wonderfully parching. The poor feed and this parching wind reduced our mules in a few days as much as two weeks' hard work would. Our lips cracked and bled, our eyes were bloodshot, and skins smarting.

We stopped for lunch at a point where the mules could descend to the river. A high terrace, or bluff, skirts the present river—that is, the plain lies from 75 to 150 feet above the present river. The mules picked some scanty herbage at the base of the bluff; we took our lunch in the hot sun and piercing wind, then drove on. We pulled off from the road a mile or so at night, and stopped beneath a bluff near the river. We had slept in the open air the previous night and did so again. It turns very cold during the clear nights, yet so dry was it that no dew fell those two nights, cold as it was! The mules found some picking where you would think that a sheep or a goat would starve.

Friday we pushed on all day, facing the wind. We met a train of seven wagons, with tents and beds—a party of twenty-five or thirty persons from San Jose going to the hot springs, some on horseback. Two-thirds were ladies. A curious way for a "fashionable trip to the springs," you say, but the style here. They will camp there, and have a grand time, I will warrant. We kept the left bank of the river, through the Mission Soledad. Before reaching it we crossed the sandy bed of a dry creek, where the sand drifted like snow and piled up behind and among the bushes like snow banks.

The Mission Soledad is a sorry looking place, all ruins—a single house, or at most two, are inhabited. We saw the sign up, "Soledad Store," and went in, got some crackers at twenty-five cents a pound, and went on. Quite extensive ruins surround the place, empty buildings, roofless walls of *adobe*, and piles of clay, once *adobe* walls. It looked very desolate. I do not know where they got their water in former times, but it is dry enough now. We came on seventeen miles farther. Here we find tolerable feed and a spring of poor water, so here is a ranch.

Sorry as has been this picture, it is not overdrawn, yet all this land is occupied as "ranches" under Spanish grants. Cattle are watered at the river and feed on the plains, and scanty as is the feed, thousands are kept on this space, which must be at least four to six thousand square miles, counting way back to the Santa Lucia Mountains. The ranches do not cover all this, but cover the *water*, which is the same thing. We could see a house by the river every fifteen to eighteen miles, and saw frequent herds of cattle. The season is unusually dry, and the plain seems much poorer than it really is. In the spring, two months ago, it was all green, and must have been of exceeding beauty. With water this would be finer than the Rhine Valley itself; as it is, it is half desert. As to the actual capability of the plain, with water, the *Pacific Railroad Reports* state that "At Mr. Hill's farm near the town of Salinas, sixteen miles east of Monterey, sixty bushels of wheat have been raised off the acre, and occasionally eighty-five bushels. Barley, one hundred bushels, running up to one hundred and forty-nine bushels, and vegetables in proportion" (VII, Pt. II, 39).

We passed through a flock of sheep, the largest I have ever seen, even in this country of big flocks. It was attended by shepherds, and must have contained not less than 6,000 sheep, judging from the flocks of 2,000 and 1,500 we have seen often before. Some of our party thought there must have been 8,000. Sheep are generally kept in flocks of not over 1,800 head.

High mountains rise on the opposite side, in the northeast, and still nearer us on the left. These latter were very rugged—from 3,000 to 4,500 feet high, black, or very dark green, with chaparral—yet not abounding in streams as one would imagine, although now only early in May. The Nacimiento and San Antonio rivers are the only tributaries of the Santa Margarita and Salinas valleys on the west side, this side of Atascadero Ranch—that is, only these two streams for a distance of 120 miles. And, from leaving the San Antonio, sixty-one miles back, we have not crossed a single brook or seen a single spring until reaching this ranch, where there is a spring.

Yesterday I climbed the ridge southwest of camp. I ascended about 3,000 or 3,500 feet, a hard climb, and had a good view of over a hundred miles of the Salinas Valley from the Bay of Monterey to above where we last struck it, or over the extreme limits of about 130 to 150 miles, with the successive ridges beyond. *Four thousand to seven thousand square miles* must have been spread out before me. I have never been in a land before with so many extensive views—the wide valley, brown and dry, the green belt of timber winding through it, like a green ribbon, the mountains beyond, dried and gray at the base, and deep green with chaparral on their sides and summits, with ridge after ridge stretching away beyond in the blue distance. Then to the north, a landscape I had not seen before, with the whole Bay of Monterey in the northwest. To the west and south of me was the very rugged and forbidding chain of mountains that extends from Monterey along the coast to San Luis Obispo and there trends more easterly—the Sierra Santa Lucia.

I have found much of intense geological interest during the last two weeks. I had intended to spend at least two weeks more in this valley had we found water or feed as we expected. Not finding it, and having four weeks on our hands before the rendezvous with Professor Whitney at San Juan, I decided to push on to Monterey, which I had not intended to visit. We are now within eight or ten leagues of there—will be there in a few days. I feel now that we are indeed working north and I long to be in San Francisco again. It is now over five months since I have attended church (Protestant) and have only had that privilege three times since I left New York.

Sunday Evening.

TODAY has been a windier day on the plain than any other day we were on it. I am glad enough we are sheltered here in camp. Clouds of gray dust, rising to the height of five or six thousand feet have shut out the view in the north all the afternoon, and even the hills opposite could not be seen at times, and all day they have been obscurely seen through this veil. If it is thus in May, what must it be here in July or August, as no rain will fall for at least four months yet! It was interesting yesterday, while on the peaks above, to watch the great current of air up the valley, increasing with the day until at last the valley seemed filled with gray smoke.

While speaking of the plain, I forgot to mention the mirage that we had. The sun on the hot waste produced precisely the effect of water in the distance; we would see a clear lake ahead, in which would be reflected the objects on the plain. This was most marked on the dry sands near Soledad—we could see the trees at the Mission mirrored in the clear surface—but it kept retreating as we advanced. The illusion was perfect. At times the atmospheric aberration would only cause objects to be distorted—wagons and cattle would appear much higher than they really were, as if seen through poor glass.

Monterey.
May 17.

WE arrived here on Wednesday. On Monday, May 13, we left Guadalupe Ranch and came about fourteen or fifteen miles and camped in a valley that turns off from the Salinas on the road to Monterey. We had hardly camped, and were eating dinner, when the stage came along. I went to the driver to get him to carry a letter to the next post office. He had to stop there to water his horses. A familiar face appeared in the stage, not at first recognized, but a mutual recognition soon took place—it was a good friend I had known on shipboard, a friend of Averill's, a lawyer from New York, a Mr. Tompkins,[3] one of the finest gentlemen and most entertaining that I have met for a long while. We were much together on shipboard. He had traded eastern property for a ranch near Monterey, on the coast, and had just been to it and was returning to San Francisco. He was the first acquaintance met since Los Angeles, or I should say, since we left San Francisco. The meeting was mutually pleasant. He tendered us the hospitality of his ranch, although he could not be with us there, but gave us a letter to his *majordomo* (head *ranchero*) to give us all attention, feed, board, horses to ride, etc. We shall go there next week. He had so improved that we did not at first know him—he was in ill health last fall, but hearty enough now. We, tanned by the sun, bronzed by exposure, without coats or vests, in buckskin pants, bowie and Colt at our belts—he said at first sight we fulfilled his beau ideal of buccaneers stopping the stage. We stopped there over Tuesday and the driver gave us a Monday's paper from San Francisco, with the latest news. That was the fourteenth, and we had news up to May 3, by Pony Express, that is, *only eleven days from New York to camp.*

We have been quite lucky thus far for news, and it has been a great item in these times. I cannot write how heavily the national troubles bear upon my mind, they are in my mind by night and by day. God grant that we may yet save the *United* States, but I fear for the worst. Newspapers from home are always acceptable, but we get the great news by earlier means. On arriving here, by a "judicious" distribution of patronage to two leading stores, we got lots of papers for reading, a dozen or more.

This sheet finishes my letter paper of thin kind—the last scrap is here—and I must use such as I can get hereafter until I get to "Frisco." Trusting that the mails will not be "seized" by pirates, it must go by next steamer. The last steamer went out fully armed, for it was currently believed that a party was going abroad as passengers to take her for the Southern Confederacy. The Union sentiment here is overwhelming.

Monterey.
Sunday, May 19.

It is a lovely evening—the moon shines brightly, the old pines and thick oaks by our camp cast dark shadows, and the quiet bay sparkles in the moonlight.

I have been to church today—attended Protestant service for the first time since last November, nearly six months ago. There is a Methodist mission station here. I heard there was to be service at 11 A.M. in the courthouse, so was on hand. The rest of the party went to Mass. I found two or three fellows loafing on the porch, and as the door was locked, a man started to find somebody who had the key. Meanwhile, a dozen collected on the porch. After much delay the key was found, and, half an hour after time, services opened. How unlike a Roman missionary—*he* would have had all ready and shown himself "diligent in business" as well as "fervent in spirit." The congregation at last numbered some twenty or twenty-five persons, not counting the few children. The clergyman was a very doleful looking man, with *very* dull style and manner, who spoke as if he did it because he thought it his duty to preach and not because he had any special object in convincing or moving his audience. His nose was very pug, his person very lean, his collar very high and stiff, and his whole appearance denoted a man entirely lacking energy, surely *not* the man for a California missionary. Yet how good it seemed to meet again with a few for divine service—it was indeed a pleasure. We have now been over a country twice as large as Massachusetts, and this is the second Protestant congregation we have seen, and both of these feeble and small. But there are Catholic churches in every considerable town.

As I came out of church and met Averill in the street, we were accosted by a man who wanted us to ride a few miles and look at a supposed silver "lead" he had discovered. We declined, but were soon beset by others, with ore and "indications" from another mine. I must take the specimens, which I did, and returned to camp and "blowpiped" them to get rid of them—found a little silver.

Monterey has about 1,600 inhabitants and is more Mexican than I expected. It is the old capital of California. There are two Catholic churches, and Spanish is still the prevailing language. Like all other places yet seen, more than half of the "places of business" are liquor shops, billiard saloons, etc.—all the stores sell cigars, *cigarritos*, and liquor. Stores are open on Sunday as well as other days, and that is the day for saloons and barrooms to reap a rich harvest. Billiard tables go from morning till midnight—cards and *monte* are no secrets. Thus it has been in all the towns. Liquor and gambling are the curse of this state. Lots of drunken Indians are in the outskirts of the town tonight.

> Pescadero Ranch.
> Monday, May 27.

AFTER examining things about Monterey for three days, we came here to Mr. Tompkins's ranch, where the feed is good. It is a ranch of four or five thousand acres, on the coast about five miles from Monterey. We pitched our tent in the yard, but a larger log house is our headquarters. Last Monday, while in Monterey, a dull day with showers, we got an "artist" to bring his camera out to camp and take a few pictures of camp on leather. He took four—not good in an artistic sense, but good as showing our camp. We divided our pictures by cutting cards for the choice, and I got the best picture.

Pescadero Ranch was formerly owned by an eccentric, misanthropic, curious man, who lived in solitude and tried to educate two boys, keeping aloof from the world and the rest of mankind. He built a large and very secure log house, for fear of robbers, just on the shore of the Pacific, by a lovely little bay. Behind rise hills covered with tall dark pines, and near the house is a field of about a hundred or more acres, fenced in, where we have fine feed for our mules. His books are still here— a strange collection on science, art, astrology, romance, infidelity, religion, mysteries, etc. Old harness, spades, implements, harpoons, etc., are stored in large numbers. I know not why he had them. He had invented a new harpoon which no one would use.[4]

By the way, Monterey Bay is a great place for whaling. Two companies are at work, and already over half a dozen whales have been taken here. On Wednesday we saw them towing one in, and on Thursday morning went down to see them cut him up. He was a huge fellow, fifty feet long. Last year they caught one ninety-three feet long which made over a hundred barrels of oil. After stripping off the blubber, the carcasses are towed out into the bay, and generally drift up on the southeast side. The number of whale bones on the sandy beach is astonishing— the beach is white with them. Hundreds of carcasses have there decayed, fattening clouds of buzzards and vultures. The whales are covered with thick black skin. The tail is horizontal. They have no fins, but a pair of huge "paddles," one on each side— oars, as it were—like great flat arms covered with skin, three or four feet wide and twelve or fifteen feet long. The ball-and-socket joint which attaches the paddle to the body is wonderful —the ball is as large as the end of a half-barrel. Barnacles grow on the skin in great numbers; I will try to collect some if they do not stink too badly.

To return to Pescadero. We came on Thursday. I had letters to several persons. Wednesday evening I had called on a prominent Monterey citizen, and spent an evening in female society, and heard a piano for the first time in many months. On Friday we rode a few miles to Judge Haight's. He is a wealthy San Francisco gentleman and has a fine ranch here, where he spends a part of the year with the whole or a part of his family. We presented our letters, but did not find him at home.[5]

We visited the old Mission of Carmelo,[6] in the Carmelo Valley, near his ranch. It is now a complete ruin, entirely desolate, not a house is now inhabited. The principal buildings were built around a square, enclosing a court. We rode over a broken *adobe* wall into this court. Hundreds (literally) of squirrels scampered around to their holes in the old walls. We rode through an archway into and through several rooms, then rode into the church. The main entrance was quite fine, the stone doorway finely cut. The doors, of cedar, lay nearby on the ground.

The church is of stone, about 150 feet long on the inside, has two towers, and was built with more architectural taste than any we have seen before. About half of the roof had fallen in, the rest was good. The paintings and inscriptions on the walls were mostly obliterated. Cattle had free access to all parts; the broken font, finely carved in stone, lay in a corner; broken columns were strewn around where the altar was; and a very large owl flew frightened from its nest over the high altar. I dismounted, tied my mule to a broken pillar, climbed over the rubbish to the altar, and passed into the sacristy. There were the remains of an old shrine and niches for images. A dead pig lay beneath the finely carved font for holy water. I went into the next room, which had very thick walls—four and a half feet thick—and a single small window, barred with stout iron bars. Heavy stone steps led from here, through a passage in the thick wall, to the pulpit. As I started to ascend, a very large owl flew out of a nook. Thousands of birds, apparently, lived in nooks of the old deserted walls of the ruins, and the number of ground squirrels burrowing in the old mounds made by the crumbling *adobe* walls and the deserted *adobe* houses was incredible—we must have seen *thousands* in the aggregate. This seems a big story, but *hundreds* were in sight at once. The old garden was now a barley field, but there were many fine pear trees left, now full of young fruit. Roses bloomed luxuriantly in the deserted places, and geraniums flourished as rank *weeds*. So have passed away former wealth and power even in this new country.

Our road to the Mission was a mere trail through the thick chaparral, crossing some deep ravines. We came on the tracks of numerous grizzlies—or, rather, numerous tracks. There are three grizzlies living in the brush near here, particularly bold and savage. One has nearly killed several people. They came here to eat a whale stranded on the beach. As we had two good Sharp's rifles, besides other guns, we concluded to watch for them that night. An Indian, an old bear hunter, entered into the project, but on examination of the ground, it was found that there was no good place—no trees to get into and watch from—for no one is so mad as to engage in a bear fight unless he has all the odds on his side. So we had to give it up.

Judge Haight came over and invited Averill and me to dinner yesterday. We rode to Point Cypress in the morning—a granite, rocky point, covered with a kind of cedar called "cypress," more like the cedar of Lebanon than any other tree I have seen.[7] Some of the trees were beautiful—and often three or four feet in diameter. I measured one that was eighteen feet eight inches in circumference as high as I could reach. Another, twenty-three feet at two feet from the ground.

Returning to camp, we took other mules and rode to Mr. Haight's, about five miles. We rode through the old Mission again—and paused a short time among the ruins. We were on hand at two o'clock, the appointed time.

Judge Haight is a fine old man, a man of much intellect,

lives in a comfortable house, has with him two daughters, most lovely young women, of perhaps eighteen and twenty-two years —pretty, agreeable, cultivated, and sensible. I don't know when I have spent an afternoon so pleasantly. The dinner was good, not brilliant—champagne was partaken of moderately. His library was well stocked with choice works. It was indeed a luxury to meet with ladies—the first time we had sat at a table with them since New Year's at Mr. Wilson's. We were decidedly pleased, and we think they were, for they are much isolated here. They had a fine piano, and one of the girls played well. We climbed a hill just above the valley, and had a pretty view of the Carmelo Valley, the sea beyond, and the mountains in the south. He has a fine ranch, keeps about twelve hundred sheep, much better animals than one generally sees here. We were so urged to stay to tea that we did, and rode home by twilight. One dared not wait later for fear of grizzlies. Where our trail ran through dense chaparral we came on fresh tracks made but a few minutes before—after a man had passed an hour before—but we were spared a sight of any animals.

El Pescadero.
Tuesday Evening, June 4.

WE were ready early Tuesday morning, May 28, for a start. Up at daylight—Averill, Peter, and a *buccaro* for a guide— saddlebags packed, and two pack-mules: Sleepy with blankets and some meat, coffeepots, and bread; Stupid with more blankets, frying pan, and more provisions. We followed a trail about three miles, then struck the road up the Carmelo Valley. We stopped at a house half an hour to wait for Charley, the *buccaro*, to overtake us. He had been to town for bread for the trip. Mrs. McDougal, where we stopped, insisted on our drinking a pan of milk, which we did, then struck up the valley.

We followed the road about twenty miles. Five ranches were passed; some barley fields along the river, and wild oats in abundance on the hills, supporting many cattle. We lunched at a stream, saddled, and were again off. Here we left the road, and for fifteen miles followed trails, now winding along a steep hillside—steep as a Gothic roof, the stones from the path bounding into a canyon hundreds of feet below—now through a wide stretch of wild oats, now through a deep canyon. We passed two more ranches, where cattle are raised among the hills, and at last struck through a rocky canyon, in which flowed a fine stream, with some glorious old trees. Before dark we arrived at a small ranch owned by a man named Finch, with whom Charley was acquainted. We camped near, and slept well, for we had been ten and a half hours in the saddle in thirteen hours. We frightened up four fine deer just as we went into camp.

Peter and Averill had each bought a "Sharp" for hunting, so on Wednesday they tried for deer. I climbed the mountain for "geology." First I passed through a wild canyon, then over hills covered with oats, with here and there trees—oaks and pines. Some of these oaks were noble ones indeed. How I wish one stood in our yard at home. One species, called *encina*, with dark green foliage, was not extra fine, but another, *el roble*, was very fine.[8] I measured one of the latter, with wide spreading and cragged branches, that was twenty-six and a half feet in circumference. Another had a *diameter* of over six feet, and the branches spread *over seventy-five feet each way*. I lay beneath its shade a little while before going on. Two half-grown deer sprang up close to me, but got out of pistol shot before I, in my flurry, had the pistol ready. Up, still up, I toiled, got above the grass and oats and trees into the chaparral that covers the high peaks. I struck for the highest peak, but backed out before quite reaching it, for the traces of grizzlies and lions became entirely too thick for anything like safety. Both are very numerous here. Finch killed three a few days before we arrived.

But what a magnificent view I had! A range of hills two thousand to three thousand feet high extends from Monterey to Soledad. It is a part of the mountains, yet there is a system of valleys behind, up which we had passed. The Carmelo River follows this a part of the way. I was higher than these hills. Over them, to the northwest, lay the Bay of Monterey, calm, blue, and beautiful. Beyond were blue mountains, dim in the haze; to the east was the great Salinas plain, with the mountains beyond, dim in the blue distance. In the immediate foreground was the range of hills alluded to, the Palo Scrito, in some places covered with oats, now yellow and nearly ripe, in others black with chaparral. Behind lay a wilderness of mountains, rugged, covered with chaparral, forbidding, and desolate. They are nearly inaccessible, and a large region in there has never been explored by white men.

I returned by the same way I had come up. There is a most beautiful tree I had not seen before, with foliage something like but even richer than the magnolia—it is a kind of manzanita. It would be splendid in cultivation in a mild climate.[9]

Averill and Peter returned without any venison, but Averill brought in an enormous rattlesnake, by far the biggest we have yet seen. He was huge, and, Averill says, decidedly savage when wounded. He was four and a half feet long, as thick as one's arm, and had twelve rattles. His head was over an inch and three-quarters broad, with mouth corresponding. I cut out one of his fangs as a specimen.

We spent an hour in Mr. Finch's house that evening. Two brothers, Americans, have a ranch, and are raising horses. Mrs. Finch seemed a meek, sad woman, with more culture and sensibility than her husband, and evidently pining for other lands and other scenes here in this lonely place, away from the world, almost away from the "rest of mankind." The house was of sticks plastered with mud, the floor, the earth. Two pretty little girls were playing upon a grizzly skin before the fire. It is a lonely life they lead there.[10]

Thursday we took a young man for guide and pushed on, over hills, through canyons, winding, climbing, toiling; our road, cattle trails; our landmarks, mountains. I saw many pretty flowers, some new to me. We struck a fine stream of water that flows toward the Salinas plain at Soledad, fourteen miles distant, but it sinks long before that in the *arroyo seco*, or dry canyon. It was a swift clear stream, and good water on that trip was one of our luxuries. It has been long since I have tasted *good* water. Here we found a little ranch, Hitchcock's. The owner was talkative, asked for papers, showed us some fine quicksilver ore, but was too shy to tell us where he found it. He only said it was back in the mountains—"A hell of a place to get to"—which I can easily imagine, if it is six miles farther in than we were, as he said it was.

Here we struck up the canyon into the heart of the mountains a few miles, now over a table for a mile, now down a steep bank and crossing the stream, up on the other side, steep as a house roof. But our mules were trusty; Old Sleepy, with his pack, proved himself equal to the occasion, and my old white mule won fresh laurels. Up this canyon the strata are bent, twisted, contorted, and broken. I never before saw finer examples of bent strata. They were less grand than the noted ones on Lake Lucerne, but more beautiful.

We saw some deer and got a shot—one was wounded, but we did not get him. All had rifles but me; my botanical box and hammer were enough for me. Soon more deer were seen. Peter and the guide started after them. We missed the trail, and in attempting to cross the stream and climb the bank came near having an accident. The bank had a slope of forty-five degrees; the path wound up it at twenty-nine degrees—I measured it. Averill's mule trod on loose stones and went down. A mule never slips, but here the path slipped. Averill got off and saved himself, but the mule went down slowly and got away. An hour and a half were spent in finding and getting her. At last all were ready again, and we took our way up the canyon as far as mules could get—and that is saying a good deal—and struck a very narrow, wild canyon leading to a little lake (*laguna*). It was a lovely spot, but a poor place to camp, so we turned back a mile, and camped on the banks of the main stream.

I wish I could describe the spot. A deep rocky canyon, with rugged, almost perpendicular sides, but green, grassy bottom, opens into the main canyon, where there is a swift stream of water of crystal clearness, grass and oats abundant for our mules, fine trees scattered around for effect and all around rise high, rugged, rocky mountains. We are now beyond all traces of human homes, but in the abodes of grizzlies and deer. A fire is built, supper (as well as dinner) got, and then we go out to hunt. In ten minutes Averill is back with a deer, and an hour later the others come in with another. I know not how many deer we saw on that trip. I took a swim in the cool stream—it was refreshing enough after riding on dusty trails and through hot canyons.

I wish you could look on such a camp at night. Scattered around are pack-saddles, saddles, bread—and oh, *such* bread as we had after sixty miles' travel on a mule's back in a bag! It needed *sifting* to get pieces large enough for mouthfuls. The mules are picketed near and around us. *They* will give the alarm if grizzlies become too familiar. Scattered on the grass around, we lie rolled in our blankets. A rifle peeps out from beneath the blankets here and there—loaded too, for, although grizzlies never molest persons asleep, it is best to have the weapons handy. The bright camp fire throws a ruddy glare on the green foliage, which shows black shadows and grim recesses back, and stately trunks and gnarled limbs shine out brighter here and there. But brighter than all, and more beautiful to me, are the stars in the deep, clear, blue sky. One is just trembling over the brow of that rugged mountain, it seems almost to touch it—others are slowly moving behind the trees, or the hills, in their majestic march to the west. The only sound to break the silence of this solitude is the murmur of the streams by us. And thus we sleep—such glorious sleep—sound and refreshing; no bad air, no close smell of feathers, no musty, ill-aired beds from which one rises in the morning with gummy eyes and heavy brain and mouth tasting as if half filled with Glauber's salts and clay.

The shadows were dark in the canyon as we rose, and some choice cuts of venison roasted on the coals were partaken of with a relish that many a hothouse millionaire might well envy. Ah, it was good! We lingered around some; I botanized an hour—and then we took our way back, following nearly our same trail. In one place the trail led along the very brink of a precipice 250 to 300 feet high; one could look down, unobstructed, *almost* perpendicularly (tourists would say *quite* so), to the rocks and water so far below. It was as steep as the north bank of Taughannock Falls, by the house, and two-thirds as high, the path scarcely a foot wide. But the mules did not hesitate—they know their own powers—and with loose rein we let them take their way, slowly, surely, now looking steadily at the path, but often swinging their heads over and looking at the abyss below. Where the path ascended a steep slope I got off, not for greater safety so much as to ease my mule, which is most too light for me. But most of them rode here, nor spoke of danger.

We got back to Finch's that night. We found some fossil bones on our way—the backbone of a large fish, not so large as a whale, yet very large. Thousands of acres of these lower hills are covered with wild oats, as thick as a poor oat field at home. These are the "live oats" or "animated oats," sometimes cultivated at home, and were introduced here from Spain by the old *padres*.

We got back safely on Saturday, June 1, after a pleasant trip, no mishaps, and much of botanical and geological interest, but well tired from the hard riding.

NOTES

1. This was not a mission, but was the chapel of Santa Margarita, an *asistencia* of the Mission of San Luis Obispo.

2. The California condor (*Gymnogyps californianus*) is now almost, if not quite, extinct.

3. Edward Tompkins, graduate of Union College, married Sarah Haight, lived in Oakland, was a state senator, worked for the founding of the University of California, and was a valuable supporter of the Survey.

4. The present Pebble Beach golf course is on the site of Pescadero Ranch. "The log house was built by a Mr. Gore, who was a brilliant scholar, kept aloof, was very proud, but was not considered eccentric by old Montereyans." (Information from Mr. Frank Doud, of Monterey.) Mr. J. Beaumont, secretary of Del Monte Properties Company, provides additional information derived from the title records. John C. Gore, the original claimant against the United States for confirmation of title, based his claim on a conveyance made in 1853, which gave him succession to the interest of Fabian Barreto, who had received the grant in 1840. Gore, on August 1, 1860, conveyed the property to Edward Tompkins, who sold it in 1862. It was acquired by the Pacific Improvement Company in 1880.

5. Fletcher Mathews Haight, a graduate of Hamilton College, practiced law at Rochester, N. Y., later at St. Louis, came to San Francisco in 1854, where he practiced law with his son, Henry, who had preceded him to California. In 1861 F. M. Haight was appointed United States District Judge. He died February 23, 1866, at the age of sixty-six. His son, Henry Huntley Haight, was governor of California, 1867–71.

6. La Misión San Carlos Borromeo was established in 1770 at Monterey, but was moved to the present site in 1771. No portion of the present structure was erected until 1793. The Mission was secularized in 1833 and the buildings rapidly fell in ruins. A partial restoration of the buildings was made in 1887.

7. The famous Monterey cypress (*Cypressus macrocarpa*) is found in an exceedingly restricted area—the Monterey peninsula and the neighboring Point Lobos.

8. The first, *encina,* is the coast live oak (*Quercus agrifolia*); the second, *el roble,* is the valley oak (*Quercus lobata*).

9. The madroña (*Arbutus menziesii*).

10. The two brothers were Charles W. and James Finch. The latter married Ellen O'Neil, daughter of Major John M. O'Neil who came to California in 1847 with Stevenson's Regiment.

BOOK II
1861

CHAPTER I

AN INTERLUDE

San Juan—A Call to San Francisco—Union Sentiment—Personnel Problems—Hot Weather—Doctor Cooper—San Juan Mission—Monterey Again—Sea Life.

San Francisco.
Sunday, June 23, 1861.

I SENT my last a few days ago from San Juan, to go by the steamer of yesterday.[1] You are surprised, no doubt, to see my letter dated at San Francisco. But, although I sent Averill up on Monday last, yet Tuesday I received a letter from the Professor asking me to come up here immediately to confer with him on some important business relating to present and future plans of operations. The next morning found me on the stage, and Wednesday night found me here. But I will continue my journal in the order of time. I closed my last letter Sunday or Monday.

On Monday, June 17, I explored alone some high sandstone hills southwest of San Juan—hills covered with wild oats, with here and there bold outcroppings of coarse red sandstone, often worn into fantastic and grotesque forms by the weather. Ledges would be perforated with numerous holes and caverns from the smallest size up to those capable of holding a hundred men. Thousands of swallows had built their nests in them, as they build in barns at home. I sat down in one of these caves, the largest I saw, opening out to the valley and commanding a lovely view. A steep ravine lay below me, brilliant and fragrant with flowers, around which swarms of humming birds were flitting, like large bees. Various species of humming birds are common, but I have nowhere seen so many as at that place.

They often came in the cave, hovering in the air near me; several times they would stand in the air so close to me that I could touch them with my hand easily, then they would dart away again.

The aspect of the scenery around San Juan is peculiar—a level valley, enclosed in large fields, hills rising beyond; all covered with oats, now ripe. The hills have a dry, soft, straw-colored look, which, in the twilight, or by moonlight, is peculiarly rich.

On Tuesday I rode a few miles to visit an asphaltum spring, or rather, several. The principal ones are on a ranch of a Mr. Sargent. He has his ranch enclosed with good fence in two fields, one of two thousand acres; the other, through which we rode, of seven thousand acres. It is hill land, and most of it is covered with a heavy growth of wild oats, in places as heavy as a good field of oats at home—a grand field of feed, that!

The mail arrives at San Juan near ten o'clock in the night, but I walked into town and waited for it that night, and with it came a letter from Professor Whitney calling me to San Francisco for consultation. I returned to camp, hastily made my preparations, left my orders, and returned to town, for the stage started at six in the morning, too early to get in from camp. Six o'clock the next morning found me in the stage for San Francisco, where I arrived at seven in the evening. The region passed through, especially about fifty miles of the Santa Clara Valley, is the garden of California—a most lovely valley, settled by Americans, with fertile fields and heavy crops, beautiful gardens and thriving villages.

There were the usual accompaniments of a stage ride—various passengers, bristling landlords running out at the stations, dust, dirt, politics, and local news. Professor Whitney had invited me to stay at his house, where I went after taking a bath and buying and putting on clean clothes—a "biled shirt" was a luxury not enjoyed for a long time before.

Thursday, Friday, and Saturday I spent in running around,

talking over plans, seeing men on business, working up geological notes, posting the Professor up on progress, etc. Every minute was occupied. How busy, bustling, hurrying, high-wrought, and excited this city seems, in contrast with the quiet life of camp!

There is much improvement going on here, even more building than usual. Business is brisk, but rates of exchange with the East are enormous—more than twice as high as they were before the War, or even three times. When I sent money home in the fall, it cost three per cent, now six to ten per cent; and so it will remain until the fear of privateers ceases.

It has been most lovely weather, but the afternoons are windy, and the air often filled with dust. No hot weather—this city is always cool—never hot, never cold.

Professor Whitney has a house several hundred feet above the city, on a hill, with a most lovely view of the city, the bay, and the hills beyond. These lovely moonlight nights the scene is surpassingly beautiful. The city below, basking in the soft light, the myriad gas lights, the bay glittering in the moonbeams, the ships, the opposite shores in the dim light, all form a picture that must be seen to be appreciated. This is indeed a lovely climate—children and women look as fresh and rosy as in England.

I meet many friends and acquaintances here; all say, "How fat you have grown," "How camp has improved you," "How stout and healthy you look," and similar assurances. Every friend I have met speaks of it.

There is a very strong Union sentiment prevailing here, although the governor is Secession, and there are thousands of desperadoes who would rejoice to do anything for a general row, out of which they could pocket spoils; yet the state is overwhelmingly *Union*. Flags stream from nearly every church steeple in the city—the streets, stores, and private houses are gay with them—but all are the Stars and Stripes—a Palmetto would not live an hour in the breeze.

On Saturday night at ten o'clock a flag was raised on T. Starr King's church. He is very strong for the Union, and this was for a surprise for him on his return from up country. A crowd was in the streets as he returned from the steamer. He mounted the steps, made a most brilliant impromptu speech, and then ran up the flag with his own hand to a staff fifty feet above the building. It was a beautiful flag, and as it floated out on the breeze that wafted in from the Pacific, in the clear moonlight, the hurrahs rent the air—it was a beautiful and patriotic scene.

Sunday I went to hear him preach. He is a most brilliant orator, his language strong and beautiful. He is almost worshiped here, and is exerting a greater intellectual influence in the state than any other two men.

Today I have been ordering a new wagon made, buying supplies, writing, etc., and must go soon to a meeting of the Academy of Natural Sciences which meets tonight (Monday). Professor Whitney has a grand scheme for erecting a great building for the state collections here; we are ventilating the matter here now. Today I saw my cousin, Henry Du Bois. He is working at his trade in this city.

Friday Evening, June 28.

I HAD expected to leave San Francisco on Tuesday morning, but Tuesday I had to meet some men to talk over matters relating to our cabinet building, so was delayed until Wednesday. We left at eight o'clock, on the top of the stage, which seat we kept all the way through, having a fine chance to see the country.

San Jose and Santa Clara are large and thriving towns, but the whole country looks dry now. The fields are all dry and yellow, the herbage on the waste lands eaten down to the very roots, the fields of grain ripe for the harvest. Many reaping machines were at work and prosperity and abundance appeared to smile on the lovely region. But *how dry it looked!* Hundreds of windmills pump water from the wells for the cattle and for irrigating the lands, but the streams are dry, and sand and clouds of dust fill the dry air. While seated on the stage we often could not see the leaders at all for the dust. Yet there will be no more rainfall for the next three or four months to revive the soil or green the landscape. The driver said the dust often became *very* fine, and eight inches deep, before the close of the dry season, filling the air with dust clouds.

Much as can be said about this lovely climate, yet give me our home climate, variable as it is. This is healthy, very healthy, lovely, but it is monotonous—four more months, long months of dry air and clear sky. The budding freshness of spring, with its resurrection of life; the summer with its flowers, its showers, its rich green; autumn with its fruits, its fall of leaves, its gorgeous forests; winter, with its dead outer world, but its life at the fireside—all are unknown here. A slowly dawning spring, tardily coming, is followed by a slower, dry, arid summer. This is the climate for a lazy man, that for labor, this for *dolce far niente,* that for *action.*

We got back a little after ten in the evening, and before eleven were in camp again. It seemed like home. The soft, luxurious bed of the city had given me quite a severe cold, which my better sleep on my blankets is improving. Ah! *camp* is the place to sleep—sweet sleep—refreshing sleep. There is no canopy like the tent, or the canopy of Heaven, no bed so sweet as the bosom of Mother Earth.

I have not had everything to my mind in camp. Men who came into it with enthusiasm, now that the novelty has worn off, abate their zeal. I had a long talk on the matter with Professor Whitney and I was thinking of either discharging Mike, or making him "toe the mark" closer. I determined to try the latter first. He has been surly under reproof and slack in his work for some time. His work was very light and his wages high (forty-five dollars per month). But yesterday I was saved all the trouble by his coming to me and asking to quit. I discharged him without further words and he went to San Francisco this morning. He has lost a good place, but I am glad to be rid of him. He, however, thought much of me, shook me cordially by the hand on parting, and said, "Mr. Brewer, I have always got along well with *you*. I have never worked under a man I liked better. You are a *gentleman*"—very emphatic. He then took his leave without saying a word or bidding good-bye to either Averill or Guirado, with whom he had not got along so smoothly. Ere this he is in "Frisco," as the metropolis is called.

We have changed plans. We keep in the field as late in the fall as we can, then disband and do our office work in the winter. This seems advisable. We have at least three men on the Survey who do not well sustain themselves. Two of these were employed because of influential political friends in this state, whom it was desirable to appease. It is not a pleasant matter to dismiss them, but by "disbanding" for the winter, stopping field work, we can throw them out. I feel sorry for the individuals, but believe it to be for the interests of the Survey. From the confidence with which I am consulted on these important matters, I feel that I am sustaining my place even better than I dared hope for, yet it is by no means impossible that policy may remove me too—but I hope not.

San Juan.
July 2.

My last letter was sent three days ago, but I fear for its safety; while the secession troubles last in Missouri the Overland may be troubled. I shall send the next by Wells & Fargo's Express. Way mails in this state are so uncertain that all important letters are carried by a private express in government envelopes. The company sends three-cent letters for ten cents, and to the states, ten-cent letters for twenty cents. Here in this state it is used very largely, the Wells & Fargo mail being often larger than the government mail. We avail ourselves of it, even on so short a distance as from here to San

Francisco, if the letter has any special importance or needs to go with certainty of dispatch. I have had letters two weeks in getting where they ought to go in two days with a daily mail. I very strongly suspect that some of the letters between here and home that were so long delayed were delayed on this side. One letter was from the first to the eighteenth of June coming from San Francisco to this place, one day's ride.

War news becomes more and more exciting; the "Pony" brings all the general news far in advance of the mails. I do not dare to think where it will all end; but I trust that in this, as in other matters, an All-ruling Providence will bring all things to work together for good in the end. The Stripes and Stars wave from the peak of our tent, the ornament of the camp.

Yesterday I climbed a high steep peak about eight miles distant, a hot, toilsome day's trip. We rode up a valley about three miles, then struck up a narrow, deep side canyon about one or two miles farther, beside a small stream. A cattle trail led up the stream, often steep and slanting, but our trusty mules managed it. Arriving at the end of this trail, we unsaddled and tied our mules beneath some trees, and mounted the ridge. It was *very* steep, of a rotten granite that was decomposed into a sand that slid beneath the feet and reflected the intense heat of the sun with fearful effect. It seemed as if it would broil us, and the perspiration flowed in streams. Often not a breath of wind stirred the parched, scorching air. A wet handkerchief was worn in the crown of the hat, as we do now all the time when in the sun, to save from a possible sunstroke.

A ridge was gained, which ran transverse to the main ridge we wished to reach. Three miles more along this ridge, sometimes down, sometimes up, now over sand, then over rocks and bushes, sometimes in a scorching heat, at others fanned by a breeze—at last we planted our compass on the highest crest of the Gabilan, a very sharp, steep, bold peak, some 2,500 to 3,000 feet high, probably nearer the latter. We had a most magnificent and extensive view of the dry landscape—the Salinas plain, Monterey and the Santa Lucia, the sea, the hills north of Santa Cruz, San Juan and its valley, the valley of Santa Clara beyond, an immense stretch of landscape beside.

Our canteens were exhausted and our lunch dispatched before we had arrived at the top, but the descent was easier, although the last slope before reaching our mules was intensely hot. A fine, clear, cold stream flowed in the canyon—and how sweet it tasted!—but it sinks before entering the valley of San Juan. A ride back to camp, a sumptuous dinner prepared by Peter, then a lounge under the large oak that stands by our camp, in the delicious breeze, then the comet in the west at night, made us forget the toils of the day.

There is a very curious meteorological fact connected with these hills. It is cooler in the large valleys, and hotter on the hills. A delicious cool breeze draws up the larger valleys from the sea, sometimes far too violent for comfort, but always cooling. It often does not reach the hills, so that rising a thousand feet often brings us into a hotter instead of cooler air. The hot weather is now upon us, much hotter than in the immediate vicinity of the sea. One does not so notice it if still or in the shade. Today it is by no means oppressive in the shade for there is a delicious breeze. It has not been above 90° F. under the tree where I write this during the day, but out in the sun or in exercise it is hot, *very* hot.

This morning at ten, when the thermometer was 80°, I laid it out in the sun and in fifty minutes it ran up fifty degrees, or to 130°. There the graduation stopped, so I could not measure the actual heat; it was probably about 140°, although it sometimes rises to 165° or 170° F. in some of the valleys of this state! Surely the state is rightly named from *calor*, heat, and *fornax*, furnace—a *heated furnace*, literally.[2]

I have not been so active today; it is too hot to enjoy work. Iron articles get so hot that they cannot be held in the hand; water for drinking is at blood heat; the fat of our meats runs away in spontaneous gravy, and bread dries as if in a kiln. You can have no idea of the dryness of the air. When I wash my handkerchief I don't "hang it out" to dry, I merely hold it in my hand, stretched out, and in two or three minutes it is dry enough to put in my pocket.

July 4.

THE "Glorious Fourth" is at hand, but I am quiet in camp, and alone. All have gone into town to celebrate. Yesterday, while hammering a very hard rock, to get out a fossil, a splinter of the stone struck me in the eye and hurt it some. It is by no means serious, but somewhat painful, and I, therefore, will keep quiet in camp today to prevent any inflammation or bad turn. With a handkerchief bound over it, it feels quite comfortable, and I apprehend nothing serious. I have one eye left for writing.

Yesterday I rode five or six miles to visit a range of hills north of the town, across the San Benito River. They were gentle, grass-covered hills, or rather *oat*-covered hills, enclosed in enormous fields and rich in pasturage. The oats were in places as heavy and thick as in a cultivated field, but generally not so heavy. They are ripe and dry now. There was some stock in, but the land here is not nearly so heavily stocked as it is farther south. There were trails up the sides, which in places were quite steep, but we rode to the summit, eight hundred or a thousand feet above the plains, an isolated ridge between the plains of San Juan and the stream on the north. That stream, tributary to the Pajaro, is in a wide valley, but the stream is a small one. The San Benito, which at times is a wide and rapid river, is now a small stream, scarcely ankle deep, filtering over its sandy bed, except where it passes through rocky hills, where it is quite a river. The view from our elevation was quite extensive and beautiful. We were back at camp before night.

Peter acts the part of cook now until we can get another suitable person. Would that he could be induced to keep the office. He is as neat and skilful as if he had served an apprenticeship in a French restaurant. We had a sumptuous dinner, but with few courses. With his revolver he had shot a large hare, which was served up in splendid style.

On our arrival in the state last fall we met a Doctor Cooper, who was very anxious to get the place as zoölogist to our Survey or as an assistant in that department.[3] He was a young man, scarcely my age, but had been over most of the United States, had crossed the plains several times, had seen much of California, and all of Washington and Oregon. He had written a large book (with Doctor Sulkley) on the *Natural Productions of Washington Territory*, was well posted in his department, was a man of more than ordinary intellect and zeal in science, but I fear not a very companionable fellow in camp. He was employed during the winter as surgeon at Fort Mohave, on the desert on the southeastern border, and Professor Whitney employed him to collect plants and make observations for us there. He was just returning to San Francisco by stage, when he stopped over night here, and we most unexpectedly met him last night. I think Professor Whitney will employ him, at least for a time. I have got many items for him, and he has collected some four or five hundred species of plants from the deserts for me.

This place is very dull for the day; everybody has gone to Watsonville, fourteen miles distant, where there is a celebration. Peter burnished his harness, harnessed the mules, and with flags on their heads, has gone to town in patriotic style. But it is hot here; the daily breeze has not yet sprung up, the flag droops lazily from our tent, and the thermometer is 85° to 90° in the tent. About noon the breeze will begin and it will be cooler.

Sunday, July 7.

YESTERDAY I rode eight or ten miles, visiting some tar springs and oil works, where oil for burning is made from asphaltum.

We received a letter a day or two ago from Professor Whitney that he would be down either Saturday or Monday, *for certain,* so I suppose he will surely be here tomorrow, after our long suspense in waiting for him. He writes, and other letters and the papers confirm it, that several shocks of earthquakes took place last week in San Francisco, and reports come in from other parts of the state. I think that I have before told you that nearly or quite the whole of this state is subject to earthquakes. There will never be a high steeple built in the state, and in San Francisco the loftiest houses are but three stories high, the majority only two. There are two hotels four stories, but all old residenters will not stop at them for fear of having them shaken down over their heads. The shocks last week (six principal shocks) were quite severe, much more so than usual, and caused much alarm. Our office is in the Montgomery Block, a high three-story stone building, full of offices, but a building many are afraid of. Professor Whitney says he thinks that all of the occupants ran out into the street except him, and says that the building rocked and swayed finely. The city was in terror, the streets filled in an instant, everyone excited. No business was urgent enough to keep a man at it; men rushed into the streets from barber shops with their faces lathered and towels about their necks; men even rushed naked from baths, but were stopped before reaching the city in that primitive costume. We have perceived no shocks, but they were reported in the vicinity two weeks ago.

I went to Mass this morning in the Mission Church.[4] It is a fine old church, with thick *adobe* walls, some two hundred feet long and forty feet wide, quite plain inside, and whitewashed. The light is admitted through a few small windows in the thick walls near the ceiling, windows so small that they seem mere portholes on the outside, but entirely sufficient in this intensely light climate, where the desire is to exclude heat as much as to admit light—so the air was cool within, and the eyes relieved of the fierce glare that during the day reigns without.

The usual number of old and dingy paintings hung on the walls, the priest performed the usual ceremonies, while violins, wind instruments, and voices in the choir at times filled the venerable interior with soft music. I never wonder that the Catholic church has such power over the *feelings* of the masses, especially when I compare its ceremonies with such as I saw in a Protestant church here. A congregation of perhaps 150 or 200 knelt, sat, or stood on its brick floor—a mixed and motley throng, but devout—Mexican (Spanish descent), Indian, mixed breeds, Irish, French, German. There was a preponderance of Indians. Some of the Spanish *señoritas* with their gaily-colored shawls on their heads, were pretty, indeed. It is only in a Roman church that one sees such a picturesque mingling of races, so typical of Christian brotherhood.

With scarcely enough Protestants here to support one church well, there are three churches, wasting on petty jealousies the energies that should be exerted in advancing true religion and rolling back the tide of vice swelling ·in the land.

Tuesday, July 9.

LAST night Professor Whitney arrived, bringing with him a topographer,[5] so today our company is quite lively again. I have returned from a trip on the Gabilan hills—quite a ride. Tomorrow the Professor and I will go to Monterey to be gone three or four days.

San Juan.
July 14.

IT is a quiet Sunday, and, although the wind blows, it is too hot to write in the tent, so I write in the shade of a fine oak by our camp. My last was sent the first of last week, by express.

Well, Professor Whitney arrived on Monday night. Tuesday was spent in arranging some small matters, and Wednesday,

July 10, we started for Monterey—Professor Whitney, Averill, and I. We were up at dawn, had our breakfast, and by half past five were in our saddles. I took one of the team mules to ride, being stronger than my little one. The early morn was clear, but soon the fog rolled in from the sea, enveloping the hills.

It was thirty-nine miles to Monterey, but a mountain trail shortened the distance some five miles, so we took that, although neither of us had ever traveled it. First up a canyon, then across the ridge about a thousand feet high, by a steep winding trail, then down on the other side. Our trail was often obscure, mingling with cattle paths, and the dense fog obscured all landmarks, but in about seven miles we emerged on the Salinas plain, where we took the stage road and crossed the plain. There was some wind, and it was cool, but the fog did not entirely obscure the hills. We stopped at Salinas, and fed both ourselves and our mules, then rode on. On striking the valley that leads up to Monterey for about sixteen miles we had hotter air, but not much dust. We arrived before night, and found the town (city, I should say—a *city* of six hundred or eight hundred population!) in much excitement over a recent discovery of silver mines in the vicinity, but which *I* don't think will ever prove of any value.

The next morning we went on to Pescadero Ranch, found no one at home, so climbed in by the window, opened the back door, and "took possession." This was the place where we had encamped so long, you recollect. I had found the geology too much for me, and I wanted Professor Whitney to see it; hence our visit to Monterey, for it was a matter of some importance to settle. Mr. Tompkins, the owner of the ranch, had tendered us its hospitalities, but his *buccaro*, Charley, was gone—all the dishes dirty on the table, and no provisions to be found, no candles, no wood cut. We spent the day looking up the objects we had come to see. Averill went into town, four miles, and got supplies, we washed up the dishes, got our dinner and supper, and made ourselves comfortable. Professor Whitney was as much interested as I had been, both in the geology and in the abundant life in the sea.

I wish I could describe the coast there, the rocks jutting into the sea, teeming with life to an extent you, who have only seen other coasts, cannot appreciate. Shellfish of innumerable forms, from the great and brilliant abalone to the smallest limpet—every rock matted with them, stuck into crevices, clinging to stones—millions of them. Crustaceans (crabs, etc.) of strange forms and brilliant colors, scampered into every nook at our approach. Zoöphytes of brilliant hue, whole rocks covered closely with sea anemones so closely that the rock could not be seen—each with its hundred arms extended to catch the passing prey. Some forms of these "sea flowers," as they are called because of their shape, were as large as a dinner plate, or from six to twelve inches in diameter! Every pool of water left in the rugged rocks by the receding tide was the most populous aquarium to be imagined. More species could be collected in one mile of that coast than in a hundred miles of the Atlantic coast.

Birds scream in the air—gulls, pelicans, birds large and birds small, in flocks like clouds. Seals and sea lions bask on the rocky islands close to the shore; their voices can be heard night and day. Buzzards strive for offal on the beach, crows and ravens "caw" from the trees, while hawks, eagles, owls, vultures, etc., abound. These last are enormous birds, like a condor, and nearly as large. We have seen some that would probably weigh fifty or sixty pounds, and I have frequently picked up their quills over two feet long—one thirty inches—and I have seen them thirty-two inches long. They are called condors by the Americans. A whale was stranded on the beach, and tracks of grizzlies were thick about it.

The air was cool, and at times fog rolled in from the Pacific, as it often does there. We found beds and blankets, and after breakfast the next day rode to Point Lobos, then over some high hills back of Carmelo Mission, but the fog obscured the fine views I wanted Professor Whitney to see.

We descended into the valley, and called at Judge Haight's, where we had visited before. Professor Whitney soon returned to Pescadero, but the young ladies pressed us so cordially to stay to tea that Averill and I did, and had a most pleasant visit. It is a very intelligent and pleasant family indeed. Our tea, and a walk we took with the ladies, detained us so that we had to ride home after dark. This would be a light matter at home, but not so here, where for three or four miles the trail led through a woods or dense chaparral as high as our heads or higher, where grizzlies sometimes dispute the right of way, and across a dark gulch with almost perpendicular sides, where none of you would trust yourselves to ride by broad daylight.

Saturday morning we had intended to start back, but were detained on our way, in Monterey, until noon, so we only reached Salinas, twenty miles from Pescadero.

An excitement in the dull monotony of the little town was occasioned by the arrival of an English brig, of only 180 tons, six months out from London. She had not seen land during all that time. She was in a terrible condition, and had put in in distress. Provisions and water very scant and bad all the way, now exhausted, men sick of the scurvy, captain dead of the same disease, second mate and boatswain lost in a storm, sailors decidedly used up—their story was a pitiful one.

As I said, we stopped last night at Salinas. This morning we were up early, and were off, crossing the plain, then over the mountain trail again, and by ten o'clock were at camp, where we found all well. We had our blankets washed during our absence. We are resting quietly this lovely afternoon after our long ride.

NOTES

1. This letter was lost. Professor Brewer's notebooks, however, show that camp was at Pescadero and Monterey June 1 to 12, at Natividad (near Salinas) June 12 and 13, and from June 14 at San Juan.

2. The origin of the name "California" has been the subject of much discussion. In 1862, Rev. Edward Everett Hale called attention to the fact that the name appeared intact in an old Spanish romance, *Las Sergas de Esplandián*, and was almost certainly familiar to Cortez and his contemporaries who first placed it on Pacific shores. That in some way the name came from this source is now the prevailing opinion. Prior to this explanation, however, there were many ingenious speculations, most of them taking the form of compounds from Latin and Greek roots; some seeking an Indian origin. The subject is discussed in Charles E. Chapman, *A History of California: The Spanish Period* (1921), chap. vi; in Ruth Putnam, *California: The Name*, "University of California Publications in History," Vol. IV, No. 4 (December 19, 1917); and in the *California Historical Society Quarterly*, Vol. I, No. 1 (1922), 46–56; Vol. VI, No. 2 (1927), 167–168.

3. James Graham Cooper (1830–1902) was the son of William Cooper (ornithologist, friend of Audubon, Nuttall, Torrey, and Lucien Bonaparte, and one of the founders of the Lyceum of Natural History of New York). He was graduated in 1851 from the College of Physicians and Surgeons, New York; in 1853 he contracted with Governor Isaac I. Stevens, of Washington Territory, as physician of the northwestern division of the Pacific Railroad Survey. In addition to his medical duties he made botanical and zoölogical collections and meteorological observations. He continued to engage in similar work up to the time of his connection with the California Survey. The results of his work under Whitney are preserved in a publication of the Survey issued in 1870, *Ornithology*, Vol. I, "Land Birds," edited by F. S. Baird from Dr. Cooper's manuscript and notes. In his own *North American Land Birds*, Professor Baird remarks: "By far the most valuable contribution to the biography of American birds that has appeared since the time of Audubon, is that written by Dr. J. G. Cooper in the Geological Survey of California." Dr. Cooper was commissioned by Governor Lowe, in 1864, Assistant Surgeon, 2d Cavalry, California Volunteers. In 1866 he married Rosa M. Wells, of Oakland. He lived in Ventura County from 1871 until 1875, when he took up his residence in Hayward, where he lived for the remainder of his life. The Cooper Ornithological Club, named in his honor, was organized in 1893. (References: "Dr. James G. Cooper. A Sketch," by W. O. Emerson, in *Bulletin of the Cooper Ornithological Club*, Vol. I, No. 1, Santa Clara, January-February, 1899; "In Memoriam," by W. O. Emerson, and "The Ornithological Writings of Dr. J. G. Cooper," by Joseph Grinnell, both in *The Condor*, Vol. IV, No. 5, Santa Clara, September-October, 1902.)

4. La Misión San Juan Bautista was founded 1797.

5. Charles Frederick Hoffmann was born at Frankfurt-am-Main, Germany, 1838, and was educated in engineering before coming to America. In 1857 he was topographer for Lander's Fort Kearney, South Pass, and Honey Lake wagon-road survey. He came to California in 1858. Next to Whitney himself, he had the longest connection with the California State Geological Survey, remaining through all vicissitudes until its discontinuance in 1874. During a hiatus in the Survey, 1871–72, he served as Professor of Topographical Engineering at Harvard. In 1870 he married Lucy Mayotta Browne, daughter of J. Ross Browne. For many years he was associated with his brother-in-law, Ross E. Browne, in the practice of min-

ing engineering, for a time at Virginia City, Nevada, later at San Francisco. He also managed mines in Mexico, and at Forest Hill Divide, California, and investigated mines in Siberia and in Argentina. During the latter part of his life he lived in Oakland, where he died in 1913. The importance of his work on the California Survey and its influence upon the development of topography in the United States have been mentioned in the introduction to this volume.

CHAPTER II

NEW IDRIA

The Road to New Idria—Examining the Mines—Hot and Dry—Back to San Juan.

New Idria Quicksilver Mines.
Sunday, July 21, 1861.

MONDAY, July 15, was spent in making preparations for leaving San Juan. All baggage that could be dispensed with was left in town; we were cut down to the shortest allowance. Tuesday we were up at early dawn, had our breakfast before five, and were soon loaded up. We passed east across the level San Benito Valley, which for eight or ten miles was as level as a floor, then struck up a side valley. We lunched at a ranch, Tres Pinos, fifteen miles, and then struck up the Canyon Joaquin Soto (*Arroyo los Muertos* of the map). The country became drier and the air hotter. Hills rose on each side, often high and steep, and in places the valley widened, showing that much more water once existed there. High terraces indicated the shores of an ancient lake. These terraces were most beautiful and made a remarkable contrast with the rugged rocky hills behind.

We at last entered a narrow canyon, and before night stopped at a spring where a Mr. Booker has a ranch. Desolate and dry as this region is, over some of the hills grass and oats grow, which although scanty are very nutritious, and a few cattle and sheep may be grazed. In this barren, hot, scorching, inhospitable, comfortless place was the ranch, and the proprietor regaled us with an account of its advantages. "The finest ranch anywheres near!" "Good water!" "The fattest of cattle!"

Around this place rise high hills, entirely of stratified gravel, in some places nearly as hard as rock, which has been washed down from the mountains and has formed a deposit of incredible thickness. It is cut into steep bluffs, in many places six hundred or more feet high, very steep; and the hills, of the same material, rise 1,800 feet high. It must have a thickness of at least two thousand feet, or nearly half a mile! Yet it is quite modern. Surely this region could not always have been so dry. Not the least remarkable fact is that a part of this has been turned up on edge by earthquakes or similar agencies in such a late geological period.

We spent the forenoon of Wednesday, July 17, examining this deposit and measuring the height of one of the hills, and in the afternoon came on twelve or fourteen miles to a well, the only water for thirty-five miles of our road. Our road lay up a canyon, rising, and our camp was near the summit, about two thousand feet above the sea, yet the day had been intensely hot. Over these hills there is scanty pasturage; a man has three thousand sheep which feed there and water is drawn by hand for them during the dry summer. The mountains here are depressed, a "gap" as it were, so the road does not rise above 2,000 or 2,500 feet. We slept in the open air, as usual. How clear the sky was! Never in Switzerland even have I seen so clear a sky of nights. The thermometer fell to 40° F. (only eight degrees above freezing) at night, yet there was no dew, so dry is the air at this season.

Early dawn found us astir, and after a scanty breakfast we were early on the way. We crossed the ridge and then descended into the valley of Little Panoche Creek. Beyond the ridge we found many pretty oak trees scattered over the gentle hills into which the chain is divided, and scattered pines were seen on the mountains at some distance, but on sinking a few hundred feet into the valley an entirely different landscape was entered. A plain extends for many miles up into the mountains, the bed of an old lake, its bottom level as a sheet of water. Lines of terraces around the margin tell unequivocally and plainly of old shores and different levels of water. This is on the map between the two branches of the Panoche River, now dry beds of sand. We crossed this plain, some ten miles or more —naked, dry, desolate. No tree cheered the landscape, no living thing, save insects, told of life. Toward the southeastern end of the valley we came to a clump of cottonwoods along the dry river bed, whose lively green looked refreshing; then we turned up a very wild, rocky, narrow canyon, the Vallecito, about six or seven miles, to a place where we found water and a house—Griswold's. The heat of the sun in this canyon was fearful.

At Griswold's the stream emerges to the surface, poor, salt, alkaline water, and here he has his ranch. We camped under some cottonwoods, bought a bag of barley for our mules, and resolved to stay the rest of the day. The place was dirty enough—dry, dusty sand, which hot gusts of wind at times blew into our eyes and mouths—there was neither cleanliness nor comfort in taking our meals and the water was very nauseous. Griswold, a hard looking customer, expatiated on the qualities of his "ranch"—squatter claim of course, for who would *buy* such a place? His cattle were "the fattest in the state" (and, strange enough, they were in good order); "*such* good water," etc. "To be sure," he said, "it does physic people when they drink much of it, but a little whiskey kills the alkali." Sitting by his "spring," a dirty mudhole, and pointing exultingly to some frogs, he exclaimed triumphantly, "Look at them toads—they can't live in poor water!" It was interesting to me to know that a frog could live in such a solution. The thermometer stood at 92° F. in the coolest place I could find, in the shade and wind. He said it was the coolest day for over a month, *too* cool for him after the last two months' heat!

The thermometer, placed in the sun, soon rose to 150° F. on this "cool" day, and that with the wind. You can well imagine what the *hot* days must have been. That night it sank to 46° F., a daily range of over one hundred degrees!—a range as great as from zero of winter to the hottest in the shade of our summer. With this daily change in temperature, no wonder that the plants of this state are so peculiar!

We found some very interesting geological facts there, so we waited over. We were up at dawn and came on ten or twelve miles to a stream, where we stopped and took breakfast. This brook looked refreshing, the first for fifty miles, and our morning ride was refreshing. Once we came on a drove of ten antelope, the first we have seen. They were very plentiful a few years ago in this state, in large flocks. They are very graceful, pretty animals, like small deer, but little taller than sheep. After breakfast, Professor Whitney and I came on ahead, leaving the rest to follow. We struck up a very wild, picturesque canyon into the heart of the mountains, rising very fast, and in six or seven miles, before noon, came to the furnace and the director's house, below the mines. We introduced ourselves, and found a place to camp, and two hours later the rest of the party arrived.

The New Idria quicksilver mines lie in the heart of the chain of mountains which runs southeast from Monte Diablo. There are three principal mines—New Idria, Aurora, and San Carlos—all in the center of the chain, with some six or seven miles between their extreme limits. The furnace and the superintendent's quarters are in a valley some six miles from some of the mines, and three or four miles from the others, at a height of 2,500 to 2,700 feet above the sea. The highest mine lies 2,500 feet higher still. Mr. Maxwell, the superintendent, is very obliging and shows us every attention. Friday afternoon we examined the furnaces and the works for preparing and reducing ore. These are on a large scale and are very complete and scientific in their arrangement. Averill and I dined with Mr. Maxwell that afternoon, and champagne was introduced in our honor.

Yesterday, Saturday, July 20, we started under the guidance of Mr. Maxwell for the San Carlos Mine. The road ran up a canyon and over ridges—steep, yet a good road. Three yoke of oxen were toiling up with an empty wagon to bring down ore. We passed a cluster of tents and cabins of the miners at "Centerville" on the way. A single frame house perched on a lofty crag at a dizzy height was seen from below, and being the only house we asked what it was. "Oh, a billiard saloon and drinking house," was the answer. "A man recently built it and I believe he has other refreshments for the miners, a load of *squaws* went up a day or two ago." We passed this cluster of cabins and continued our way, and in due time reached the San Carlos Mine.

The mine is almost on the summit of a mountain about five thousand feet high. The ore is diffused in streaks through the rocks and is wrought extensively. Diggings, galleries, shafts, and cuts run in every direction, wherever the richest ore may be found. The rock is a very remarkable one, a sort of altered slate, acted on by heat and hot water, and the brilliant red ore is diffused through it. We spent five or six hours there, visiting every working. We planted our barometer on the summit. I had carried it up, and we got our observations for altitude, to be calculated after our return. The miners are mostly Chileans, a hard set, and their quarters are in shanties covered with bushes, in huts, and even in deserted workings. We went into one to get a drink of water; it was a deserted gallery, and a squaw was in attendance, the wife *pro tem* of two or three miners.

The view from the summit is extensive and peculiar. It is to the north and east that the view is most remarkable—the Panoche plain, with the mountains beyond—chain after chain of mountains, most barren and desolate. No words can describe one chain, at the foot of which we had passed on our way— gray and dry rocks or soil, furrowed by ancient streams into innumerable canyons, now perfectly dry, without a tree, scarcely a shrub or other vegetation—*none*, absolutely, could be seen. It was a scene of unmixed desolation, more terrible for a stranger to be lost in than even the snows and glaciers of the Alps.

Beyond this lay the San Joaquin, or Tulare, Valley, wide and dreary. It is fifteen to twenty-five miles wide, without trees, save a green belt along the river—all the rest dry and brown. Dust rose from it, shutting out the mountains beyond, but in places we could see the snows of the Sierra Nevada glittering in the sun through the veil of dust that hung between us. They looked grand and sublime in the faint outlines we could see and appeared ten or twelve thousand feet high. Although we were so high, five thousand feet, yet the temperature was about 80° F.

New Idria Mine.
Wednesday, July 24.

SUNDAY, July 21, seemed hot enough but it was only the beginning. We remained three days longer, the thermometer each day rising from ninety-five to one hundred degrees in the coolest place we could find. God only knows how high in the *hottest* places!

Monday we visited the largest mine, the New Idria Mine proper. We were on hand early and three of us went in, accompanied by the superintendent and mining captain. We spent the day under ground. For six hours we threaded drifts, galleries, tunnels, climbed over rocks, crawled through holes, down shafts, up inclines, mile after mile, like moles, sometimes near the surface, at others a thousand feet from daylight.

The distribution of ore through the rock is very capricious, and where a thread of it can be found it is followed up, so the workings run in every conceivable direction, and being mostly mined by Chilean and Mexican miners, the work is more irregular by far than the burrows of animals. Sometimes we climbed down by a rope, hand over hand, bracing the feet against the wall of rock, sometimes on *escaladors*, sticks merely notched. But the trip was interesting, and as they wanted our professional advice, we saw all, the two men devoting the day to us.

Iron pyrites occurs in the rock, which decomposes on exposure to the air, causing heat, and the temperature of some of the galleries was near 100° F. Most of the mine, however, was deliciously cool. Although so deep in places, there is no water, save a very little in the very lowest drift, and the rock has been so broken by volcanic forces and cracked in every direction and the galleries are so extensive that the air is perfectly pure and so dry that the miners use rawhide buckets.

The effect was often picturesque indeed—the brilliant red ore contrasting with the dull color of the rock, the miners, naked above the waist, their lithe forms and swarthy skins shown by the light of our candles, the broken walls, the occasional sound of a blast, like heavy, dull, underground thunder. We emerged at 4 P.M. into the hot, dry, scorching outer world and took our dinner at five o'clock—the air heated to near 100° in our cool camp, where there was a brisk breeze.

Tuesday we rode to the Aurora Mine, which is not good for much, then visited the top of a ridge 4,500 feet high. Here there was a very extensive vein of chromate of iron, the black heavy ore occurring in immense quantities. This was *supposed* to be silver ore of the richest kind! A company was formed, many tons got out, a furnace built, thousands of dollars expended, and then it was found that there was not a particle of silver in it. But this led to the discovery of the quicksilver mines.

But oh, how hot it was, even at that height! The sand burned our feet through our boots, and a stiff breeze, dry and hot as if from a furnace, played over the ridge. We saw much of geological interest and got back to camp before night.

Last night I returned on foot to a cragged ridge, 1,500 or 2,000 feet above camp, for specimens. It was a toilsome walk, but I was repaid. The sun set while I was there, coloring with orange light the barren mountains north and east and even showing plainly the snowy Sierra in the distance. The view was glorious but desolate as a desert. A few clouds curled over the distant snowy peaks, crimson in the rays of the setting sun. But the shades drew on, the valleys grew darker, and I took my way back. It was cooler, but still hot. I stopped often with my load of specimens. How still it was!—no sound of a bird in the evening twilight, no chirrup of insect, but silence, deathly stillness reigned.

Wednesday we spent in making another visit to the New Idria Mine, and getting ready to be off. It was even hotter than before. We took double sets of astronomical observations that day, for latitude and longitude. In such cases I have to "mark time" with the chronometers, while Professor Whitney observes with the sextant. He says that I mark time more accurately than many old astronomers who have practiced all their lives—making long sets agree to the tenth of a second.

Before leaving, a few words as a summing up. The mines have been profitable to the stockholders, and are still. They own several miles in extent, have a store, sell goods to the miners at great prices and profits. There are 250 or 300 persons employed in the various departments. The miners work by the job—the average wage is about three dollars a day, but often less than two, but in rich luck sometimes as high as twenty or twenty-five dollars a day. Such streaks of luck are profitable to the company as well as to the miners, and can only take place on finding unexpectedly very rich ore. The yield of quicksilver now is about nine hundred flasks of seventy-five pounds each per month, or about 67,500 pounds, and it sells at thirty-five to forty cents a pound. It is sent to San Juan, thence to Alviso on the Bay of San Francisco, and there shipped.

The work at the furnaces is much more unhealthy and commands the higher wages. Sulphurous acids, arsenic, vapors of mercury, etc., make a horrible atmosphere, which tells fearfully on the health of the workmen, but the wages always command men and there is no want of hands. The ore is roasted in furnaces and the vapors are condensed in great brick chambers, or "condensers." These have to be cleaned every year by workmen going into them, and many have their health ruined forever by the three or four days' labor, and all are injured; but the wages, twenty dollars a day, always bring victims. There are but few Americans, only the superintendent and one or two other officials; the rest are Mexicans, Chileans, Irish (a few), and Cornish miners.

I can hardly conceive a place with fewer of the comforts of life than these mines have—a community by itself, 75 miles from the nearest town (San Juan) and 135 from the county seat, separated from the rest of the world by desert mountains, a fearfully hot climate where the temperature for months together ranges from 90° to 110° F., where all the necessities of life have to be brought from a great distance in wagons in the hot sun. As might be expected, little besides the bare necessities of life is seen, and if any luxuries come in, it is only at an extravagant price.

Such is New Idria and by such toils and sufferings do capitalists increase their wealth!

San Juan.
July 29.

THURSDAY, July 25, we were up at three o'clock. We loaded up by the bright moon, and at four o'clock, just as the first streaks of morning light began to be seen in the east, we were on our way. The twilight in this latitude is shorter than it is farther north, and it soon became light. As we emerged from the canyon into a wider valley six miles below, the sun came upon us, now in all its force. We stopped at Griswold's, fifteen miles, for breakfast, and to feed. Our mules had had no hay or grass, merely dry grain (barley), and but little of that, since we got into this region. Our breakfast was eaten with the thermometer at 95°, although but 8° in the morning; but our appetites were good after seventeen hours' fast.

We were now on our way again, along the foot of the most

barren chain of mountains, or rather between two such, in the Vallecito Canyon. Then we struck across the plain of Panoche. I wish I might describe that ride that you might realize it, but words are tame. The temperature was as high as any traveler has noted it (so far as I know) on the deserts of Africa or Arabia. Hour after hour we plodded along—no tree or bush. A thermometer held in the shade of our own bodies (the only shade to be found) rose to 105°—it was undoubtedly at times 110°, while in the direct rays of the sun it must have fluctuated from 140° to 150° or 160°. I think, from other observations, it must have risen to the last figure!

Imagine our feelings when we found that after our first drink a hole had been knocked in the canteen and our water was all gone. But the plain was crossed, and at its edge we stopped for a few minutes under the shade of a large oak. Here it was cooler—the mercury sank to 100°. It was the first tree we had seen for twelve miles of the road—I mean near the road—and we were still ten miles from any water. Here the road takes over the hills, dry and barren, but with scattered trees, oaks. None of the trees there (although some of the oaks are large) have foliage enough to make a complete shade—the leaves are too small, scarcely an inch long, and too few to shut out the sun. So, too, the pines, although their leaves are very long, cast only a very scanty shade. Many shrubs have the leaves with their edges turned to the sun, like the trees of the deserts of Australia—a most curious feature!

As we climbed the hill, panting and thirsty, we met a wagon with some *watermelons!* A man visiting the mines had bought, at San Jose, 110 miles distant, some fine melons. Needless to say, we bought two, although the price was the modest sum of *two dollars each.* We stopped under a tree near by and ate them. They were large and fine ones, and I never knew before how deliciously refreshing a watermelon could be. He will sell the rest at two and three dollars each at the mines; a large muskmelon he valued at four dollars. These facts are significant as showing how rare such luxuries must be there to command such prices. I surely felt satisfied with my expenditure of a dollar for a third of one of them.

One of our mules, Old Sleepy, had been tied behind the wagon, his rider riding in the wagon. He had pulled back some, got a little choked, the dust and intense heat affected him like the blind staggers, and when discovered, grew suddenly very bad. Professor Whitney and I rode on ahead. It took several hours for the rest of the party to get him on to camp—the sheep wells before alluded to—the only water within many miles. Here we camped after our terribly hot day's ride. Old Sleepy grew worse, a bottle of brandy poured down his throat partially revived him, but he was very ill.

Friday was another hot day. It was impossible to move Sleepy, so after breakfast the rest of the party went on, leaving Hoffmann, Guirado, and me with him, the rest taking all the baggage to the next camp, fifteen miles distant, at Booker's.

Leaving Guirado with the mule, Hoffmann and I started to visit some hills in the vicinity. We toiled up—temperature 98° to 102° all the time—crossed a ridge, then a deep canyon, then on to another ridge, where we found a worse canyon between us and the peak we desired to scale. But, oh, how hot it was! At times there was a wind more scorching than the still air.

One observation will explain. You all know that evaporation produces cold. If we take two thermometers and keep the bulb of one wet it will sink below the other, the drier the air the more the difference between the wet and dry bulb. In our eastern climate a difference of ten degrees F. is considered very high. I think it is very rarely as much as that in our driest times. I had a thermometer along, and when it had stood at 99° or 100° (the latter heat was observed at the time of the experiment), by merely wetting it with saliva from my tongue, it sank in a few seconds to 64°—or, it sank thirty-five degrees F. I dare say, an ordinary "wet bulb," where the bulb is covered with cotton and wet with water, would have sunk much lower. Such is that climate! The deserts of Africa are not hotter nor drier than this region at this season, the difference being that here there is rain in the winter. You cannot conceive the thirst that this dry air rapidly creates.

We were very dry, but it was desirable that we scale the peak, so we pushed on. I began to regret that we had attempted to make the peak, when we suddenly and unexpectedly came upon a little spring; the only one, we afterwards learned, within several miles, and this will be dry soon. We stopped there half an hour, and then, refreshed, started on and soon planted our barometer and compass on the peak. It commanded a magnificent view as regards extent, but a desolate one. It was about 2,000 feet above camp, and 3,500 or 4,000 feet above the sea, yet the thermometer stood 98° and 99°, at times 100°, on the summit, and in the breeze!

We were back to camp at three or four o'clock, took our dinner of bread, out of which all moisture had dried, and fat bacon, which we roasted by holding slices on a stick over the fire, and washed the whole down with water neither very pure nor cool—not a luxurious meal, but our appetites proved a good sauce.

We sent Guirado on to the camp where the rest were. We stayed. We had only one saddle blanket, which was insufficient, and the gravel on the rocky ground made a poorer bed than usual. But how clear the sky was at that height! The myriad stars shone with more than the splendor of a winter's night, and at midnight the moon came up and lit up the scene. We were out of food, but the man who herded the sheep most hospitably offered us his fare, which we thankfully partook of— fried beans, dry bread, and poor coffee without sugar, but it was sufficient.

Before eight o'clock, Saturday, July 27, Peter came back, with provisions to last him three days, to stay with the mule. Hoffmann and I pushed on for San Juan, forty-one miles distant. The first thirty miles was a hot ride, but before ten at night we were at San Juan, having stopped twice, and got something to eat at a Mexican ranch house.

Here we are in San Juan, and are preparing to leave in the morning for Santa Cruz. Professor Whitney left for San Francisco immediately on his arrival here. I did not see him after leaving him at the sheep wells, but he will join us at New Almaden, near San Jose, in about two weeks.

The delightful temperature here at San Juan, of 75° to 85°, is now too cold. I left my coat with Peter and shivered all day yesterday (Sunday) as I lay in camp resting after our hard trip.

CHAPTER III

NEW ALMADEN

*Pacheco Peak—Santa Cruz—Mountain Charlie's—
Governor Downey—New Almaden Mine—A Visit
to San Jose—Social Life at New Almaden—Climbing Mount Bache—Enriquita and Guadalupe Mines.*

Santa Cruz.
August 4, 1861.

We had intended to leave for Santa Cruz on Tuesday, July 30, so on Monday made our preparations, but that afternoon Professor Whitney telegraphed to me to wait until Thursday for letters and orders. This gave us three days on our hands. I resolved to visit Pacheco Peak, about twenty-eight miles northeast of San Juan, so Hoffmann and I started on our mules. Averill in the meanwhile went to Monterey for a horse he had bought.

I will describe our trip. First a ride of eighteen miles across the dead-level plain, tedious and monotonous. The Gabilan Range on the south, the Monte Diablo Range on the north, to which Pacheco, Santa Ana, and other peaks belong. To one who has never tried riding on a level plain, no description is adequate to cause a full realization of its tediousness.

The distance seems near, very near, to begin with. Pacheco Peak rises in such full view from our camp, its rocks and ravines every one so distinct that any of you would estimate its distance, without California experience, as five or six miles at most, instead of the near thirty we must ride to reach it—twenty, at least, in a straight line. The road is crooked, several miles are lost by its windings; the air is hot as we plod across it; league after league is passed without the aspect of the scene changing; the Pacheco seems no more near, nor the

Gabilan no more distant, than it did an hour before. But at last a belt of scattered oaks is entered. Then we strike up a canyon, on the Pacheco Pass, through which the Overland made the crossing of this chain. Eight miles up this canyon, by a good road, brought us to Hollenbeck's tavern, in the heart of the chain at the foot of Pacheco Peak and at some elevation above the plain.

We had ridden the twenty-four miles from camp, but were not too tired to climb the peak, and as time was valuable and as it was still only two o'clock, our mules were put out, our dinner was got, and we started. There is a good and easy trail most of the way up, three or four miles, made by cattle and sheep. The wind had been high during the middle of the day and still shrieked through the canyons; it had annoyed us much on coming up the pass by the dust it raised.

Up, up, we toiled. The peak rises like a very sharp cone, forty to forty-six degrees' slope on each side, about six hundred feet above the mass of mountains around—a sharp, steep, conical peak, towering far above everything near. It has generally been put down as a volcano (extinct) and the rock on top as lava, hence it was important to visit it. All wrong, however. The rock was the same as the rest of the chain, metamorphic, only a little more altered by heat. The ascent of the last cone was steep and laborious, but it was accomplished, and we stood on the highest rock of the summit—about 2,600 or 2,700 feet above the sea, and 2,500 feet above the plain on either side.

Strange, that the strong current of wind that sweeps through all the valleys scarcely reaches us here, only a gentle breeze plays over the summit. Below, to the south lies the great San Juan plain, its division into two parts, one running to Gilroy, the other to San Juan. It is four or five hundred square miles in extent, but it seems not one-fourth of that. A cloud of dust rises from it, raised by the high wind like a gauze veil two to three thousand feet high, but the Gabilan Mountains, the Salinas plain more obscure through the gaps, and the Santa Cruz Mountains to the west, are all visible.

The chain we are on is about thirteen or fifteen miles through the base, rising in an innumerable number of ridges to the height of about two thousand feet, furrowed into countless canyons, the whole elevated mass running northwest and southeast. To the east is the great Tulare, or San Joaquin, Valley, but a dense cloud of dust rises from it and forms an opaque white veil that shuts out the view of the Sierra Nevada beyond. We have a view of eighty or a hundred miles of the chain we are on, with the higher peaks to the east of us, some of which rise about three thousand feet, but are without names.

When you go off in rhapsodies over the *grandeur* of Mount Tom or Mount Holyoke, around which cluster so many poetical associations, think that the former is less than 900 and the latter but 1,200 feet high, with all their poetry! Here, peak after peak raises its grand head against the sky, and the addition of either of the Massachusetts heroes to the height of any one of them would scarcely be noticed; yet these peaks are not only unknown to fame, but are even without names. I was about to say, "Born to blush unseen, and waste *their grandeur* on the desert air."

The mountains are covered with oats, now dry, giving them a soft straw color. We lingered on the peak until the sun was nearly set, the shadows long and dark on the plain, the canyons dark and gloomy, the sunny slopes bathed in the softest golden light. We returned to our tavern, ate supper, talked in the barroom on Secession at home, then retired—Hoffmann to fight fleas all night, I to congratulate myself that they do not bite me but only crawl over me in active troops. I wished for my blankets that I might go out and sleep.

Wednesday we returned to camp—a terribly windy day—and found Peter back. He had run out of money and left Sleepy thirty miles back. Michael, also, was back; he had left us, you recollect, some weeks ago, but came back so penitent that I concluded to try him once more.

August 1, Hoffmann and I again climbed Gabilan Peak, about seven miles from camp, for the sake of measuring it and getting fresh bearings from its summit. It was 2,932 feet above camp and I suppose about 3,100 feet above the sea—quite a peak. I dispatched Guirado, meanwhile, after Sleepy, and to our surprise he got him into camp the next forenoon, having traveled all night in the cool. The mule was much better, but we had to leave him at San Juan.

Friday, August 2, we intended starting early. We were up and had our breakfast at five o'clock, but Averill did not arrive. We packed up and waited all the day, or at least until 1 P.M., then started on without him. We came on about eighteen miles, first over the hills, then across the most lovely Pajaro Valley, a little bottom five or six miles wide and eight or ten long—a perfect level—an old lake filled in as is shown by its position and by the terraces around its sides. On the river is Watsonville, a neat, thriving, bustling, American-looking little town. The country around is in the very highest cultivation, divided into farms, covered with the heaviest of grain, or a still heavier crop of weeds. Several threshing machines were seen at work in the fields, and the hum of industry in the little town sounded American-like.

I ought to mention a little item. The squirrels were very thick around our San Juan camp—they came out of their holes to eat the barley near our mules. Guirado killed seven at one shot with the shotgun, and Peter, one morning, shot twenty-one in four shots. The morning we left, while we were waiting for Averill, as I sat making some calculations, I kept my revolver by my side to shoot at the squirrels when they came out. I killed five in thirteen shots, so you see I am getting to be quite expert with the "instrument."

We camped about three miles from Watsonville and came on here to Santa Cruz yesterday, Saturday. Santa Cruz is a pretty little place. We camped at a farmhouse about a mile from town, near the seashore. The church bells on Sunday morning reminded me of home, and as I had been in a Protestant church but twice since last November, I resolved to go.

We went down to the beach for a bath and preparatory wash, and I took the barometer along for observations at the same time. Climbing down the rocks at the beach I slipped and fell about ten feet, and in trying to save the barometer I received a terrible bruise on my left knee. The barometer, however, was saved, but a thermometer was smashed, my pants and drawers were torn and ruined, and my knee was quite seriously damaged. I stayed in camp quietly that day, and then the next went to work again.

Monday, August 5, I visited some limestone quarries and lime kilns near town. Lime is burned for the San Francisco market. All the arrangements are very fine and complete, and about five thousand barrels per month are burned. The proprietor, Mr. Jordan, was very kind and showed me around. These people own a large ranch, raise their own cattle, keep their own teams, cut their own wood, etc. They have schooners for shipping the lime, and the wagons they use in drawing it to the wharf are enormous. These wagons draw from 90 to 150 barrels to a load, each barrel weighing 200 to 250 pounds—surely quite a load.

The mountains back of Santa Cruz are partially covered with magnificent forests of redwood, pine, and fir—tall, straight, beautiful.[1] Many of the redwoods are from ten to twelve feet in diameter; one is nineteen or twenty feet in diameter. You cannot appreciate how tall and straight these Californian evergreen trees are. Mr. Jordan cut a fir for a "liberty pole" that was 14 inches in diameter at the butt, and 2½ inches where it broke off at the top, perfectly straight, and *171 feet long*. In falling it broke in the middle and was spliced. I saw some magnificent sticks cut from these fir trees. The redwood is by far the most valuable timber of California. It grows only on the mountains near the coast. It is between a pine and a cedar—the wood is as coarse grained as pine, but of the color and durability of red cedar. The trees grow of enormous size, *very* large and tall, with a habit something like our hemlock, only the trees swell out more at the base. The "giant trees" of the Sierra Nevada are of the same genus, and much like it in all respects, only larger.[2] The wood splits well and all the cabins in the mountains are built of it—fencing, timber, everything. Northward they grow still larger. Chester (Averill) tells of seeing in Humboldt County a house and barn —roof, timber, boards—all—and several acres fenced—with *one tree*. Such is the redwood.

Tuesday, we went to see some noted rocks called "The Ruins," six or seven miles back in the mountains. We found them a humbug—nothing near what we expected—some were small outcrops of sandstone, which had weathered in curious forms, some were in tubes like chimneys two or three feet high. These had been regarded as artificial works—"chimneys of furnaces"—and at one time a company was actually formed to dig for treasures that might be buried there. So inflamed is the public mind here on *hidden treasures!*

The town stands on a terrace about sixty feet above the sea, a table that runs back a mile, where another terrace rises— an old sea beach about 180 feet higher, which may be traced for miles. Wednesday we spent in examining these terraces and measuring their heights. The town is prettily situated on a clear stream, the San Lorenzo River. The place looks quite American—neat homes, trees in the yards, gardens, flowers, and American farms around. Even the old *adobe* mission church is nearly torn down and a neat wooden church stands by its side, the old *adobe* walls and ruins telling of another race of builders. There are some Spanish people and Indians left yet, but the town is American.

Thursday, August 8, we left for the New Almaden Mines, striking north through the mountains, where a fine road runs to San Jose and Santa Clara. It is the most picturesque road we have yet traveled. We struck back into the mountains directly behind the town, wound up valleys, through and across canyons, over ridges, and along steep hillsides. Sometimes the way lay along a bare hillside, giving a fine view, but oftener in dark forests, where enormous trees shot up on every side; sometimes singly, their giant trunks like huge columns supporting the thick canopy of foliage above; at others groups of trunks started from a great rooted base, half a dozen or more trees together, their spreading tops forming only one head of immense size. Laurels, or bay trees, with their fragrant foliage, firs, pines, oaks, mingled in that picturesque scene, which continually changed with every turn of our ever winding road.

We at last struck a ridge and followed it some miles, and at 2 P.M. camped on the summit at "Mountain Charlie's," about two thousand feet above the sea. Evergreen oaks grew around, a lovely place to camp, and a little pond overgrown with sedge and rushes lay beside the house. After dinner we went on a hill near camp, a few hundred feet above it, for bearings and observations. The view was both extensive and lovely, one of the prettiest I have yet seen. Around us was the sea of mountains, every billow a mountain; deep dark canyons thread them, the form and topography very complicated, the geological structure *very* broken. Redwood forests darken the canyons on the west toward the sea, chaparral covers the ridges on the east. Some of the peaks about us are very high—Mount Bache[3] rises in grandeur nearly four thousand feet, Mount Choual and others over three thousand feet. To the south lie the whole Bay of Monterey and a vast expanse of ocean.

The beautiful curved line of the bay seems like an immense semicircle, thirty-five miles across, but every part of the arc distinct. Monterey is obscured by fog, but the mountains rise above it in the clear air. Fog forms at the head of the bay and rolls up the great Salinas Valley as far as we can see, but we are far above it and looking on the top of it—it seems like a great arm of the sea, or a mighty river, stretching away to the distant horizon. The mountains beyond Monterey also rise above it, in the clear air. Each ridge is distinct—the range of

Point Pinos, the Palo Scrito hills, while in the blue dim distance rise the high peaks of the center of the chain, ninety or a hundred miles distant from us. Southeast, over the lower ridges, rises the Gabilan, thirty-five or forty miles distant; nearer, the high ridges of our own chain.

The "chain" is here a series of ridges and we are near the middle. Through a break we can see the lovely Santa Clara Valley in the north, and the towns of Santa Clara and San Jose, eighteen or twenty miles distant, are in full view—the separate houses can be distinguished with our glasses.

Some plants were collected, barometrical observations made, supper eaten, and in the twilight we sang songs beneath the branches of the trees until we turned in and watched the stars twinkling through the leaves in the clear, blue mountain air.

Friday found us astir early, and we came on. The valleys and canyons and the distant bay were all covered with dense fog, but we were above it, looking down on its top. As the sun came up this fog seemed a sea of the purest white, the mountains rising through it as islands and the tall trees often rising above its surface. It was now tossed into huge billows by the morning breeze, and as the sun rose higher it curled up. The scenes gradually shifted as the curtain rolled away, and the mountain landscape was itself again.

We came on, first down a gradual slope for about nine or ten miles, then down a canyon that breaks through the outer ridge, until we emerged into the Santa Clara Valley. We passed over the tableland at the base or among the foothills, about twelve miles, and camped near the mines of New Almaden.

New Almaden Mines.
August 17.

On the way to the New Almaden Mines we met Governor Downey[4] and his wife and sister in a carriage with driver. Guirado was ahead, and as I came up he introduced me. I was decidedly in a "rig" for introduction to such society—buckskin pants, leggins, dirty gray shirt, without coat, vest, cravat, or suspenders, begrimed with the dust of our very dusty road—yet it was all the same. We stopped and talked a short time, then passed on.

He is a wealthy man, Irish, has a ranch worth $300,000, has risen from the ranks, has a Spanish wife, is a zealous Catholic—so has the elements of political popularity here. President Lincoln has made a requisition on this state for mounted volunteers to protect the Overland mail route, and the Governor has offered Guirado a first lieutenant's commission, decidedly a fine opening, so he left the party yesterday morning.

Well, we arrived. A most lovely little town has sprung up by the furnace—neat houses on a long street, with a row of fine young shade trees, green yards, pleasant gardens, etc. The superintendent, Mr. Young,[5] is absent. He lives in most magnificent style, like a prince, has a wife half Spanish-Californian, half Scotch, who has a lot of single sisters, the Misses Walkinshaw, lovely girls, of whom more anon.

Mr. and Mrs. Whitney were at the house of Mr. Day, the mining engineer, and head man in Mr. Young's absence. We were introduced to the ladies in the same unpresentable costume, which excited much mirth. It was the first time I had seen Mrs. Whitney since taking the field. She scarcely knew me in my bronzed, burnt skin, robust looks, and un-Parisian costume, and she failed entirely to recognize Averill, so has he changed by exposure and the growth of a tremendous beard. We were soon camped below the town, by a stream, in a pretty spot. Another rig donned (colored shirts, of course, but clean), and I returned to talk about affairs with Professor Whitney. He had to go to a party, so I went up late in the evening to talk with him after his return.

Saturday, August 10, we visited the principal mine of New Almaden, with Mr. Day to guide and conduct us. It is probably the richest quicksilver mine in the world, and is worth one or two million dollars. The pure red cinnabar (sulphuret of mercury) is taken out by the thousands of tons, and the less rich by the hundreds of thousands of tons. The mine is perfectly managed and conducted. The main drift is as large as a railroad tunnel, with a fine heavy railroad track running in. Six hundred feet in and three hundred feet below the surface of the hill is a large engine room, with a fine steam engine at work pumping water and raising ore from beneath. The workings extend 250 feet lower down. We went down. The extent of the mine is enormous—miles (not in a straight line) have been worked underground in the many short workings, and immense quantities of metal have been raised. We were underground nearly half a day, then came out and had a sumptuous dinner at a French restaurant near, then climbed the hill over the mine.

A ridge runs parallel with the main chain of mountains, about 1,700 feet high, in which are three mines, of which this is the principal one. The view from "Mine Hill" is perfectly magnificent—over the region east and north, with the bold, high peaks back of it.

Mrs. Whitney went to San Francisco that day. Professor Whitney returned on Sunday, August 11, when I rode down twelve miles to San Jose with him.

San Jose (always pronounced as if one word, *San-ho-zay'*) is a pretty town in the Santa Clara Valley, on a level plain. Orchards, gardens, pretty houses, fruit, fertile fields are the features—a bustling, thriving town, of probably six or seven thousand inhabitants. The roads at this season are dusty beyond description and the town looks accordingly.

I bade good-bye to Professor Whitney and then went to church—Mr. Hamilton's church (of Ovid).[6] He was absent, but Mrs. Hamilton (formerly Miss Mead, you remember, of Gorham and Ovid) was there, and on my going up to her after service she was as much surprised as if I had really dropped from the clouds, as she asked me if I had—although no *clouds* had been seen here for months. I went to the parsonage, a neat house next the church, where they are living very neatly and comfortably indeed. She has become thoroughly Californian, is delighted with the country, climate, people, etc. A pretty little boy with light hair and fine gray eyes was running around, the pride of his mother's heart. Mr. Hamilton had been absent six weeks in Oregon, but was daily expected back. I need not say that we had a pleasant chat until it was time for me to return.

We were down with the wagon. I ran around town but very little, for the climbing up and down so many hundred feet of ladders the previous day in the mine had told fearfully on my lame knee, and I began to be anxious. I bought some medicine for it, and we returned to camp at nightfall. Monday I was so lame that I resolved to stay in camp quietly and recruit my knee before it grew worse, as I had some mountain climbing to do during the week.

In the evening I rode up to Mr. Day's.[7] He has two daughters now home. A young lady was visiting, and half a dozen came in, and a lively time we had of it. We were invited to a horseback ride the next afternoon with some of the ladies. Mr. Day is a son of President Day of Yale College, is a fine man with a very fine family, has been here twelve years, and many a chat we have had about New Haven and old Yale. The room I occupied in college was his old room at home.

Tuesday, August 13, I went to the mines and collected specimens. The mines are about two miles from the furnaces, on the hill. We collected two or three boxes of specimens, then returned. The furnaces are complete, and about three thousand flasks (seventy-five pounds each) of quicksilver are made each month. More might be made if desired, but that is enough for the market. An old furnace has been taken down, and the soil beneath for twenty-five feet down (no one knows how much deeper) is so saturated with the metallic quicksilver in the minutest state of division, that they are now digging it up and sluicing the dirt, and much quicksilver is obtained in that way. Thousands of pounds have already been taken out, and they are still at work.

No wonder that there has been such legal knavery to get this mine, when we consider its value. Every rich mine is claimed by some ranch owner. These old Spanish grants were in the valleys; and, when a mine is discovered, an attempt is made to float the claim to the hills. Two separate ranches, miles apart and miles from the mine, have claimed it, and immense sums expended to get possession. The company has probably spent nearly a million dollars in defending its claim—over half a million has been spent in lawyers' fees alone, I hear.[8] The same at New Idria—it was claimed by a ranch, the *nearest edge of which is fifteen miles off!*

And this is only a sample of the way such things go here. Were I with you I could relate schemes of deeply laid fraud, villainy, rascality, perjury, and wickedness in land titles that would entirely stagger your belief, yet strictly true. The uncertainty of land titles and the Spanish-grant system are doing more than all other causes combined to retard the healthy growth of this state.

But to our horseback ride. Mrs. Young's Scotch father, Robert Walkinshaw, having died, the children now live here. There are several girls left, all beautiful. Professor Whitney thinks one the most lovely lady he has seen in the state. I hardly go to that length. They ride every afternoon and invited us to ride with them. Three of them, with Miss Day, a Miss Clark of Folsom, a young Englishman, Averill, and I, made up the party. It was the pleasantest time I have had in the state. We started at five o'clock and rode five or six miles east to the top of a hill that commands a lovely view of the Santa Clara Valley and the opposite mountain chain.

I wish you could see those Mexican ladies ride; you would say you never saw riding before. Our American girls along could not shine at all. There seems to be a peculiar talent in the Spanish race for horsemanship; all ride gracefully, but I never saw ladies in the East who could approach the poorest of the Spanish ladies whom I have yet seen ride. I cannot well convey an adequate conception of the way they went galloping over the fields—squirrel holes, ditches, and logs are no cause of stopping—jumping a fence or a gulch if one was in the way. The roads are too dusty to ride in, so we rode over the hills and through fields, sometimes on a trail, sometimes not. We took tea at Mrs. Day's, then were invited to Mr. Young's, where we spent a pleasant evening. My lame knee was better, but still bad enough as an excuse for not dancing.

Wednesday I went to the Enriquita Mine, about six miles distant. It is a poor mine, but yields some quicksilver. It is twelve miles or more distant by the road, but not half that distance by the trail we took across the hills. My knee was much better, but I refrained from using it much.

It was desirable to ascend the mountains about five miles back, in a direct line, from the mines. Professor Whitney was anxious to have them examined. They form a long ridge, covered with chaparral, rugged, forbidding, but not so steep as some we have seen. The highest point, Mount Bache, is the highest point of this chain, and as it was a Coast Survey signal station its height had been carefully ascertained. Because of its known height, we wanted to test our aneroid barometer.

A good trail had once been cut to the top, and the view was so grand that several ladies wished to make the ascent with us; Mr. Day and several other gentlemen also wanted to go. A man was sent the day before to reconnoiter part way. He declared it to be impossible for ladies, so they were left behind. Five were to go, and five o'clock named as the hour of starting. I suggested a later hour, as previous experience had taught me the uncertainty of getting the outside party out so early, although *we* could start at any time; but five was the hour set.

Thursday, August 15, we were up before four, ate our breakfast by the light of the stars, and at five were on hand at the rendezvous, punctual, as we always are. Our aneroid had not arrived the previous evening by express, as was expected, but we found the expressman—he had it—returned to camp, unpacked it, got our other barometer, got back—the rest not ready. Two long hours were consumed before we got the party off, eight persons instead of five—Averill, Hoffmann, and I, of our party, Mr. Day, Wilson (a young Englishman who had never climbed a mountain), a Doctor Cobb, a Mr. Reed, and our guide. Mrs. Day insisted on putting up our lunch, so we could not well carry along any other. We were to ride to the foot of the chaparral, about 1,500 feet above camp, leave our horses by a spring—then, two hours' climb to the summit, 2,000 feet above. A single gallon canteen was to be amply sufficient for the ascent—since there was such a fine spring on the summit! To be sure, none of the party had ever been there, but we had the minutest directions.

Well, we reached the water, unsaddled, found good water but very scanty; it took an hour to get enough for our horses and mules. I hung up my barometer for observations, when we found the aneroid smashed—a total wreck. I had not examined it in the morning; Averill had carried it, and it must have been broken in the express on the way from San Francisco.

It was ten o'clock when we started from there, the time we were to have been at the summit, at the very latest. Our visitors "pitched in" vigorously on the start. I remonstrated, told them they never could stand it. It was hot, the hill steep. Mr. Day had once or twice said he "knew more about mountain traveling than any of us." Five hundred feet were gained, some began to lag, the perspiration streamed, our green climbers pitched into the water, scarce as it was, rested, pushed on, some fell behind, and before we had ascended a thousand feet the intense heat, dense chaparral, and hard climbing told so fearfully that two gave out, drank up part of the water, rested, then went back. Averill went back with them, foreseeing a hard time, and fearing he would give out. The rest of us pushed on exultingly—hotter, more chaparral, greenies hotter, so they drank up the water.

At last the summit of Mount Choual, 3,400 feet, is reached. All are terribly thirsty, it is most intensely hot, all pant, all are dry. Here we lunch, for it is noon, in the heat, in the sun, of 136°, and no shade, no air, no water—and our lunch such as was fitting for a picnic. The butter had melted and streamed over everything—of course, *we* stood the best chance, being used to all kinds of fare.

Here two more gave out, as the hardest of the work was still ahead. Mr. Day's "knowledge of mountain climbing" was hardly sufficient; he lay panting under the bushes and I feared for his safety. Four of us pushed on, intensely thirsty, but water lay ahead at the summit. We descended a few hundred feet through bad chaparral, then over a knob four or five hundred feet high, then descended again as much more. Here we got into a terrible chaparral. We toiled, tugged, worked, and lucky were those without instruments. We got so tired that, finding a shady spot under the dense brush, we lay down and rested or slept half an hour—quite refreshing. Strange, *all* dreamed of *water*. Our guide was hardy—he stood it well. Poor Wilson vowed that a "pleasure trip" would never get him on a mountain again. Our last ascent was a toilsome one, but soon after three o'clock we stood on the summit, 3,800 feet high, as nearly used up a set of men as you ever saw.

Water would have readily brought five dollars a quart, and sherry cobblers would have been above price—but there was no spring, no water. We must return; some feared they never could get back. I have not suffered so before with thirst—it was terrific. It was half past three when we got the instruments set; no one admired the landscape. One who gave out behind, pushed on and gained the summit; but, finding no water, all started back except Hoffmann and I, who stayed until after four o'clock taking our observations. Hoffmann's eyes looked as sunken, his visage as haggard, as if he had been on hard fare for two weeks. I suspect I looked worse, for I had carried the barometer, level, etc. He carried the compass, tripod, and sketch book—but the fine view was left unsketched—only the topographical bearings and barometrical observations were taken. As before remarked, the rest started back, and straggled to the spring, one by one, haggard, used up, faint—some staggering, so faint—disgusted in general with mountain climbing.

Hoffmann and I took it back more slowly, often lying down to rest, and it was after sunset when we got back to the spring. Averill came up a few hundred feet and met us with a canteen of water. We soon drained the gallon between us, and were quite fresh when we reached the others. We soon saddled up, rode back by the bright light of the moon, and at half past nine were in camp. Supper, drink—two bottles of wine—and we sat and sang until eleven o'clock in the clear light of the moon, for it was Averill's last night with us. I sent him to Monterey today to join Doctor Cooper, our naturalist.

Thus ended our trip. Had *we* the management of it all the suffering would have been avoided; as it was, it was a day of *intense* suffering. Today I am in camp, writing, packing specimens, etc.

New Almaden.
August 17.

It is a lovely Sunday—above 90° in the tent where I write, but less outside in the cool breeze, beneath the trees—there the air is delicious. Our camp seems small and quiet, but although my party is small, it is effective.

Yesterday (Saturday) we were up in good season, and Hoffmann and I rode to visit the other quicksilver mines of this vicinity. There are three mines within a region of six miles, all in the same ridge, which is about 1,700 feet high, lying parallel with the high chain of mountains behind, and separated from the Santa Clara Valley by a still lower chain of foothills.

We rode to the Enriquita Mine,⁹ about six miles from camp, by a trail over the hills. It is the poorest of the three, and lies about midway between the other two, the three being in a direct line. We introduced ourselves to the superintendent and engineer, Mr. Janin, who showed us every attention, going into the mine with us. It is much like New Almaden in character, but vastly poorer. Its owners are one set of disputants for the title of New Almaden also, so of course there is much feeling between the two mines. Four sets of claimants are lawing for New Almaden, and two more wait behind, to claim of these claimants should the latter be successful—a pretty "kettle of fish," to be sure.

The mine of Enriquita has been going about two years, has its great works, deep cuts, long tunnels, furnaces, steam engine to work machinery, etc. Mr. Janin is a very young man, New Orleans birth, but has spent several years in Europe at mining schools, mostly at Freiburg in Saxony.¹⁰ He knew many of my friends, and we had a capital time.

The third mine, the Guadalupe, is two or three miles farther to the northwest. As he had never been in that, Mr. Janin offered to go with us, which he did, and introduced us to Doctor Mayhew. The mine is more rich, extensive, and profitable than the Enriquita, but vastly poorer than the New Almaden. It is mostly in a valley between the "mine ridge" and the main chain, and its workings extend down about 450 feet below the bottom of the valley. As it is so much lower it is much more wet and dirty than the other mines here. There are also workings in the hill, all of which we visited, spending some hours in climbing ladders, crawling through passages, threading tunnels, examining the character of the rock and ore everywhere.

We came out dirty, but were taken to Doctor Mayhew's house, introduced to a lovely little wife, with two fine babies four months old. Doctor Mayhew is a rough Baltimore man, probably forty or forty-five, his wife a lovely girl of perhaps sixteen or seventeen. His story was brief but comprehensive:

"Well, I wandered many years, concluded a year ago last spring that I ought to marry and settle, went home from Salt Lake, married, returned, and in nine months and two days was astonished with a pair of daughters. And here I have, in about a year and a half, traveled, stopped, married, settled, and have a smart and growing family—California is a prosperous state!"

His wife laughed heartily at her mistake, for seeing three men in red woolen shirts, belts, Spanish hats, saddles, spurs, etc., she had exclaimed, "There come some fine looking *Mexicans!*"

I loaded my saddlebags with specimens, returned to Enriquita, spent an hour or so more with Janin, then returned here at evening. We are having lovely moonlight nights now—cool, clear, bright. I am sorry that I must so soon leave this place and the acquaintances I have made. We spent the evening at Mr. Day's. It was pay day yesterday, and last night the town rang with the sound of violins, guitars, dancing, *fandangos*, singing, and mirth.

NOTES

1. The redwood (*Sequoia sempervirens*) is found as far south as the Santa Lucia Mountains, in Monterey County, and extends north barely beyond the Oregon boundary. In its natural state it is found nowhere else. The pine referred to here is the yellow pine (*Pinus ponderosa*); the fir is not a true fir, but the Douglas fir (*Pseudotsuga taxifolia*).

2. The big tree (*Sequoia gigantea*) is found in its natural state only in the Sierra Nevada.

3. This name, given in honor of Alexander Dallas Bache, Superintendent of the United States Coast Survey from 1843 to 1867, is no longer in use; the mountain is now called Loma Prieta.

4. Governor John G. Downey, born in Ireland, 1827, came to America in 1842 and to California in 1849. He was elected Lieutenant-Governor in 1859 and succeeded Governor Milton S. Latham when the latter resigned to become United States Senator in January, 1860. Downey was succeeded by Leland Stanford in January, 1862. His first wife, daughter of Don Rafael Guirado, was killed in a railroad accident in 1883. His sister, Eleanor, married Walter H. Harvey in 1858 (d. 1861) and Edward Martin in 1869; she died in San Francisco in 1928 at the age of 102 years. Another sister, Annie, married Peter Donahue in 1869; she died in 1896.

5. John Young, a native of Scotland, came to California in 1844 or 1845. He was a trader and master of vessels on the coast for a time. He died at San Francisco in 1864. He married a daughter of Robert Walkinshaw, a Scot who had been a resident of Mexico before coming to California in 1847. He managed the New Almaden mines prior to his return to Scotland in 1858.

6. Laurentine Hamilton, born at Seneca, N. Y., 1827, was educated at Hamilton College and at Auburn Theological Seminary, N. Y. He married Miss Isabella Mead, of Gorham, Me., and Ovid, N. Y. In 1855 he came to California as a missionary and went to the mining town of Columbia, where he built the Presbyterian Church that still stands. Later he became pastor of a church in San Jose until called in 1865 to the First Presbyterian Church in Oakland. He was an advanced liberal and was charged with heresy. Resigning, he established the Independent Presbyterian Church, which later was the Independent Church. On Easter Sunday, April 9, 1882, he dropped dead in his pulpit. His sermons and addresses were published in a volume entitled *A Reasonable Christianity*. Mount Hamilton, now famous as the site of Lick Observatory, bears his name. The little boy mentioned by Brewer was Edward H. Hamilton, a well-known journalist now (1930) living in San Francisco.

7. Sherman Day was graduated from Yale, A.B., 1826, and received also the degree of A.M. His father, Jeremiah Day (1773–1867), was president of Yale from 1817 to 1846. Sherman Day was born in New Haven, Connecticut, 1806; he died in Berkeley, California, 1884. After his graduation from Yale he lived in New York and Philadelphia for a time as a merchant. For several years he was in Ohio and Indiana as an engineer. In 1843 he published *Historical Collections of the State of Pennsylvania*. He came to California in 1849 and engaged in civil and mining engineering at San Jose, New Almaden, Folsom, and Oakland. In 1855 he made for the state a survey of wagon-road routes across the Sierra; he served in the State Senate, 1855–56; was United States Surveyor General for California, 1868–71; was one of the original trustees of the University of California, and for a time was Professor of Mine Construction and Surveying. He married Elizabeth Ann King, of Westfield, Mass., in 1832. The two daughters at New Almaden in 1861 were Harriet (Mrs. Charles Theodore Hart Palmer) and Jane Olivia (later Mrs. Henry Austin Palmer). The Miss Clark, of Folsom, was either Mary or Annie Clark; a sister, Harriet, had recently married Roger Sherman Day, son of Sherman Day. (Information from members of the Day family.)

8. The contest for title to the mines is summarized in Bancroft's *History of California*, VI, 551.

9. This mine was opened in 1859. J. M. Hutchings, in *Scenes of Wonder and Curiosity in California* (1860), pp. 168–172, quotes an account of the dedication, which describes the naming of the mine for the little daughter of the manager, Mr. Laurencel, Henrietta (Spanish, *Henriquita* or *Enriquita*).

10. Louis Janin (1836–1914) was a native of New Orleans, where his father was a prominent member of the bar. After completing his sophomore year at Yale, in 1856, he went to Europe and studied mining engineering at Freiberg, Saxony, for three years. He also studied at the École des Mines, at Paris, before returning to America in 1861. His engagement as

superintendent of the Enriquita Mine was a brief one. He was succeeded there by his younger brother, Henry, whose career up to this time closely paralleled his own. Louis Janin went to the Comstock mines in Nevada, where he advanced rapidly as a metallurgist, becoming superintendent of the Gould and Curry Company in 1864. After a period as manger of mines in Mexico he was employed for a year by the Japanese Government and then entered upon a long and distinguished practice as consulting engineer and mining expert with offices in San Francisco. During the last twenty years of his life he resided at Gaviota, California. (Biographical notice by R. W. Raymond in *Transactions of the American Institute of Mining Engineers* for 1914, XLIX, 831.) It was through Louis Janin that Herbert Hoover received his start in the profession of mining engineering.

WILLIAM H. BREWER
1859

JOSIAH DWIGHT WHITNEY
State Geologist of California, 1863

IN CAMP NEAR MONTEREY, MAY 1861
From a photograph taken on leather. The cracks appear to be due to shrinkage of the emulsion, not of the leather

HOLLENBECK'S ROCK, PACHECO PASS
By Charles F. Hoffmann

THE BEER KEG
By Charles F. Hoffmann

THE SURVEY PARTY IN CAMP NEAR MOUNT DIABLO, 1862

GABB   WHITNEY   AVERILL   HOFFMANN   BREWER   SCHMIDT

JAMES G. COOPER

WILLIAM ASHBURNER

CHESTER AVERILL

WILLIAM MORE GABB

THE OBELISK GROUP (MERCED PEAKS) FROM PORCUPINE FLAT
By J. D. Whitney

MOUNT DANA AND PEAKS AT HEAD OF TUOLUMNE RIVER
By J. D. Whitney

UNICORN PEAK, FROM TUOLUMNE MEADOWS
By Charles F. Hoffmann

CATHEDRAL PEAK AND FAIRVIEW DOME,
FROM TUOLUMNE MEADOWS
By Charles F. Hoffmann

CATHEDRAL PEAK, FROM TUOLUMNE MEADOWS
By J. D. Whitney

THE SUMMIT OF MOUNT LYELL
By Charles F. Hoffmann

LASSEN PEAK
By Clarence King

THE CALIFORNIA STATE GEOLOGICAL SURVEY, DECEMBER, 1863
AVERILL GABB ASHBURNER WHITNEY HOFFMANN KING BREWER

JAMES T. GARDINER

CHARLES F. HOFFMANN

RICHARD D. COTTER

CLARENCE KING

WILLIAM H. BREWER
San Francisco, 1861

MOUNT BREWER
By Charles F. Hoffman

MOUNT SILLIMAN
By Charles F. Hoffman

MOUNT WHITNEY, IN DISTANCE, FROM
SLOPE OF MOUNT BREWER
By Ansel F. Hall

RECORD FOUND ON THE SUMMIT
OF MOUNT BREWER IN 1896

CLIFF NEAR CAMP 181 ON CHARLOTTE CREEK,
HEADWATERS OF KINGS RIVER
By Charles F. Hoffmann

CANYON OF THE UPPER SAN JOAQUIN RIVER
By Charles F. Hoffmann

CHAPTER IV

APPROACHING THE BAY

Santa Clara Valley—A Camp-Fire Scene—San Jose —Mount Hamilton—Mountain View—A Democratic Barbecue—Mission San Jose—A Rattlesnake —A Skunk—Financial Disturbances—Amador Valley—San Ramon Valley—Oakland—Sacramento and the State Fair.

Camp 48, San Jose.
Saturday, August 24, 1861.

Monday I made my preparations to leave, packed and sent off specimens, etc. In the evening I called again on our lady acquaintances. The Misses Walkinshaw were even more lovely and agreeable than usual. We had a pleasant time.

We visited a famous soda spring near the mines—a copious spring highly charged with carbonic acid and other mineral matter. A house is built over it. It is as highly charged and as sparkling as the Saratoga Springs. It is delicious, and the gas is so abundant that it fills the spring that is walled in.

Tuesday, August 20, we left camp early, crossed some low hills by a byroad, then down the valley of Llagas Creek, a pretty, picturesque valley, and in about fifteen miles emerged into the Santa Clara Valley and camped at the "21 Mile House," twenty-one miles below San Jose.

The Santa Clara Valley (San Jose Valley of the map) is the most fertile and lovely of California. At the point where we came into it, it is about six miles wide, its bottom level, a fine belt of scattered oaks four or five miles wide covering the middle. It is here all covered with Spanish grants, so is not cultivated, but near San Jose, where it is divided into farms, it is in high cultivation; farmhouses have sprung up and rich fields of grain and growing orchards everywhere abound. But near our camp it lies in a state of nature, and only supports a few cattle. One ranch there covers twenty-two thousand acres of the best land in the valley—all valuable. This Spanish grant land-title system is one of the great drawbacks of this country. One man will make an immense fortune from that ranch, but the public suffers.

We camped under some beautiful oaks, near a house, where we got hay and water. Two days were spent examining the hills to the east of the valley, from the summits of which (near two thousand feet above the valley) are to be had most magnificent views. One sharp peak rose near camp, on the west, conspicuous from every direction. It was *very* sharp, and rose very steeply (over thirty degrees on each side), more than eleven hundred feet above the valley. The view from its top was superb. It has been burned over this summer, and its black cone is a grand object, whether seen by day or by the clear moonlight of these lovely nights.

Peter had been sent to San Juan to see about Old Sleepy, the mule that had been left there. He found him still unfit for traveling, so he sold him for twenty dollars, which was five less than he has cost us since he gave out. He cost us a hundred last December.

We camped under the trees and did not pitch our tent. A camp scene may repay writing. Thursday evening Mike resolved on a "treat," so he bought a keg of lager beer and some cigars and brought them into camp. After supper I went to the house to read the news and when I returned a fine fire was lighted, the beer tapped, the moon was bright, and all were happy.

I only wish you might have beheld the scene. Five large oaks, their branches festooned with lichens, are our canopy. The bright fire lights up their trunks and foliage and the group around. The moonlight lies soft on the plain and lights up the black mass of the peak, Ojo de Agua de la Coche, that rises back of the camp, its black outlines sharp against the blue sky.

But it is the group near the fire that demands our attention. The baggage and equipage are piled against a tree—saddles, axes, instruments, provisions—back stands the wagon in the shadow, the harness .hanging on the pole. A Sharp (rifle) is leaning against a tree, while from the trunk of another are suspended barometers, thermometers, etc. Some piles of blankets lie on the sod, ready for their occupants at bed time. The bread chest stands modestly in the distance. The light is reflected from the bright tin canisters of tea, coffee, etc., and the grim kettles and gridiron stand against them. The water pails and washbasins are near too, and a few towels hanging on a limb flutter in the gentle breeze. The table stands under the largest tree, a few notebooks and maps on it.

But these are only the *background* of the picture. Nearer the fire are the group. Beside that tree, on a box of specimens, a beer keg is poised, a pitcher and four glasses standing on the ground near it until distributed; soon this takes place. That man on a camp stool, his California hat on one side, his legs crossed with ease, his plaid overshirt brilliant in the firelight, but not entirely concealing the luxurious "biled shirt" which he has on today, puffing a cigar with the dignity of a senator, is Peter. Between him and the beer keg sits Michael, the host of the evening. His red shirt is doubly brilliant by the bright firelight, his face beams with more than his usual good humor, it even seems to me that his light hair curls tighter—he evidently enjoys it, and he puffs his cigar with gravity worthy of the occasion.

Beyond the fire, on the *manta* (saddle-cover), with the grace of a Turk on the divan in his harem, reclines Hoffmann, our topographer. His well used red pipe lies beside him—a cigar has taken its place—and maps, bearings, and topography are forgotten as the smoke curls up lazily, only interrupted by taking another glass from time to time. His black shirt looks somber, in keeping with night. That great mass lying between him and the fire is not an outcropping of metamorphic rock, as it might at first be taken for, but his mountain shoes, from the soles of which those stupendous nails loom up and glitter grandly in the firelight.

That demure, modest looking individual on the ground, leaning against a tree (but close to the beer keg) is the humble botanist. His face is indistinctly seen, as it is modestly hid behind a huge stone pipe, a native "California Meerschaum" from which occasionally curls up a blue column of smoke as from the crater of a half-sleeping volcano. His last pair of pants are a little torn, and a flag of truce is displayed in the rear—emblem of peace, even if not of plenty.

Song follows story, and laughs follow both, until the oaks echo again with their ring. The keg is finished, the cigars are smoked, the embers have ceased blazing, the moon is higher and its shadows shorter, the lights are out at the house near, and the owls are hooting among the trees as we turn in to our blankets and sleep closes and covers the scene.

Friday, August 23, we came here, twenty-one miles down the valley, and camped just on the edge of the town. The day was warm, and the roads very dusty—*à la California.*

San Jose.
Wednesday, August 29.

WE arrived here Friday night. Saturday morning we found we had lost a roll of maps. Here was a loss truly. The maps had cost us months of labor, were partly compiled from the Land Office surveys, ranch surveys, coast surveys, and our own labor—the compilations and results of immense labor—and we had no copies. One was, moreover, of the field we were to work up for the remainder of the year. Five hundred dollars would have been a trifling loss compared with the loss of these few sheets of paper. They were in a roll in a tin case and had dropped out of our wagon. I wrote some notices, offering ten dollars reward for their recovery, and sent Hoffmann back. He rode back to our last camp, twenty-one miles—no tidings. Then, recollecting that he was not sure of seeing them since leaving New Almaden, he took the back trail and, by a streak of good luck, found them about twenty-five miles from our camp in a place where we had lost our trail and they had rolled out on the rough ground *five days before*.

Sunday I went to hear Mr. Hamilton preach. It seemed like old times to hear him again. He has changed much, and has improved. He stands very high in this state as a preacher. Hearing him carried me back to the old times and other scenes and other hearers. He is pleasantly settled here. Both he and Mrs. Hamilton have grown old fast since they left the East six years ago. I took tea in town on Sunday evening at a Doctor Cobb's.

Professor Whitney is on his way to Washoe, or is there now. I got a letter from him from Placerville a few days ago. I shall be glad when he gets back, which will be in three weeks now.

Camp 49, Mountain View.
Sunday, September 1.

NEARLY east of San Jose, some distance in the mountains, is a high peak[1] we wished to reach, being the highest in that part of the Diablo Range. As near as we could judge from our maps, we supposed it nine miles distant in a straight line. It proved over fifteen. Mr. Hamilton went with us. A ride of six miles across the plain brought us to the foot of the ridge. All this is enclosed, in farms, and under good cultivation. Farmhouses, orchards, etc., give it an *American* look. We then struck the ridge, and on rising, had a capital chance to see this part of the Santa Clara Valley. It is perhaps twelve or fourteen miles wide at San Jose, an almost perfect plain, very fertile, a perfect garden, and much of it in higher cultivation than any other part of California.

This first ridge was about 1,000 or 1,500 feet high. Then we crossed a wide valley, then up another ridge. We had attained an altitude of nearly three thousand feet, when we came upon another deep and steep canyon cutting us off from the peak. Here we left our mules and proceeded on foot about three miles and reached the peak after 4 P.M. The view was very extensive and the day very clear. It was about 4,000 feet high—we made it 4,200 feet—but that is doubtless too high.[2] We could see various portions of the Coast Range, from far above San Francisco to below Monterey, probably 140 to 150 miles between the points, and the Diablo Range for about a hundred miles.

It was five o'clock before we left and after sundown before we got on our mules, with at least fifteen miles to ride. Night closed on us among a labyrinth of hills and canyons twelve miles from camp and at least six from any road. We gave our mules the bridle and let them find the way back, which they did with a sagacity beyond belief, over steep hills, along ridges, through canyons, to the road at the foot of the hills at the edge of the plain. It was near midnight when we reached camp. It is at such times that I realize how healthy we are in camp. While others must bundle up and put on extra clothes for fear of catching cold, we never have colds. Not anticipating any such delay we were without either coats or vests. We were wet with the perspiration of a six mile walk and climb, the last three miles very vigorous, then a ride of that distance in the cool night air, much of the way against a chilly wind—yet no cold or symptoms of any. Averill writes me, "Since I have taken to living in a house, full of rats and fleas, haunted by tom cats (or the devil), I have taken an abominable cold." He never had a cold in camp.

Tuesday, August 27, we went to examine a hill east of the head of the bay and north of San Jose.[3] It was both farther (14 or 15 miles) and higher (2,500 to 2,800 feet) than we expected, so it took us all day. The valley looked like a map, and the head of the bay, with its swamps intersected and cut up with winding streams and bayous crossing and winding in every direction, made by far the prettiest arabesque picture of the kind I have ever seen. It was wonderful.

Wednesday afternoon we took dinner with Mr. Hamilton, then rode to some sulphur springs and rocks that produce alum, about eight or nine miles east of town, returned and took tea with him. We had a pleasant time.

Thursday I spent in trying to sell mules, could not, so gave up, and dined out. Friday I had resolved to put up my mules at auction on Saturday, so had the day for leisure. In the morning, with a young lady who was visiting at Mr. Hamilton's, I went out of town a mile and visited the residence of a wealthy citizen, Mr. Belden.[4] He and his wife had come here early (1841), poor, had got rich, visited Europe, bought many works of art, etc. He lives here very comfortably on his money, has a fine house, pretty grounds, etc. We spent two or more hours most pleasantly in looking over pictures, photographs, etc., which he had brought from Europe. He was absent, but his wife appears a very fine and pleasant woman.

Saturday I sold the mules, got as much as I expected—$71 for two mules we gave $130 for—then came on here, seventeen miles northwest of San Jose. This is on a farm of Mr. Putnam,[5] a brother-in-law of Professor Whitney. He is in business in San Francisco, but has a farm here, where his family spend the summer. He comes down every Saturday night and returns on Monday morning. It is in the foothills, at the base of the high mountains, a lovely, quiet, secluded, beautiful spot.

One event of the week must not be forgotten—a grand *barbecue* of the Breckenridge Democrats (Secession), in a grove about a mile from camp. The Breckenridge party is quite large in this state and is much feared. Some of its men are open and avowed Secessionists, but the majority *call* themselves *Union* men, *Peace* men, most bitterly opposed to the Administration and opposed to any war policy—in fact, are for letting all secede who wish to. They are making great exertions just now, and hope to carry the state at the election next Wednesday. If they do I fear this state will be plunged into the same condition that Missouri is in. There are many more Secessionists in this state than you in the East believe, and many of them are desperadoes ready for anything in the shape of a row.

But to my story. From quite early in the morning a stream of carriages and horses poured into the grove—men, women, children. After dinner Hoffmann and I rode out. Such a *political* meeting I never before saw. It seemed a cross between a camp meeting and a German May picnic. There were as many women and children as men, some listening to fierce political speeches, but more loitering in the shade of the large sycamores. All were well dressed, as if for a festival, and all good natured.

Dinner was announced. A long ditch or trench had been dug, a fire built in it, and spitted over it on iron rods laid across were immense quantities of mutton, beef, and pork, finely roasted. These were served up at long tables, with bread, peaches, etc., and if poor Lincoln's army is assaulted as vigorously as was that pile of eatables it stands a narrow chance, and if Secessionists fight as valiantly as they eat, then the Union is indeed in danger.

It was a scene for a Hogarth or a Cruikshank. Here a youth with a huge bone in one hand and a chunk of bread to match in the other—there a rustic beauty, her cheeks distended with juicy meat—another maiden with countenance equally indicative of bread—children with faces, from their eyes down, daubed with pie, happy and greasy—men with fingers distended and hands elevated, greasy, and afraid to touch anything because of it—old ladies in agony for fear the gravy would get on their best frocks—matrons attending to the wants of a numerous band of rising and growing, but youthful, Democrats—young men helping their sweethearts—family groups, friendly groups, crowded spots, solitary eaters.

Then, the scene of desolation over the tables half an hour later, as women begin to wipe their fingers, children ask if there are any more peaches, young men and large boys begin to parade, each with a long nine cigar stuck in one cheek, men begin to talk of crops, mules, horses, hard times, or politics, children to play, and women to look up acquaintances and inquire why on earth they had stayed away so long and not been to see them, or talk of family cares and domestic duties.

Then the speeches commenced again. Men and women were seated, and the eloquent speakers told of the horrible designs of the other parties, of their infamous doctrines, of their wonderful inconsistencies, of the scoundrels who were the leaders; and they pathetically told of the cruel persecutions and slanderings their own party had received, of its patriotic leaders and pure principles, of its innocence and the immaculate purity of its office seekers.

I sat and listened for a while, and as I gazed on the scene around I felt sad that so pure a party should not have all the offices, and the scoundrels of the other parties could not all be instantly hung.

Two or three women near me, who were feeding their infants in the natural way, impressed me deeply with the productiveness of the state, and its capacity for feeding the rising population.

There was much good humor, no fighting, some faint cheering for the Union, some equally faint for Jeff Davis and his cause.

We left and went to camp, through a dust the like of which you never saw. The wind was high, and the dust of a thronged road in this dry climate, where not a drop of rain has fallen for three or four months, can never be appreciated until it is seen and felt. The fiercest snowstorm is not more blinding.

That evening I heard Conness, the candidate of the Douglas Democrats for governor. I then learned that it was the other party that were plotting the downfall of the state and general ruin of mankind, which terrible catastrophe could be averted only by voting the ticket of this patriotic and moral party.

Tuesday, September 3.

YESTERDAY was a most lovely day. We started early on foot to climb a high mountain that rises behind the camp, over three thousand feet.[6] It was hard to get at. We were over eleven hours at it, and had no water. We had a canteen along to fill at a spring on the way, but we found it dry. We took it up moderately, however, and did not suffer much from thirst. We found tracks and traces of grizzlies, more abundant than we have seen them before—we were in paths where their fresh tracks *covered* the ground, but we did not meet any.

Today has been a very fine day, very clear. We took a ride of five or six miles over the hills west of camp, pretty rolling hills, covered with wild oats, and stocked with much better cattle than one sees in the southern country. We see everywhere here the evidence of *American* enterprise in the farming and stock. There is a thousand dollar Durham bull on this ranch.

This whole valley abounds in the best of fruit: peaches, apples, pears, melons, etc. Peaches are very abundant and cheap. I saw a *steam* threshing machine at work in a field last week—it worked well, and easily threshes 1,500 bushels of barley in a day. A thousand bushels a day they call light work.

Formerly there was a lack of water here for stock and for irrigating. The large streams that run into the valley either sink or dwindle away to mere pools. So hundreds of artesian wells have been bored. Sometimes water is struck within a hundred feet, but many wells are three or four hundred feet deep. Many of these overflow, often with a large stream, but with the majority the water only rises to near the surface without overflowing. It is then pumped up by windmills, and hundreds of these may be seen in motion every afternoon when there is a strong breeze up the valley. Many of these are very ornamental, costing from three hundred to six hundred dollars, and impart a very peculiar feature to the landscape.

Camp 50, near Mission San Jose.
Sunday, September 8.

IT is a lovely, warm, quiet day. I have been to Mass at the Mission, have done my Sunday's allowance of "plain sewing," and will now drop a few more lines.

Wednesday, September 4, we left our last camp and crossed the plain *via* Alviso and Milpitas to the east side of the bay, and camped about a mile from the old Mission of San Jose.[7] It was election day, and much excitement existed at the several polls passed. This place is Secession.

There was more excitement in this state than there has been since the days of the Vigilance Committee. But the state has gone overwhelmingly Republican. There was much fear on the subject, from the fact that the Secessionists were united while the Union men were divided into Republicans and Douglas Democrats. But California is still for the Union, one and undivided.

Thursday we examined the hills north of camp and Friday visited some noted hot springs near. These have quite a reputation for the cure of sundry diseases, and the houses, grounds, etc., are better fitted up for comfort and luxury than any of the mineral springs we have before seen here. The water is somewhat sulphury, contains various salts in solution and has a warm temperature. There are five principal springs, with temperatures varying from 87° F. to 95½° F. The baths are really luxurious.

Yesterday we started for a high mountain some twenty miles southeast, in the center of the range. We followed up a canyon several miles, and when we left it we struck up the wrong ridge and came out on another peak twelve miles distant from the one we started for. It answered our purpose, however. We got the topography of a region on which maps were blank. We attained a height of over three thousand feet, but the distant view was obscured by smoky air.

An incident of the day may be noted, as it came very near costing the Survey a valuable mule, if not an assistant also. We stopped on the summit, took our bearings and observations, unsaddled, and fed our mules some barley we had brought, then sat in the cool shade of a fine oak and took our lunch. Then a smoke while waiting for an hour, hoping the air would clear and enable us to obtain distant bearings. It was useless, so we saddled up. Just as I finished, I felt something squirming under my feet vigorously, looked and found I was on a snake, holding him fast under my well nailed boots while he was writhing to get free. I stepped off from him and saw that it was a large rattlesnake! He ran between the forefeet of the mule, over one hoof, then through between the hind feet. I took the tripod of our compass and caught him, held him fast and cut his head off with my bowie. He was between two and a half and three feet long and had nine rattles. Had he shown as much fight as the others we have seen, both I and the mule would have got our share. But my buckskin pants and my boots would have been pretty good armor, for buckskin absorbs the poison from the tooth so much that but little enters the wound, it is said.

He had three fangs, two on one side, side by side. They were as sharp as needles. I took them out as trophies. A cool breeze blowing over the summit probably accounts for his stupidity. He struck the tripod vigorously, however, when I overhauled him. He had come out from under the tree, from a hole less than two feet from where we were sitting. Here was a danger *seen,* how many *unseen* we pass only our Protecting Providence knows.[8]

Later.

ALL are in bed. I cannot sleep as much as the rest do, so go to bed later, and have a quiet hour when all is still. We all sleep outside now. Here, near the bay, the nights are much cooler, the sky is clear, with sometimes a fog in the morning, and a heavy dew falls every night. Often our blankets are quite wet. We are in a cooler region than we have been in most of the time for the last two months.

This is a little old mission town—a large dilapidated church, old *adobe* houses with tile roofs, a few dilapidated walls and gardens, and new American buildings springing up around and among them. The very houses show the decay and decline of one race and the coming in of a superior one.

The old church is large, gaudily painted on the inside, but dilapidated; the congregation a mixture of Indian, Spanish, mixed breeds, Irish, with a few German, French, and American. There are a few stores here—it is a little village, one that will never be a large one. As we work north the decay of the native and Spanish element becomes more and more marked.

A camp incident this morning: We had just finished breakfast when we saw walking leisurely along the road the largest kind of a gentleman skunk. Peter started with revolver, but fired at such a respectful distance that he missed four shots, all he had loaded, and his skunkship started up the hill. Pete got the shotgun and Mike and Hoffmann joined pursuit. Hoffmann shot five revolver balls at him—all missed. I got my revolver from under my pillow—four barrels were loaded—but I ran so hard that my shots met with like success. Then an Irishman and dog joined in the pursuit, and something might have been smelt for some distance.

I was mending my stocking at the breaking out of the *mêlée*. I returned and finished the work while the battle raged over the hill. I had just loaded up my revolver when the party returned, the skunk still ahead, coming into the field near our tent. I again rushed to the battle, drove him from the field into the road to get him away from camp, then finished his career with two balls through him. We covered him with earth. A fragrance pervaded the valley, decidedly rank, but the hot sun and fine breeze entirely dissipated it in a few hours. Thus ended the tragedy. It made ten times the excitement of yesterday's rattlesnake adventure.

We all sleep with loaded revolvers here, and Mike and Pete sleep with the mules in the corral, with a double-barrel shotgun extra, for horse thieves are thick and bold. A fine horse was stolen from this very spot last week, valued at two hundred dollars. We have been fortunate with our mules thus far, considering their value. But then we have been vigilant, and loiterers who come into camp can easily ascertain that each man has a navy revolver handy and is expert in its use. This may seem to many a superfluous caution, but I am convinced that it is judicious, considering our occupation and mode of life. It insures the "respect" of the class of gentry most likely to covet our property.

            Sacramento.
         Wednesday, September 18.

MORE than ten days have elapsed since I have written anything—days of too much anxiety and labor to allow any time for letters or journal. But I will continue my story in order. Monday morning, September 9, we left camp at Mission San Jose and moved up to Haywards, fifteen miles, along the east side of the bay, near San Leandro. The road led through a lovely region. A slope from four to eight miles wide lies between the hills and the bay, of beautiful land and of extraordinary fertility. It is all under cultivation, and enormous crops are raised on it. The fields are fenced, the houses American, and all tells of American enterprise. On arriving at Haywards I found a dispatch calling me to San Francisco immediately. I got in camp, left my party, and at night arrived at the city, twenty-three miles distant. I spent Tuesday, September 10, there with Professor Whitney trying to devise "ways and means" of continuing the work, and then returned to camp. I will give the main features of our trouble.

There was an appropriation of $20,000 before we came. We expected to use but $10,000 or $15,000 this year, so got an additional appropriation of $15,000 to carry on our work this and the first part of next year. We drew immediately in advance $10,000, and afterwards got $5,000. There was, on our arrival, over half a million of specie in the state treasury. We apprehended no difficulty in getting the money as we needed it after the opening of the new year's accounts. But at last session a transfer of over a quarter of a million was made from one fund to another, which has thrown trouble into all the machinery of state finances, depleted the treasury, and crippled us. We went without our salaries, but were promised $5,000 certainly the first of this month. The time has arrived but the money can't be got. We have seen the Governor, the Treasurer, and others, but our only answer and consolation is: that there is no money in the treasury; that they can't pay us until the first of December, and then but $5,000; that we are no worse off than the rest of the state officers; that the Governor and all the officers have received no salaries for months; that we get none of our last $15,000 appropriation until next March; and that we must retrench and cut down our party and wait.

I have held long consultations with Professor Whitney and the present plan is to curtail expenses, dismiss one or two assistants, do without our salaries until next March, except what is absolutely necessary for the direst necessities, keep in the field say six weeks yet, then withdraw from the field and dismiss all but the three assistants in the first departments, and run on borrowed money, *if we can borrow*. We borrowed $2,500 for present use.

To continue my story. Tuesday, September 10, I returned to camp. Wednesday I visited some of the hills in the neighborhood, and among the rest, a "coal mine" where much money had been expended and not a particle of coal found, and where a *very little* geological knowledge would have saved the money. Then a "copper mine" just as bad and more expensive. I went down a shaft a hundred feet, hanging on a rope, then into a drift in rock where it is impossible for a mine to occur—money thrown away. I told the man to stop digging, and I think he will—after sinking a few more hundreds.

That night Professor Whitney joined us in camp. He had just returned from Washoe. I wish I could recount his tales of that country—it seems a fable—a desert region, inhospitable, but with mines of fabulous richness—$700,000 taken from one vein in a short time—the largest steam stamping mills in the world erected, where the freight amounted to sixteen cents per pound ($320 per ton)—for a short distance of the road *boilers* sent by *express* at thirty cents per pound—tales of money being both lost and won by the hundreds of thousands, of a large town springing up in that desert in two years, etc.

We were camped on a lovely spot by a stream, under a stupendous sycamore, on a rich bottom. The land was recently sold for one hundred dollars per acre, as it had produced over ninety bushels of wheat per acre. But pretty as it was, a heavy fog rolled in from the sea by night and dripped from the trees like rain, wetting our blankets as we slept under them—reminding me of rain once more.

The next day we moved on to Amador Valley, northeast, at the south side of Mount Diablo, then climbed a high hill after we got in camp, where we had a fine view of the region. The San Ramon Valley, west of Mount Diablo, lay at our feet, the richest and most lovely I have yet seen in the state. It is all held in farms, where wheat is grown, and crops of over sixty bushels per acre are expected—they sometimes rise to over ninety—such crops does this state produce! The premium crop of wheat last year was nineteen acres, accurately measured, which *averaged* ninety-five bushels per acre over the whole, or over 1,800 bushels on nineteen acres! well authenticated—and so very dry that each 100 bushels would be at least 105 to 108 in the eastern states.

I have many things to write about the agriculture of this state, but every letter I don't do it, for I have so much else that I want to tell. Were I with you it would take me a month to "talk out."

We camped at the farm of a Major Russell, who had been with the Mormons. He sat in camp during the lovely evening and told us much of Mormon life. The universal testimony about the Mormons is the same; those that know the most of them give them the worst name.

The sky was very clear, the stars and moon bright, as we went to bed under some lovely live oaks by a little brook. The brook had "broken out" after the earthquake in June last—it is good water, and Russell says is worth $5,000 to his farm. The ground had cracked quite extensively near our camp, and a number of good springs had broken out in the valley at that time.[9]

Game was once very abundant—bear in the hills, and deer, antelope, and elk like cattle, in herds. Russell said he had known a party of thirty or forty to *lasso* twenty-eight elk on one Sunday. All are now exterminated, but we find their horns by the hundreds.

Friday, September 13, we went up the San Ramon Valley about twelve miles, and left our party to camp, while we pushed up the valley, then climbed the hill, 2,500 or 2,800 feet high, where we had an extensive and comprehensive view. Mount Diablo was the grandest object in the landscape. I will not attempt a description of the view, but cannot pass over some geological facts.

The strata here are all filled with shells and are of enormous thickness. They are turned up at high angles and much broken. The whole country is of mountains 2,000 to 3,500 feet in elevation, made by the broken edges of the strata. We saw sections of these strata over a mile in thickness, yet full of shells through their whole thickness. I think the Tertiary rocks of this region are two or three miles thick! Who shall estimate the countless ages that must have elapsed while they were being deposited in that ancient ocean? While these myriads of animals were called into existence, generations lived and died, and at last the species themselves became extinct. Each day reveals new marvels in our labors, teaches us new truths in the world's history.

We had camped in a "wind gap" and the air drew through fiercely. We expected it to die down with the night, but it did not. We "retired," scattered here and there on the ground, some under the trees, but I out in the clear open air, for should fogs come on it is drier, as the sequel proved.

How the wind howled round my head and played with my hair and shrieked through the trees, as I lay and watched the stars before I went to sleep! The wind continued through the night, and with it came a dense fog from the sea, which wet our blankets, searched out every crevice, probed every rheumatic corner of our bones, and dripped like rain from the trees. The morning was dark, but we got out, got breakfast, packed up, and started on our way before the fog cleared and the sun came out. The stream by that camp, like the last, opened and sprang up at the last earthquake.

We traveled all day and reached Oakland that night, opposite San Francisco. We camped and the Professor and I went into town. Sunday I returned to camp to see if all was right and to make preparation for leaving for a week. The ferryboats were crowded by the thousands who were going over to see a parade at the camps near Oakland that day.

Oakland is a pretty little place, springing up with residences of San Francisco merchants. It is like Brooklyn from New York, only it is farther, the bay being some seven or eight miles wide there. Pretty oaks are scattered over the sandy flat.

I returned to the city that night. The next day, Monday, September 16, I transacted various business affairs and then came up here along with Professor Whitney.

Sacramento.
September 20.

WE left the city by steamer for Sacramento, 120 miles, at 4 P.M. and did not get into the river until after dark. The sail up the bay is very fine. The islands and the shores of hills are bare and brown now—I mean bare of trees—only dried grass. The effect of the setting sun, illuminating this with its mellow light, was most beautiful indeed. Mount Diablo stood up, a grand object, in the landscape.

The Rev. T. Starr King, the celebrated orator and clergyman, was on board with us. I got an introduction and had a pleasant time with him. He is as agreeable in conversation as he is eloquent in the rostrum. Night closed in on us before we entered the Sacramento River, and when I got up in the morning we were lying quietly at the wharf of that new city, the capital of the state, the "Albany" of California.

The State Fair is being held here. The noise and bustle distracts me. I feel nervous and excited and long for the camp again, with its clear air, calm still nights, simple life, and its loneliness, rather than this bustle and crowd. I took cold when in San Francisco ten days ago, and again now—had I my blankets here I would be tempted to sleep out on the fairground.

The Fair is like other fairs—hundreds of big cattle, horses, etc. (the horses the finest)—many more Durhams than I expected to see, few Devons (in fact, none at all), some few sheep, fewer hogs, some mules and jacks. The grounds are fine, over twenty acres enclosed with a high brick wall with ten entrances, a fine track, etc. The stalls for cattle are finely arranged around the outside, and a promenade is to be built on the flat roofs of the stalls. There is a large stand for two thousand spectators, and a fine track. The races were received with California *gusto*, where horsemanship is such an accomplishment.

Indoors the Fair was more peculiar—no flowers at all, but fine fruits. These latter were more remarkable for *size* than any other characteristic. It is too late for the best plums and peaches. I will give you some items of pears and apples. Numbers of apples which *I measured* were over 15 inches in circumference—one $16\frac{1}{2}$ inches! Three pears on one plate, I measured, both around them lengthwise and around the largest part crosswise, and their measurements in circumference were 17 inches by 14, $16\frac{1}{2}$ by $14\frac{7}{8}$, and $18\frac{7}{8}$ by $14\frac{3}{4}$ inches, respectively, the three weighing $6\frac{1}{2}$ pounds! Numbers of pears were seen measuring over fifteen inches around, and proportionately larger if measured around from stem to blossom end. In two instances of three pears on one stem, each cluster weighed together over five pounds. There were grapes of four-, five-, six-, and even seven-pound clusters! Yet, I must say candidly that I think the quality of all the fruit, except pears, to be inferior to the same kinds in the eastern states. Pears grow peculiarly well here.

The park is about a mile outside of the city, the pavilion for indoor show is in the city and was built earlier. The park was only fitted up this summer. The pavilion is an enormous brick building, has cost already over $30,000, and it will take $10,000 more to complete it according to the plan. It is lit by gas, and the greatest crowds are there in the evening, when the beauty and fashion of the city are on hand.

It is unnecessary for me to speak of two laborious and "borous" days I spent on a committee to make arrangements for sending things from the Pacific Coast to the great World's Fair at London next year. Governor Nye, of Nevada Territory, was our chairman, but most of the work devolved on a small sub-committee of seven, of which I was one. Governor Nye had as "delegates" to the Fair from his territory (Nevada) three great ingots of silver from the Washoe Mines, each of over sixty pounds avoirdupois weight—nice pocket pieces.

A noticeable feature of the Fair was the gambling. Besides the usual sideshows of live snakes, big cows, fat hogs, fat women, etc., there were hundreds of *fan, monte,* and other gaming tables, each with their piles of silver and gold, often to the value of hundreds and even thousands of dollars, in full blast, with the crowds around. Music, females singing or dealing cards to draw the custom, liquor, noise, swearing, etc., were the accompaniments. Yet the whole Fair was orderly. I never saw a Fair in the East where the crowds were more orderly or so well dressed as at this.

NOTES

1. This is the earliest account known of an ascent of Mount Hamilton. Professor Whitney vetoed a proposal to name the mountain for him (Brewster, *Life and Letters of Josiah Dwight Whitney* [1909], p. 238). It was thereupon named for the San Jose clergyman, and is cited as Mount Hamilton in the Whitney Survey, *Geology,* I, 43, 50, etc.
2. The altitude of Mount Hamilton is given by the U. S. Geological Survey as 4,209 feet.
3. Monument Peak (2,647 feet).
4. Josiah Belden was a member of the first party to cross the plains to California for the purpose of settlement, known as the Bartleson-Bidwell party of 1841. The story of this party is well known through John Bidwell's articles in the *Century Magazine* (November and December, 1890), and his *Echoes of the Past,* first printed at Chico, reprinted by the Lakeside Press (Chicago, 1928), and *A Journey to California, 1841 . . . The Journal of John Bidwell,* published by The Friends of the Bancroft Library, 1964. Belden settled in San Jose about 1848 and was its first mayor (1850). He married Sarah M. Jones, a pioneer of '46. *Josiah Belden, 1841 California Overland Pioneer: His Memoir and Early Letters.* Georgetown, California: The Talisman Press, 1962.
5. Samuel Osgood Putnam, one time of Milwaukee, married Elizabeth Whitney. He came to California in 1850 and was connected with the California Steam Navigation Company.
6. Black Mountain (2,787 feet).
7. La Misión del Gloriosísimo Patriarcha Señor San José, founded 1797.

8. In later years Professor Brewer frequently told this rattlesnake story to illustrate a peculiarity of human fear. A portion of the story omitted in the Journal is supplied by his son, Arthur Brewer. After cutting off the rattles and head and wrapping them in a piece of paper, he put them in his pocket and proceeded down the mountain. He had traveled but a short distance when he decided to go back and measure the rattler. As he started to straighten out the snake, in order to measure it with his tape, the body coiled and struck him on the wrist with the stump where the head had been. Although he knew there was absolutely no danger, as the head, fangs, and poison sac were in his pocket, he was so unnerved by the incident that it was some time before he could remount his mule and resume his journey. Professor Brewer was accustomed to point out by this incident how one's fear could not always be controlled by one's reason.

9. An earthquake, July 4, 1861, "cracked open the earth, started a new spring, threw the water out of an old one, and cast down men standing up in the field" (*Daily Alta California*, August 8, 1864).

CHAPTER V

THE MOUNT DIABLO RANGE

Camp at Clayton's—Measuring the Height—Raphael Pumpelly — War-Time Shudders — Geology of Mount Diablo Region—Corral Hollow—Rumors of Coal—Return to Mount Diablo.

San Francisco.
Sunday Evening, September 29, 1861.

For over a week I have been entirely too busy to write to anyone. I arrived back here from Sacramento, Friday, September 20, and the next two days were spent here, active and busy. Monday I rejoined my party with Professor Whitney. They were camped at the foot of Mount Diablo, on the north side. We found a noted German traveler, a Mr. Jagor,[1] from Berlin, who has been on a tour of four or five years to Asia, the East Indies, the Philippines, etc. He spent several days with us. He was a quiet, inoffensive, and unassuming man. He has taken elaborate notes and sketches of the countries visited and has over 1,700 drawings.

Mount Diablo is a bold peak nearly four thousand feet high, which rises quite abruptly on the north side, and is one of the most extensively seen objects of the state. Because of its central position and the great distance that it is visible from every direction, a point on the top is the starting point of all the surveys of the state. The Coast Survey determined its position—latitude and longitude—and the land surveys started from the same point, a base line and meridian being run through it.

Tuesday, September 24, we ascended it. It was the easiest mountain to climb we have yet had. A pretty good trail runs to the top, so we could ride our mules most of the way. We carried to the top a new barometer, one made for the government topographical engineers, and an aneroid barometer of the finest construction, to test the accuracy of the other instrument. We found it far less accurate than the mercurial barometer, and it required just as much care to carry it.

The view from the summit was remarkably fine, but the day was not clear and the distant views were shut out. There were no clouds, but a haze or smoke in the air shut out everything over fifty miles distant. But the bay, the valleys of San Joaquin and Sacramento, and the mountains around were fine indeed.

As an instance of how dry this climate is, I found a four-bladed knife on the top that must have been lost by some previous party weeks, or possibly months, ago—there was not a speck of rust on the new and polished blades!

Wednesday Professor Whitney and I climbed on foot a side spur, very steep and rugged, some 3,200 or 3,500 feet high, a much harder day's work than the last, and a much hotter one.

Thursday Averill and I visited the hills to the north and ran a line of barometrical observations across a number of hills and valleys in order to get an accurate section across from the mountain through the coal mines that are just now attracting much attention here. It was a laborious day's work, making many miles of hot walking and climbing.

Coal on the Pacific Coast is a great desideratum, where anthracite coal for ocean steamers has sometimes sold for fifty dollars per ton, and *often* for over thirty dollars, and is now, I think (not sure), over twenty dollars. Bituminous coal is cheaper, of course, but brings ten to twelve dollars per ton, if good. The true "coal measures" do not exist, in any probability, west of the Rocky Mountains. But in various places a kind of soft coal known as "brown coal" and "Tertiary coal" is found, especially up in Russian America. It is now being discovered in this state, the most valuable deposits thus far found being near Mount Diablo. I will reserve a description of the mines until I pay them another visit.

Friday we visited the hills east of the mountain. One of the localities visited was a so-called "borax spring," for which we were told the owner had refused $17,000. We examined it, found it only a *salt* spring, and a poor one at that, not worth 17,000 cents at most.

The region north and northwest of Mount Diablo is a beautiful one—pretty valleys scattered over with oaks, many of enormous size, with wide branches often drooping like the elm. The rugged mountain rises against the clear sky, and when illumined by the setting sun is an object of peculiar beauty. Our camp was in a very pretty place, with great trees around, and the mountain in full view.

Friday evening we had a little incident. We were camped near a Mr. Clayton's house. His dog treed a large coon close by camp, up a very large and wide-branching oak. Mike and Pete built a fire under it, we all got around but no coon could be seen. Pete climbed the tree, no easy feat as the trunk was four or five feet in diameter. Up in the branches—still no coon—built another fire—it lit up the foliage, and the scenery around—caught a glimpse of him—Pete followed him out to the end of a lofty limb sixty or eighty feet from the ground and shot him with his revolver. It was exciting, as he got higher and higher, shooting and missing, in the dark, until he had but one more load left, then following him out to near the end, and that ball brought him. He was a huge fellow, much like our eastern coon, but rather larger, and of a different species.

While in the city I had caught cold and I had slept badly, but the clear sky and ground soon brought sound and refreshing sleep. This is no affectation, but plain truth. I have now caught cold *five times*, each time very hoarse and with more or less sore throat, *each time* that I have slept indoors for the last ten months, and these are the *only colds* I have had, and each vanished as soon as I took to the open air again! I have slept in the open air, under the open sky, for the last three months—sometimes with fog like rain—I have slept in the rain itself, slept in wet blankets for night after night, and not taken cold, yet go into a house, sleep in a bed, and I am sure to take cold! And my case is not peculiar. It is the general experience here—need we stronger proof that colds are owing to a close and artificial mode of life? It has long been known that those who sleep in small or close and warm rooms are more subject to colds than those who sleep in large and well-ventilated rooms.

Camp 58, near Martinez.
October 1.

I RETURNED here last night and have spent the day among the hills and the evening in closing up the last month's accounts. For six weeks I have kept the accounts of the party, but as Averill has returned to camp he will keep them hereafter. It is quite a job—I have written six large "cap" pages, added the long columns, and balanced all.

It was necessary to have barometers. We bought one new one, and hearing that we could get tubes, the Professor and I went to San Francisco Saturday to fill some for the old cases. Sunday we went to work—the tubes were so poor that only after trying and breaking three did we at last fill one, taking us nearly all day; but it broke of itself before Monday morning. So much for Sunday work!

Monday I got stores, etc., and returned here. I met Blake,[2] an old classmate, in San Francisco. He has accepted a place to go to Japan, to aid those people in developing their mines there.

We go again to Mount Diablo tomorrow. I am glad to get in camp—but the season flies—as I lie in my blankets at night and see the Pleiades rising so high in the clear sky I am reminded that winter is at hand—and what a terrible winter for our country—I tremble to think of it. I have been anxious and low spirited much of late over our unhappy troubles—the end looks dark. I would rather see the nation reduced to poverty and a million men perish than see the Union broken—but what will be the end? God only knows, and in Him we must put our trust.

Clayton, at the foot of Mount Diablo.
October 4.

THIS is my first chance to write in camp during the day time for a long while. We made Mount Diablo higher than it is marked on the maps, so wished to make another and more careful measurement. We sent Averill to the top with barometer this morning, to observe today, while I stay here in camp and observe another barometer. I have been making calculations all day and will now write—with the necessary interruption of having to note observations every fifteen minutes.

We have a pretty camp, on the north side of the mountain, about five miles from the summit in direct line. Fine oaks shade our camp, and the grand mass of the mountain looms up in front of us. When lit by the evening sun it is a magnificent object.

The Californians tell us that once in olden time they had a battle with the Indians here; it was going hard with the Spaniards, when the Devil came out of the mountain, helped the Spaniards, and the Indians were vanquished.[3] I cannot vouch for the truth of their story, but the story gave the name to the mountain, and the rocks certainly do look as if the devil had been about at some time. There is a breaking up and roasting of strata on a grand scale.

We are having lovely weather here now—days and nights clear, not a cloud to be seen for week in and week out, days warm (70° to 80° in shade, but the sun is scorching) yet with a delicious breeze every day from the bay. The nights are cool (it has been down below 40° the last two nights) but without dew—glorious nights to sleep under the clear sky, only a little cool.

As I have read home letters telling of "dry weather," "very dry spell," "no rain for six weeks," I often compare that with this climate. There was a heavy rain *here* last April, but none since. *I have seen no heavy rain since early last January.* We had slight showers later, but not a drop since a slight shower at Monterey late in May (possibly the first week of June). Think and try to conceive, if possible, *how* dry it must be— everything, except trees, parched and sere, watercourses but dry beds of sand, roads two to eight inches deep of the finest dust, soil everywhere cracked to the depth of two to six feet, the cracks often wide enough for the foot to slip in when walking, and indeed, the whole surface fissured with cracks one to three inches wide. "Dust thou art and to dust shalt thou return" extends to the soil here as well as to its inhabitants.

A young man of whom I have long heard much has just returned from Arizona. His name is Pumpelly,[4] originally from Owego, New York, then in the Scientific School at Yale after I left, then at Freiburg and other European schools. A year ago he went to Arizona as engineer in one of the mines of that country. You have seen enough in the papers of that region— a region of vast mineral wealth, but the most inhospitable part of America—desert, hot, parched—producing only thorns and cacti—with here and there a fertile valley—cut off from the world—inhabited by the Apaches, the most treacherous and bloodthirsty of American Indians—with Mexicans, more treacherous but less honorable than the Indians. Twice in the history of that country, before its purchase by the United States, the Apaches expelled and exterminated the Mexican race from the territory, and now they have expelled the Americans—the first place on the continent where our race has had to resign territory once occupied. But the treachery of the United States officers, the withdrawal of troops, the inciting of the Indians to murder by the Southern Confederacy, has inaugurated scenes of horror in that country for which the early history even of the eastern states shows no parallel for cruelty and atrocity.

Every superintendent of mines in the country has been killed, men, women, and children murdered, their dead bodies always mutilated, many tortured, others burned, and no means of redress for the present. Pumpelly, with some others, escaped, as if by miracle, traveling six hundred miles across deserts with only *panoli* (roasted corn or wheat, pounded) to eat, and is now safely here. A terrible responsibility rests on the heads of those who inaugurated this war on our country. It makes me lose my patience to hear them excused or even palliated.

I must fear trouble in this state. I know that the state *as a state* is loyal—it has shown it at the last election, it has shown it at the recruiting offices. But we have a large desperado population, most of whom belong to the Secessionists—men ready for anything, who care nothing for the cause of either North or South in the abstract, but who would inaugurate war for the sake of its spoils. Then there are others of southern birth and southern sympathies to lead them. A large Mexican population, but semi-civilized at best, and who, as a class, hate the Americans with an inveterate hatred, is being incited by the Secessionists, especially in the southern part of the state. Already, over a large region life is very insecure, unarmed men stand no chance, robberies are daily committed by armed bodies calling themselves Secessionists. This does not extend here. It is mostly in San Diego, San Bernardino, and Los Angeles counties—an immense region, sparsely peopled, and containing much desert. It is the worst in San Bernardino, and while I hope for the best, there is just cause of apprehension for a terrible state of affairs yet in this state.

We came here from Martinez day before yesterday. A fair (County of Contra Costa) was in progress at the village of Pacheco, and we stopped for a few hours. It was the poorest fair I have ever seen—some poor fruit, half a dozen poor cattle, and some good horses made up the exhibition, for which a dollar admittance was charged. The town was overrun with gamblers—*faro* and *monte* tables and similar establishments abounded everywhere, with all their attendant scenes of drunkenness, swearing, and vice.

We then came on here, and yesterday Averill and I took a tramp of twelve or fifteen miles over the coal mine region—a long and hot walk, over steep hills and through hot canyons.

Pumpelly goes up with Ashburner, working without salary for the rest of the season, when he will probably be employed as an assistant.

Clayton (formerly Deadfall).
Sunday, October 6.

STILL here, and a lovely day it is, but warm. I am writing this in a lovely "bower." By our camp runs a watercourse, the bed of a considerable stream in winter and spring, but all dry now. The bed is three or four rods wide, sunk a few feet, covered with gravel, and shaded overhead with large oaks and here and there willows and grapevines. A pretty place here is shadowed and festooned by vines, under which I am writing, a gentle breeze plays through, very pleasant, but it blows my papers —the thermometer in this cool place is 80° F., but in the sun, laid on the dried grass, it is 120°. Pete is making a couple of pies for dinner, the rest are lounging about in the shade, birds flit and sing overhead, and quail trip around near me.

We have had delightful nights lately—I wish you might be with us once. As I sleep less than the rest, and the evenings are getting longer—they go to bed at eight or eight-thirty—I sit in the tent and read until cold, then go out and sit by the fire, warm myself, gaze into its embers and reflect on distant scenes and distant friends, take a quiet smoke (for I smoke in camp), then retire. The brilliant shooting stars, so common in August, have almost ceased—but here the sky is clearer, like our clearest winter's night, and the stars twinkle as brightly. The oaks are grouped around with their drooping branches, and the stars twinkle through them—while in the southern sky loom up the bold and grand outlines of the majestic old mountain.

The strata about Mount Diablo are of most enormous thickness, in all probability not less than one and a half or two miles! I think even more than the latter number. With the elevation of the mountain these were broken up—the central mountain mass roasted and baked, yet perfectly stratified often, but on all sides the strata only broken—the broken edges stuck up, forming ridges 1,500 to 2,000 feet high, the strata dipping at a high angle, often entirely perpendicular. Scattered through these are many fossils, and in this great mass is a bed of coal over four feet thick. The bed, like the strata in which it is found, is inclined about forty-five degrees. Several mines are opened, and companies have formed with capital to the amount of some three or four millions of dollars. They are now getting perhaps a hundred tons per day and making preparations for more extensive work.

We took a piece of wood only partially decayed from a stratum of clay under the coal. The stick was some six feet long. We got out one piece fifteen or sixteen inches long and four or five in diameter, quite perfect, with the knots in it—it seemed like a piece of wood that might have lain a few years in wet mud, partially rotted, the rest sound, yet this lay in a stratum that must have had nearly or quite a mile of rock deposited over it after it was placed there, then thrown up at the raising of the mountain! The mind vainly tries to grasp the ages that stick must have been thus buried, now to be dug out by moderns. It was about a hundred feet from the present surface.

I rode over the hills, leaving the party, to a hill several miles northeast, nearer the bay. Here I found some pretty fossils, mostly shells; but the most interesting was fossil wood, trees silicified as hard as flint, but with the whole structure preserved in its minutest details. The grain was pretty. I cannot compare it with anything but curled maple, yet finer than that.

I collected as much as I could carry, then rode back to camp, arriving after sundown. As I rode up the valley the old mountain never looked so gorgeous before, tinged in purple with the setting sun.

We expect a "Pony" today, and are anxious for the news.

Corral Hollow.
Sunday, October 13.

IF you look on the map, southeast of Mount Diablo, you will find a valley, "Corral Hollow," watered by a curved river, enclosed in the mountains. If you are posted in newspapers you have heard of the "rich coal mines" in said Corral Hollow. Well, here we are! As distance lends enchantment to the view, just believe it a lovely spot; but as we are here, we find it a most Godforsaken, cheerless, inhospitable, comfortless region.

I will not anticipate, however, but keep on the even tenor of my story. Monday, October 7, we sent Averill to Martinez to take barometrical observations, to determine the height of our camp at Clayton, the basis of all our observations about Mount Diablo. Professor Whitney remained to take observations at camp. I took my mule, to visit the hills eight or ten miles to the north and northeast of camp. A very lonely ride, first through the Kirker Pass, then among rounded hills, almost bare of grass or herbage, in places entirely so—no trees to cheer the eye, no water in the many canyons and ravines. I found much of geological interest, quantities of fossil wood— of the hardest flint, yet the finest grain of the wood preserved in the minutest details—fossil shells of more than ordinary beauty, immense beds of sandstone, and thick strata, over a hundred feet thick, of volcanic pumice stone and lava. I had no lunch along, and found no water for self or mule, except some alkaline springs which neither mule nor I could drink. The day was hot, and it was long, too, being without either water or lunch. I got so many specimens that I packed them on my saddle and led my mule, walking eight or ten miles back to camp, where I arrived just at sundown. Didn't my supper taste good, and a drink taste better!

A few clouds were in the sky during the day, light fleecy clouds, and the barometer fell rapidly and all prophesied rain, for it is now getting time for it. Night came on and the clouds vanished. No rain, but instead, a windstorm. Whew! how it blew! The wind just shrieked—clouds of dust—dried leaves—pieces of grass—etc. It was hard work to keep one's blankets on, and the wind blew through the blankets, cold and rheumatic. The ground is so dry that a wind raises much dust, and in the tent it is much worse, for the tent flaps in the wind and raises an "infamous" air. Professor Whitney slept in the tent and came out in the morning looking decidedly grimy. He reported a miserable night inside.

Tuesday we left that camp, Averill remaining to observe barometer. We went on ten or twelve miles, crossing a pass on the east side of the mountain, and camped in a deep canyon, where we found good water, a stream that came to the surface for two or three miles. The wind was very high during the day, and the air so filled with dust that at times nothing distant could be seen. The old mountain was at times obscured, at others stood out faintly in the thick gray air. We slept among a few trees that grew near the water in the canyon, but the wind howled throughout the night, blankets seemed but a partial protection, and every rheumatic joint would almost creak as one turned and turned again to keep his blankets over him.

Wednesday the wagon returned for Averill and to bring the rest of our things. Professor Whitney returned with it, much to my regret, for we were now on our way to a hard region, its geology complicated and its "accommodations" horrible, but he thought it best.

Hoffmann and I visited a high ridge southeast of Mount Diablo, a ridge over 2,600 feet high. The wind was high, but it died down about noon, and in two hours more we had a fine view of the surrounding country for many miles—not clear, however—but we got the bearings we desired from the summit. To the south of Mount Diablo the range is depressed in the middle and there is a great valley basin, of probably two hundred square miles or more—a great plain, once probably an inland lake. It is the Livermore Valley. Livermore Pass is on the east side. The valley is not marked on the maps. It is a great basin, branching off into several valleys—Calaveras Valley, Amador Valley, San Ramon, etc.—but the streams do not follow them at present, so great have been the changes in modern geological times. This great valley was spread out beneath us. We had seen it before from various points, and had been in its western edge from Haywards, but this was the best view.

Thursday, October 10, we came on to Camp 60, in the entrance of Livermore Pass. First, down the canyon about eight miles to a Mr. Marsh's ranch. He has a fine stone house, by far the finest in this whole region. As this was the last water, we stopped two hours, lunched, and I visited a "quicksilver lead" of his, which proved to be no quicksilver at all, but a red clay. He gave us some fine grapes.

We had expected to get across through the hills, but found it impossible—we must take the plain. It is about thirty-six miles to the coal mines, water to be found in only one place on the way, sixteen miles on.

We strike out on the plain—oh! what a tedious plain—league after league stretches away—it seems as boundless as the sea—we go slow, for it is sultry—but we pull up into the hills at the place directed and find a little tavern at a spring, with a few stunted willows around, the first trees we have seen for over a dozen miles.

The San Joaquin (pronounced *San Waugh-keen'*) plain lies between the Mount Diablo Range and the Sierra Nevada—a great plain here, as much as forty to fifty miles broad, desolate, without trees save along the river, without water during nine or ten months of the year, and practically a desert. The soil is fertile enough, but destitute of water, save the marshes near the river and near the Tulare Lake. The marshy region is unhealthy and infested with mosquitoes in incredible numbers and of unparalleled ferocity. The dry plain on each side abounds in tarantulas by the thousands. These are spiders, living in holes, and of a size that must be seen to be appreciated. I shall try and catch some to send home, but I have seen them where two would cover this page, as they stand, their bodies as large as a half-grown mouse, their hairy legs of proportionate size, their fangs as large as those of a moderate sized rattlesnake. Pleasant companions! We never think of pulling on our boots in the morning without first shaking them, for fear of tarantulas—but luckily they seldom travel by night. They bite vigorously when provoked, and their bite is generally *considered* fatal, although I have heard of but one well-authenticated case of death resulting since I have been here; but the bite generally proves a painful and serious affair.

The Diablo Range is skirted on the east side (I think its whole length, certainly for 150 or more miles) by ranges of barren hills, sinking into the plain, and sometimes rising to the height of 1,500 feet, mostly rounded, nearly destitute of water. Barren, very barren, few trees—often one will have a prospect of a dozen or even twice that number of square miles without a tree or shrub large enough to be seen, the ground either entirely bare or with a very scanty vegetation of stunted grass and low weeds.

A few cattle feed among these hills or on the adjacent plain, knowing where the water is to be found in the ravines, or going to the river. Hundreds of trails here lead to the river. Cattle go down in the forenoon, linger near the water until near night, then start in droves, single file, for the hills. They will thus go six or eight miles for water each day, going to the hills to feed and to keep away from the mosquitoes. The streams that form in these hills in the spring all sink when they enter the plain, and as summer advances they dry up farther and farther until they all disappear. Such is an immense region, such it must ever remain, supporting a scanty population.

Our camp (No. 60) was at Zimmermann's Mountain House at the entrance of Livermore Pass. The hills near had been extensively prospected for coal, but nothing of importance found. We spent Friday there, and Hoffmann and I visited a peak about eight miles distant, 1,700 feet high, while Averill went to the river to observe barometer at the same time.

Saturday morning we started for our present camp. We were told that it was seventeen miles. First out on the plain, hot, sultry, tedious, four weary leagues were made, when we struck up the canyon of Corral Hollow. Here we followed up the dry bed of the creek; sand and gravel deep, often dusty, the air close, no wind, hot and sultry. It was but 85° in the shade, but it seemed much hotter. We were heavily loaded, for we had, besides our regular baggage, barley for our mules, for nothing can be got here for them to eat. Our seventeen miles proved over twenty, which took us seven hours to accomplish, with no water on the way save that in our canteens, which was a little salt and alkaline and got warm, say 80°, and was insipid and nauseous enough.

We arrived at the mines, and an hour was spent looking for water before we came to camp. Water to drink had to be carried by hand from a canyon a mile distant. Our mules could drink the water that ran from the mine, a little stream where the teamsters watered their horses. It was half a mile from camp, and it was *awful*, contained alkali and sulphur, and the poor animals refused it until driven by keen thirst. There was a deep well at a house, but it was insufficient in quantity to supply the people and too alkaline to drink. The woman told us: "It is good water, we can cook some things with it and make coffee, but it *spiles tea*."

Under these disadvantages we camped and got our dinner

at sunset after the day's fast. We concluded to move camp this morning down the valley again to a spring. So after breakfast we packed up and moved down the canyon three miles. Here is a spring, tolerably good water, but not enough. We will stay here, for this will do for cooking and drinking, although we must take our mules three miles to the mine to drink.

A few trees grow in the bottom of this canyon. We are camped under some oaks, a good enough place if we only had water. We have spent a quiet afternoon, a quiet Sunday, but with more fasting than prayer, I am afraid. Our mules drank the muddy spring so dry that we could get no dinner until night, and then the tea was gritty with the mud. But we are used to going without dinners as well as to other discomforts.

Alas, how little we appreciate the blessings we enjoy amid the comforts of a home, where we have food when we are hungry, water when we are thirsty, shelter in the storm, and beds when we are weary; where we can sit and talk with those dearest to the heart when the day's work is done, can be cheered by friends when we feel sad or lonely, comforted by them when we feel troubled, advised when we are perplexed. Ah, my dear friends, you little appreciate or know what you enjoy—they are so much matters of everyday life that they cause no thought.

All are asleep but me—my "claim is located," the stones picked out of my bed, my blankets spread on the rough gravel and invite to sleep. The sky is unusually murky, the wind howls down the canyon, it feels like an unpleasant night. A year ago yesterday I left home.

Corral Hollow.
Tuesday, October 15.

I FIND what I wrote about the San Joaquin plain may be misunderstood. There *is* water in the *river* that runs through it, but from the river to the hills on each side, especially the west side, a distance of four to fifteen miles, there is no water—fifty or sixty miles might be passed on the plain between the river and the hills without crossing a stream of water, for those figured on the map are all dry now.

Corral Hollow runs up west into the mountains, then suddenly turns southeast, the canyon much narrowing at the same time. The coal mines are near the curve, about nine miles up. The sandstone that forms the hills is broken and thrown up, and there a few seams of poor coal are found. There is but one mine of any account or that has as yet *sold* any coal, and that not over three hundred or four hundred tons at most. I question if any mine here will ever prove profitable. But there are several companies, and many thousands have been expended, as well as much money spent in prospecting. One company spent $11,000 I hear, and got no coal worth speaking of—not a ton of workable coal. I have spent most of today in the mines.

Yesterday Averill, Hoffmann, and I visited a hill to the southwest, over three thousand feet high and over two thousand feet above camp. We rode up the canyon, then mounted a ridge and crossed several knobs, but the air was so filled with dust that we could see but a few miles, scarcely ten in any direction. The first hills crossed were over the sandstone, but the soil is clay, dry and cracked.

A fine rattlesnake sounded his alarm and then retreated into one of the cracks in the soil. I punched him with the tripod of our compass, the only stick we could get, until he ceased rattling, but we could not get him.

We soon struck other hills, where the sandstones had been twisted and baked by volcanic heat for miles, and here the scene changed—some trees scattered here and there, canyons more narrow, and hills more sharp.

We took a circuitous route up, which we thought we could shorten several miles on the return by descending into the Corral Hollow canyon above its curve and following it down. We descended into it, a narrow gorge more than a thousand feet deep, down a *very* steep slope, our mules sliding and getting down as best they could—it was too steep to ride them—a slope of thirty degrees or more—then struck down the stream. We got into a fix.

The gorge got narrow, huge rocks had fallen in and choked it up in places, but we got our mules down nearly to the road, when the route became absolutely impassable. We spent two hours in getting them about a mile through the rocks, and then had to get them out by making them climb a slope having an average incline of forty-seven degrees, and in places over fifty degrees, for five or six hundred feet. Think of that! But they did it, and we got out safely.

We were without lunch yesterday and got no dinner until dark, and far from a sumptuous one then. Our coffee has given out, the last "fresh" meat, in an advanced state of blueness and beginning to have a questionable odor as well as color, was eaten for breakfast, but bacon yet remains. We get no good pork in this state. Our sugar gave out this morning, but as bacon and beans are very nutritious, there is no danger of starving.

In the beds of sandstone a mile or so north of our camp we found today the finest fossil leaves I have ever seen. The rock was filled with leaves of several species of trees, most minutely preserved, as fine as the finest paintings, black in the light colored sandstone, with many hundred feet of sandstone resting on them. They were in a deep canyon. Much fossil wood abounds. We found one tree, or rather stump, erect, its stony roots still in the bed containing the leaves, once soft mud, the stump sticking up into the sandstone strata above—all now flint, but with the finest markings of the grain of the wood.

All this shows that *true* coal cannot exist here, only "Tertiary" coal, which must be, of necessity, an inferior kind; while the way the rocks are broken and tossed about must make following the beds very precarious and uncertain. Most of the coal stands nearly perpendicular, all at an angle over forty-five degrees. To follow such seams far one must go very deep, and the beds are cut off by the breaks or faults in the strata.

Friday Night, October 18.

EVERY night now is windy, the days warm (about 80°–84° F.), hazy, or dusty—I cannot say *smoky*—but the air is thicker than in our thickest Indian summer. It interferes with the work of our topographer very seriously.

We will leave tomorrow. I sent Averill on today for barometrical observations to connect with this camp. I have been hard at work every day, and provisions getting lower. Peter shot some quail and rabbits, and we have had two or three "potpies" that vanished before our attacks like dew on a summer's morning. But small game is scarce (except tarantulas) and this morning we bought half a deer of a hunter.

This hunter, by the way, is an old companion of "Grizzly" Adams.[5] His cabin is near our camp. We are camping by Adams' spring, the ruins of his cabin are within a hundred yards of me. This man came here and lived with Adams before he left, and has hunted ever since, but he complains that civilization has interfered seriously with his sport. "We had good times before the settlers came," he says, and he bears terrible scars, the trophies of contact with grizzlies. He told me this morning that he had killed seven or eight hundred deer here since he came, but they are getting scarce now. He was so badly used by a bear last spring that he has hunted none all summer and is just beginning again now. His venison was very acceptable, for our "table" has not exhibited a great variety of late—tea without sugar or milk, bread, pork, and beans. We have tea for only two days more, and water too bad to drink alone.

For the last few days we have been hard at work here, exploring the hills, and will be off in the morning. I shall be glad when we can get stores again, hear the news, and get a drink of *good* water. I assure you the last is no little item. But such things are incidental to our work and are most cheerfully borne.

Much of the region around here is practically a desert, not

called so, but really so. The bed of this stream tells of a large stream at times, but often years pass without any water flowing down it to the plain, much less to the river beyond.

The region is thrown up into hills from one thousand to three thousand feet high. These on the north are all rounded and furrowed with canyons, but almost destitute of springs, and at the present time are dry beyond anything you can picture to your mind. Sometimes the soil is cracked, in other places dry sand, and in others a dry clay-loam, as dry as ashes, into which the mules sink to the fetlock at every step. There is scanty herbage here and there, but large patches are as bare as a dried summer fallow. I have been on hills today where there were such soils, with here and there scattered bushes—artemisia, sage, etc.—living, yet the leaves so dry that they crumble with the slightest touch.[6] They are reduced to dust in the hand in an instant if you rub them, yet they are alive, and with the first rain these same leaves will show that they are alive. They are not shed every year, only dry up. I picked some low green herbs—small, to be sure, but perfectly green—in one place, on a soil as dry as if it had been dried in an oven, a soil that had been exposed to this scorching sun for many months without either rain or dew. There is no dew here now, the nights are as dry as your dry days, and things will dry as fast.

I wished to preserve some tarantulas—I will send you one when I can—so a day or so ago I caught a couple. This incited the boys—yesterday they caught some and made them fight. I tried it this evening. One of the boys went out and caught four near camp, huge fellows, and placed two near each other and irritated them. Soon they closed in—such a biting—they clasp each other firmly, then bite until one or the other dies. You can see the poison exude from their jaws. Pleasant fellows to find in one's boot or coat sleeve! They live in holes in the ground, and, on the whole, are not dangerous.

There is a large blue wasp with orange wings, a wasp two

or three times as large as the largest hornet of the East, that is the natural enemy of the tarantula. I have never seen a field battle, but we caught one yesterday and made him fight a tarantula in a box. In this instance the tarantula was victorious. In the field the wasp is—he lies down on his back, and as the tarantula pounces on him, he stings him and suddenly glides out, and soon kills his bigger foe.

The wind is roaring down the canyon, a stiff breeze, and not comfortable for sleeping.

San Ramon Valley.
Monday Evening, October 21.

AFTER I finished writing on Friday evening and went to bed, the wind howled all night—a tremendous wind. I had to pile boots, saddlebags, etc., on my blankets while I rolled myself up in them to keep them, and then the wind blew through them. It was by no means a comfortable night, and I often thought of a bed at home. The moon shone bright, the sky was clear, the wind in its fury. When we got up in the morning—with eyes and faces full of dust, hair and beards full of sticks, pieces of grass, and leaves, blankets in the same fix—could you have seen us you would have thought us a rough set, to say the least. We packed up, and I sent the wagon on, clouds of dust following it down the canyon.

A pass leads over the hill into the Livermore Valley, and could we get over it it would be but six miles to Livermore's. As it is, it is over thirty, and we must take the long road. Buggies can get over the pass, and light empty wagons. A hundred dollars would fix the hill. There is but one hill, about 1,760 feet high, or 890 feet above the valley. It rises this 890 feet in about three-quarters of a mile—somewhat steep, surely, but all the rest is a good road.

Hoffmann and I stayed in camp until noon, observing barometer. I took the time to visit a prospecting shaft near, where they are looking for coal. The shaft is a miserable hole, scarcely larger than a well, very insecurely timbered, and 150 feet deep. To hang on a poor rope, much worn, stranded in several places, and thus be let down that distance into such a hole was decidedly suggestive of accidents, but I concluded that if they could trust it for themselves, I could. I stood on the bucket and they often stopped to give me a chance to examine the strata as I went down. The worst was in coming up, for the bucket would catch against the timbers and would have to be lowered a little and tried again. Their work is folly —they never will get a profitable mine there. One seam of coal two feet thick was passed, standing nearly perpendicular, and I think that it is all they will find. The ground is so broken up by the forces that have upheaved and twisted the strata, that even if they find more, mining there must ever be risky.

Another uncomfortable night at Zimmermann's. The wind swept through the pass, not a tree or shrub to break its force, everything dusty. The wind died down in the night, and in the morning heavy clouds hung overhead and enveloped the mountains. All prophesied rain. We were in a sorry plight to meet it there, out of provisions and no wood—we used dried "buffalo chips" for fuel, but as there are no buffalo here that means cow chips, or in camp parlance "counterfeits." Money was reduced to less than twenty-five dollars, so I ordered a start, although it was Sunday. The whole camp was decidedly in favor of it. It is the first Sunday that we have traveled all day, although twice we have traveled a few hours, and the previous Sunday had moved our camp three miles.

First our way lay through Livermore Pass, about eight miles, among rounded hills over a thousand feet high, then we emerged into the Livermore Valley. We crossed the plain about fifteen or sixteen miles, a tedious ride. At Amador we stopped, fed our mules, and got our dinner. Here are two taverns, a grocery, and about two houses besides. A horse race was coming off in the afternoon, and a mixed crowd of fifty or a hundred Americans, Mexicans, and Indians had assembled—decidedly a hard looking crowd—drinking, swearing, betting, and gambling. After dinner we came on here, where we camped some five weeks ago—at Major Russell's.

Today Hoffmann and I have been up about eight or ten miles on the ridges at the foot of Mount Diablo. We passed over the edges of perpendicular strata, standing perfectly on edge, for two or three miles, showing that these strata have that enormous thickness. They contain shells in abundance at intervals through the whole of that immense thickness—oyster, clam, and other shells. We were on a ridge over two thousand feet high, with these shells on the top and in the rock. Last week we were on a ridge 2,200 feet high, where wagon loads of immense oyster shells might be picked up. Today I found also the joint of a whale's backbone! These are some of the marvels of Californian geology.

NOTES

1. Friedrich Jagor (1816–1900).
2. William Phipps Blake, graduated from Yale Scientific School, 1852; geologist and mineralogist for the Pacific Railroad Survey, 1853; a candidate for the position as State Geologist ultimately given to Whitney; Professor of Mineralogy and Geology, College of California, 1864; later at the University of Arizona.
3. See also Bret Harte's "The Legend of Monte del Diablo," one of his *Spanish and American Legends.*
4. Raphael Pumpelly (1837–1923) made extensive geological explorations and was in charge of important surveys and mining enterprises in many parts of the world. In *Across America and Asia* (1870) he describes his Arizona experiences and his work in Japan, referred to by Brewer. On page 67 he says: "In preparing for this journey I became indebted to many kind friends, especially to Professor J. D. Whitney, of the State Geological Survey, and to his Assistants, Messrs. Brewer and Ashburner, as well as to Messrs. Louis and Henry Janin, of the Enriquita mines." Further references are found in *My Reminiscences,* by Raphael Pumpelly, 1918.
5. *The Adventures of James Capen Adams, Mountaineer and Grizzly Bear Hunter, of California* (1860), by Theodore H. Hittell, is one of the classics of California literature. Adams captured and tamed wild animals, specializing in grizzlies, several of which—Ben Franklin, Lady Washington, and Samson—were especially famous. They were exhibited in San Francisco, 1856–59. In 1860 he took his animals to New York and exhibited them under contract with Barnum. Adams came to Corral Hollow in 1855, where he made a bargain to hunt with a man named Wright. The experiences in Corral Hollow are told in *The Adventures,* chap. xiv.
6. Sagebrush (*Artemisia californica*) and sage (*Salvia mellifera*) are both found here (Jepson).

CHAPTER VI

NAPA VALLEY AND THE GEYSERS

Overland Telegraph—Benicia—Judge Hastings—Suscol—Napa—Yount's—St. Helena—McDonald's—Pluton River—Pioneer Mine—The Geysers—Rain—San Francisco and Winter Quarters.

Camp 62, near Benicia.
Saturday Evening, October 26, 1861.

ON Tuesday last we came on from San Ramon, intending to come a few miles and stop over one day, but finding no place to camp we came on to Martinez.

On Wednesday night Professor Whitney came and brought with him a tremendous package of letters for us all. Peter got the sad news of the death of his mother, who lived in Illinois (or Iowa). He had not seen her for two or three years, but often spoke of her in terms of the tenderest affection. He is much depressed by it.

I had expected that Professor Whitney would remain with us, but Pumpelly has accepted a place on the Japan affair, so the Professor has to go up with Ashburner. He spent but one day with us.

That day, Thursday, October 24, was a most memorable one for California. The last piece of wire was put up on the Overland Telegraph, and dispatches were received from the East, Salt Lake, etc. You cannot appreciate the importance of this, but great as it is, it made but little excitement here. A dispatch leaving New York at noon may now be received in San Francisco at a quarter before nine of that morning, or over three hours ahead of time!

This line is built by two companies acting in concert and meeting at Great Salt Lake City—the Overland Telegraph Co., owned in San Francisco, on this side, and the Pacific Telegraph Co., owned in the western states mostly. The latter has put up 1,600 miles in four months! News may now come by telegraph from Cape Race, seventy-two degrees of longitude east, 3,500 miles distant in a direct line, or 4,500 miles by the telegraphic route—surely a most gigantic circuit—the difference in time being nearly five hours. The tariff to New York City is now a dollar per word, and for the first two days the office has been crowded with dispatches.

Friday morning we crossed the ferry to Benicia and camped four miles north of the town. As an item of California prices, we paid *nine dollars* to be ferried over from Martinez on the regular (steam) ferry, and had our whole party been there it would have been ten dollars. Hoffmann and Professor Whitney were absent. The former we had sent to San Francisco. He returned that evening (last night) and brought papers with news from New York up to the previous evening.

This morning a cold heavy fog hung over everything, and many shrugged rheumatic shoulders. Tonight I shall take to the tent and sleep there after this.

Camp 63, at Suscol, five miles from Napa.
November 1.

SATURDAY night, October 26, was foggy, but it cleared up in the morning and the day was most wonderfully clear. It was the loveliest day for a long time. Two gentlemen called at camp —Judge Hastings and a Mr. Whitman. Judge Hastings invited us to dinner on Sunday. After dressing, Averill and I went into town and attended the Episcopal church, the first service of that denomination I have attended for a long time. The church was small but neat, the attendance good, and very "respectable." At the close, we were met at the door by Mr. Whitman, one of the vestrymen, who invited us to his house near *to take a whiskey cocktail*. As it was too early to go to

Judge Hastings we went in, he being an old friend of Averill's. We sat an hour, walked into his garden, the most noticeable feature of which was an abundance of most delicious almonds, now just ripe. The cocktail story may *seem* a large one, but it is literally true—you have no idea how prevalent drinking is in this state—one scarcely ever goes into a house without being invited to drink. If you go to dinner you are asked to drink before, but a refusal to drink is always courteously received—at least, I have always found it so, as I quite seldom accept the invitation except wine.

At three we went to Judge Hastings' and were most cordially received. The Judge is quite a noted man here, has made much money, is a man of influence, was once one of the supreme judges of the state.[1] He has a pretty wife and several children; some of the latter very pretty, others not.

Don't ask me to describe dinners, that is out of my line. The dinner was quite ordinary—two or three courses. The waiters were three Digger Indians, of the homeliest kind, two young squaws, and a boy—far from neat, yet tolerably handy. After dinner Mrs. Hastings bragged of her Indians, told me all their merits and demerits, admired them as servants, but not as cooks—she has a Chinaman cook. So are the races mixed up here. Judge Hastings is a convert to the Roman church; his wife is a leading Presbyterian here. He, like all proselytes, is very zealous, is probably the most influential man in that church here. He had traveled abroad, and gave us a most interesting description of his presentation to the Pope.

Benicia is noted for its schools. There is a college, and there are two large female schools. One of the latter is Roman Catholic, in charge of nuns (sisters, rather) of the order of Ste. Catherine. They have recently made great enlargements, put up a new building at an expense of thirty or thirty-five thousand dollars, and have made proportionate improvements. Judge Hastings asked if we would like to visit it, so we did, were introduced to the "Mother" (Mother Mary) at the head, and went through many of the rooms. About thirty-five sisters have charge, dressed in the most untasteful white garbs of their order. They are very kind and the pupils are greatly attached to them. There were about a hundred girls (besides the day scholars), who, of course, are kept very secluded. It being Sunday there were no studies, but we met them all coming out of the chapel and saw most of them there or in some large recreation rooms. A peculiar feature was an examination hall, apart from the other buildings, in the grounds. It is a building with roof, lattice sides, and floor, built in a hollow, with the ground rising on each side forming a natural amphitheater. It was an excellent arrangement, but of course only suited to a Californian climate.

Our visit there so prolonged our stay that we rode back after dark in the chilly night air. The night was glorious, but cool. The day had been of the most lovely kind, the sky intensely blue, and the mountains across the bay, twenty miles distant, seemed scarcely three miles away.

Monday, October 28, was another magnificent day, clear as the previous one, harbinger of the fall rains they say. Myriads of wild geese flew over our camp, as they have for several days, their numbers incredible. At this season of the year they come from the north to winter in this state. They congregate on the plains, and at times hundreds of acres will be literally covered with them. I believe I wrote last winter of the immense numbers we saw near Los Angeles.

We raised our camp and moved up the Napa Valley to Suscol Ferry, five miles from Napa. The roads were dusty almost beyond endurance. There is much travel, and every team moved in such a cloud that it was impossible to see it at any distance —you only saw its cloud of dust.

We camped by a pretty brook, near the Suscol House. On our way we passed the pretty little village of Vallejo (pronounced here in the Spanish style *Val-lay'-ho*), where the United States has a navy yard. We passed through a fertile region, fine farmhouses at frequent intervals, farms fenced, young orchards growing, everything with an American look.

After camping, Peter shot a fine wild goose near camp. A flock came down in a stubble field, and he stole up and killed one.

Camp 64, Sebastopol.
Sunday, November 3.

TUESDAY, October 29, was a cloudy morn, but soon partially cleared up. Hoffmann and I started for a sharp rocky peak some six or seven miles east, and, as the fields were fenced, we took it afoot—a long walk. We reached it, a sharp rocky knob 1,332 feet high, from which we got bearings of all the surrounding country. A cold wind swept over the summit, but although the sky was cloudy, the air was clear, and the prospect extensive. The arms of the bay, with the winding bayous (called here "sloughs") in the swamps around them, were very beautiful, the effect heightened by the rugged mountains on the north and northwest.

As the wind was so high and raw, we crept behind some bushes on the lee of the hill and hung up the barometer behind them and waited there half an hour for observations, taking a quiet smoke and enjoying the lovely prospect beneath. The wind shrieked over our heads. The pipes smoked out, observations taken, barometer put up, as we emerged from the bushes we saw a dense mass of clouds sweeping down from the north with rain. The whole aspect of the sky in that quarter had changed while we were in the bushes. We struck for the camp "double quick" pace, but had not made over one or two miles when the clouds came sweeping over us. Our way here lay along the crests of a series of ridges for about three or four miles—the ridges in places quite sharp, in other places branching off in spurs—easy enough to see in the clear air, but blind enough in the dense gloom of the thick fog that shrouded us.

With this cloud and fog came a fine rain, sweeping with the

wind. But a few minutes were necessary to wet us to the skin, cold and uncomfortable, but the smell of the rain, the damp air, the smell of wet ground, was delightfully refreshing. It carried me back to days of summer showers, to green grass, fresh air, and no dust—this feature was positively delightful, in spite of the discomfort of the wet.

We pushed on our blind way, facing the stiff wind, and as we neared the end of the ridge where we were to descend into the valley the rain ceased. In a few hundred feet we got below the clouds, and the lovely Napa Valley, with its neat village, pretty farms, green trees, and pretty sites, lay below us, lit by patches of sunshine here and there. A spot of sun lay directly on the village of Napa, six or seven miles distant, producing a delightfully beautiful effect.

We found it had rained but little at camp, not enough to lay the dust. The rest of the afternoon was quite fair, but not clear, and dense masses of clouds hung over the ridges we had left. We changed our clothes for dry ones. Mike soon had the goose roasting, and at half-past four it was served up in good style and partaken of with an appetite any epicure would envy. The skeleton was dismembered, the bones polished, and every vestige that was eatable soon disposed of.

The evenings are getting longer. The weather drove us all to the tent to sleep, and after all the preparations for rain were made, the long evening was whiled away with reading and euchre, varied by a game of "seven-up." But we had no more rain.

Wednesday there was no rain, but a dense fog hung over everything. Averill went to Napa for letters. I lounged down to the tavern to read the news. While there, a rough but intelligent-looking man entered into conversation and invited me to his house a few rods distant for a "glass of *good* cider." I went, got the cider, the best I have tasted in the state, and went into his house. I found him an intelligent man, quite a botanist, and even found that he had some rare and expensive illustrated botanical works, such as *Silva Americana,* worth sixty to eighty dollars—the last place in the world I would have looked for such works. He does not own the ranch, is merely a hired man, having charge! There is an orchard of ten or twelve thousand trees and a vineyard—he makes wine and cider and sells fruit.

It cleared up, and Averill and I took a tramp of ten or twelve miles over the hills—visited a knob over 1,700 feet high and got a magnificent view. The day was most delightful and clear after eleven o'clock.

Thursday, October 31, another foggy morning. Mr. Beardsley came to camp and invited us to his house for more cider. We went, spent an hour, when it cleared up, and we started for a peak seven or eight miles northeast. We got on the wrong ridge, got up about two thousand feet, saw much of geological interest, and got back long after dark.

The hills of the ridge to the east are of strata derived mostly from volcanic products—lava, pumice stone, volcanic ashes—which appear to have been thrown up in water and deposited in strata many hundred feet thick. After being thus deposited, the strata have been upturned, much twisted and broken, and again baked by volcanic agencies—a most complicated affair. These rocks are of all colors, red as brick, gray, black, brown.

Friday, November 1, I sent Averill and Hoffmann to a point we failed to reach the day before, while I visited some points nearer camp, collected and packed specimens, observed barometer. The day was very clear, after a cloudy morning, and they had an all-day's ride of it, getting back after dark. They reached the point, 2,200 feet high, and got good bearings and had a most extensive view, reaching to San Francisco and Stockton on the southwest and southeast and far into the mountains in the opposite direction.

The swamps bordering all the rivers, bays, or lakes, are covered with a tall rush, ten or twelve feet high, called "tule" (*tu'-lee*), which dries up where it joins arable land. On the plain below camp, fire was in the tules and in the stubble grounds at several places every night, and in the night air the sight was most grand—great sheets of flame, extending over acres, now a broad lurid sheet, then a line of fire sweeping across stubble fields. The glare of the fire, reflected from the pillar of smoke which rose from each spot—a pillar of fire it seemed—was magnificent. Every evening we would go out and sit on a fence on the ridge and watch this beautiful sight, some nights finer than others.

Saturday morning we started, first to Napa, five miles, then on to this place, nine or ten farther. We had gone but a short distance when we came on a large flock of geese, several hundred feeding in a stubble field close by the road. They are very sagacious, always keeping several on watch while feeding, and never allowing a man to approach on foot. But they are not so afraid of horses and wagons going along the road. We stopped and loaded our double-barrel shotgun. Peter walked behind a mule to within eighty yards, then shot. The geese rose, but three fell; two we got, the other fell in the tule where we could not get him.

The various tricks that hunters resort to, to kill these geese, are ingenious, and the sagacity of the geese is as marked. The hunters lie in the bushes and shoot the geese as they fly over, but the geese learn in a few days to fly high over these bushes. Sometimes they train a horse, but the geese soon learn to avoid a horseman. On the San Joaquin plain, where there are multitudes of cattle, which can approach the geese, an ox is trained so that a man can walk behind into the very flock and bang away with both barrels and kill several, but the geese soon learn which ox is the suspicious one. The geese bring a good price in market. (N.B.—Mike is now roasting a goose outside the tent—I smell the savor thereof—I wish you could dine with us.)

To go on—Napa is a pretty, American town, on a stream large enough for a small steamer to ply to San Francisco, and is a place of much trade. We stopped a few minutes. I got letters, one from Professor Whitney, one from home—which I read after mounting, as I rode along.

We came on nine miles farther up the valley. Pretty farms, neat farmhouses, fine young orchards, lined the way. The bottom is three to five miles wide most of the way, very fertile land, and the fields have scattered over them many most grand oaks, which would be an ornament to any park with their broad spreading branches, drooping at the ends like those of an elm—majestic trees.

But *the* feature of that nine miles was the dust, *dust!* DUST! The road has been much used, hauling grain, and from fence to fence the dust is from two to six inches deep, fine as the liveliest plaster of Paris, impalpable clay, into which the mules sink to the fetlock, raising a cloud out of which you often cannot see. Each team we met was enveloped in a cloud, so that often you could not see whether it was a one-, two-, four-, or eight-horse team we were meeting; the people, male and female, were covered with dust—fences, trees, ground, everything covered. Need I say that on our arrival to camp a wash was one of the first performances?

San Francisco.
November 17.

SAFELY back here again in the city. The last two weeks have been such laborious ones—one week in the rain—that I have done no letter writing. I will now bring up my journal as I have time.

This was left off at Camp 64, at Sebastopol, the "Yount's" of your maps.[2] And so it should now be called, in justice to the settler, Yount, who settled there twenty-eight years ago.[3] His story seems a romance. He was a western man who wandered across the Rocky Mountains, lived with the Indians, hunted, and trapped. While plying the trade of trapper he entered San Francisco Bay, pushed up its northern arm, entered Napa

Valley, and found this lovely spot, inhabited by savage Indians and a few semi-savage Mexicans.

In his youth, in a western state, a fortune teller had predicted to him his future home, settled in a lovely valley, etc. Here seemed to be the place—a fertile valley, enclosed by high mountain ridges, a rich bottom with grand trees, a stream rich in fish. He did not stay, however, but the prediction of the sibyl so often came up to him that he returned a year or so later and got a grant of the Mexican Government of two leagues of land in the valley. He built him a cabin—at once fort, fortress, and home.

By his force of character and kindness he overcame the Indians and made them such warm friends that to this day many live on his ranch. With his rifle he compelled the submission of the treacherous Mexicans. His exploits and adventures smack of the marvelous, but he held his place, fully determined that Fate had destined that spot to be his home. He raised cattle, had a village of Indians on his ranch, and lived that patriarchal life for fifteen years before the discovery of gold here and the immigration. His Mexican grant was confirmed by the United States Government after much delay and difficulty, and now he is surrounded by thriving and valuable farms, fine orchards, above and below him. We camped on his land, but missed seeing him. Here are the "heads" for a "romance" or "tale" for some future author.

One evening while at this camp I attempted to go up to Yount's and call on the old man as he had sent us an invitation. It was but a mile or so from camp. We had had a hard climb that day and the others declined going, so I started alone. I passed the Indian village on the ranch. It was after dark. The homes were mere sheds covered with bushes or rushes, the front side entirely open to the air. These seemed their "sitting rooms." A bright fire lit up the hut. Standing, lying, squatting around the fires were the Indians—some with bright red blankets around them—squaws doing various work, dressed in skirt and chemise, the latter quite scanty in the neck, and in many of the middle-aged women showing enormous and flabby breasts —children playing about. The whole, lit up by their bright fires, made a most picturesque and peculiar picture. The Indians of this state, except some of the tribes in the extreme north, are perhaps the ugliest looking in North America—they are certainly much inferior to the other tribes I have seen. I missed the way, for the night was very dark, and I could not find Yount's house, so I returned to camp. I thus missed seeing him, as I had no other opportunity.

November 4 we climbed a high rock ridge east of the valley—a rough craggy mountain 2,147 feet high, its sides furrowed by canyons, and very picturesque as seen from the valley. Although not very high it was quite a rugged climb and the view from the top decidedly fine. We had no lunch along, but a roasted goose about sunset answered for dinner and supper both, and some prophesied for breakfast too. But the prediction proved untrue—we ate it all, polished the bones, and left not even a "wreck" of its trimmings, and morning found us hungry again, of course. By the way, we came near losing our geese. The night after our arrival, just after dark, a stranger approached camp. We did not *see* him, but his presence was unmistakable. Mike, notoriously careless, had tied the geese in a bush, quite too near the ground. We "heard something drop," rushed out, cautious, for we had a very formidable thief to contend with, but by our infernal yells, shouts, pounding on a board, etc., he left the geese and sneaked away through the bushes. We then tied them so high that no marauding skunk could get them, but he hung around during the night, loading the air with the odor of his presence.

November 5 Hoffmann and I climbed a ridge west of camp, about five or six miles distant. A steep sharp ridge rises from the valley, very steep, its top a very sharp point, 2,400 feet high, with no rocks in sight and but few trees. Its steep sides and sharp top show a smooth surface of dark green chaparral

—beautiful to the lover of scenery, but forbidding to our experienced eyes. We have seen too much of that treacherous surface, which looks so inviting in the distance but is so difficult and laborious to penetrate.

A ride of about four miles up and across the valley, then back through the fields, and then we strike up a ridge. A few hundred feet are gained, when suddenly a board fence bars further progress. But we are ready for the difficulty, as it is one we often meet. I have some nails in my saddlebags, my geological hammer is the right tool—I soon knock off the top boards, jump our mules over, replace the boards with fresh nails, and ride on. We soon get into a cattle trail that carries us up until it gets too steep to ride farther. We tie and make the rest on foot, find the chaparral not tall, and less difficult than was anticipated.

The view from the top is finer than any we have had since crossing the bay, more extensive and more grand. San Pablo Bay gleams in the distance; the lovely Napa Valley lies beneath us, with its pretty farms, its majestic trees, its vineyards and orchards and farmhouses. Its villages, of which three or four were in sight, the most picturesque of which is St. Helena, are nestled among the trees at the head of the valley. Bold rocky ridges stand across the valley, a bold broken country around us. To the northwest lies the distant valley of the Russian River, one of the finest in the state, many thinking it even finer than the Napa and the Santa Clara valleys. But *the* feature is Mount St. Helena, rising to the north of us, over four thousand feet—steep, bold, and rocky—an object of sublimity as well as beauty.

We stayed for nearly two hours on the peak. We were tired and hungry. I thought we had no lunch. Hoffmann had told Mike, however, to put up something. We searched my saddlebags, and lo! six quails, finely broiled (cold), with bread, salt, etc! How we feasted! and the scene looked even more beautiful after it. As the peak was without a name we called it Mount

Henry, as it was something like Mount Bache across the bay, and we thought it well to honor the distinguished man of science.[4]

Our return was without especial interest. Found a goose roasted for dinner, the last of the season for us—last but not least. It went the way of its predecessors.

Professor Whitney came to camp the evening of the fifth. It had been our intention to visit the new quicksilver region of Napa County and the Geysers, but we had already exceeded the time set for withdrawing from the field. The rain had kept off unusually long, and it was feared that when it would once begin there would be no stop. The Professor was more than half inclined to turn back as it began to look like rain, but I was anxious to get up into that country, even if it were only for a week, and I carried my point.

The next day we started on, following up the valley eighteen miles. The bottom grew narrower, the hills on each side more rugged, and trees more abundant. The pretty little village of St. Helena, with its fifty or more houses, many of them neat and white, nestled among grand old oaks, was very picturesque. We got meat and vegetables and pushed on eight miles farther and stopped at Fowler's Ranch, Carne Humana ("Human Flesh Ranch") but we could not find out the origin of the name —near some quite noted hot springs. These last are curious— a number break out around the base of a low but sharp conical hill that rises in the valley. The springs vary in temperature from 157° to 170° F. The waters smell quite strongly of sulphur and have some considerable reputation in the cure of diseases. There is a fine public house with bathhouses, etc.[5]

The weather continued to look like rain, but November 7 we pushed on. The trees grew more numerous, not only oaks, but fir or spruce, pines, the majestic redwood here and there with trunks towering perhaps two hundred feet high, and a lovely madroña tree growing finer than we had seen it before. This is a beautiful tree, has leaves like the magnolia, rich dark green, sheds its outer bark every year, and is very peculiar as well as beautiful.

Here let me state that your maps will give but little satisfaction for the region I am to describe. All north of the Napa Valley is guess work—much totally wrong—only a few of the main features are correct. There is a Mount St. Helena (not St. Helens), but no Mount Putas (Whore Mountain). The latter is probably the one now called Mount Cobb, as there is a high Mount Cobb near where the map puts Mount Putas. But all the region through to Clear Lake and on to the Humboldt is a *very* rough country, of which there are as yet no maps anywhere near correct. We, of course, are getting as many details and bearings as we can, and will eventually, I hope, get a tolerably good map. But do not wonder if my letters and the maps disagree during the remainder of this trip.[6]

We passed up to the head of the Napa Valley, then over a low divide toward the northwest and descended into Knight's Valley, a lovely valley watered by a tributary of the Russian River. The divide between these valleys on the south side of Mount St. Helena is very low, not over five or six hundred feet high. We passed down Knight's Valley a few miles, then across by an obscure road, over low hills to McDonald's, on a creek of his name, a tributary of Knight's Creek. Here we camped—Camp 66.

McDonald is a quiet, fine man, and what is rare in such regions, a pious man. He settled here twelve or fifteen years ago, then the remotest settler in this region between San Francisco and the settlements in Oregon. His wife, then but twenty years old, is still pretty, an intelligent and amiable woman. It must have required courage to settle here at that time, surrounded by Indians, so far away from civilization.

As this was the "headwaters" of wagon navigation we made our preparations to go on with mules. To the north of this lies a region now creating much excitement from the discovery of many quicksilver "leads." This we wished to hastily visit.

We had a cold night, and November 8 the temperature was 37° F. in the morning—but the sky was intensely clear. We started, Professor Whitney, Averill, and I, each with a mule, and Old Jim carrying our blankets and some provisions packed on his saddle. The Professor and I carried saddlebags on our mules, while tin cups, pistols, knives, and hammers swung from our belts. We agreed to ride the two mules by turns, but it proved in the end that it was little riding that any of us did. Mr. McDonald went with us on foot as guide the first day. First we went up the valley of the little stream two or three miles, then struck up the ridge and crossed Pine Mountain, some 3,000 or 3,200 feet high, commanding a glorious prospect from its summit. The ascent was steep, the trail very obscure, and we would never have found it alone, even with all our mountain experience.

The summit of the ridge has some scattered pines, but on descending its northern slope we passed through a forest—firs, pines of several species, oaks, madroñas, etc., and with these the strange "nutmeg tree"—a curious tree something like a pine, more like a hemlock, but bearing fruit in size, shape, and taste very like a nutmeg![7]

We descended a few hundred feet into a canyon, where a cabin had been built beside a fine brook. It was now deserted, but here we camped, unloaded, and picketed our mules to feed, then descended into another wild canyon to a claim that was being worked.

On our return we found two other men had followed us, with pack-horses, having "interests" in some of the leads. A fire was built in the cabin, a dirty coffeepot found, which we cleaned up, coffee made, bacon roasted, and we had a comfortable supper, after which we sat by the bright fire and listened to the tales of pioneer life and adventure in these wilds. The rest slept in the cabin, but the Professor and I spread our blankets out under a majestic fir tree. The night was clear, the stars very bright, twinkling through the foliage; the sighing of the wind through the pines carried me back to recollections of other years and far distant scenes. The night was cold, but we were tired and slept well. To be sure, we were not over ten miles in a direct line from camp, probably not that, but distance in such places is not to be reckoned by miles but by the hours required to accomplish them.

November 9, by early dawn, we were astir and before the sun had gilded the mountain tops had breakfasted and packed up. We had but four miles to make that day, in direct line, but *such* miles! On our way we visited several leads, some quite rich. But *such* a trail—across steep ridges, zigzag, up and down, through deep gulches, over ridges—one of which was three thousand feet high and two thousand feet above other parts of our trail—through chaparral. In one place we were two hours making one mile in a direct line. We struck a trail, however, and descended into the valley of the Pluton River near its head—a canyon rather than valley—where the furnace of the Pioneer Mine is situated.

Here are the furnace, smith shop, and the homes of the workmen, among the trees, in a most picturesque spot, and, although in a canyon, still over two thousand feet above the sea. No wagon road leads here—everything must be packed on mules, even the fire brick for the furnace, tools, even an anvil; a wagon had been taken apart and packed in over a trail that crosses ridges over three thousand feet high.

We were most cordially and hospitably greeted and welcomed by Mr. Wattles, the foreman of the mines. We spent the rest of the afternoon in examining the furnaces and in visiting the "Little Geysers," some remarkable hot springs near. The furnace was new and but two charges had yet been burnt in it. Great hopes are entertained of its eventual success. The ore of the Pioneer Mine is remarkable for its being native quicksilver, or the metallic quicksilver mixed with the rock. In places the rock is completely saturated with the fluid metal. It appears in minute drops through the whole mass—shake a lump and a silver shower of the glittering metal falls from it. It sparkles in every crevice, and sparkles like gems on the ragged surfaces of the freshly broken rock. Sometimes a "pocket" will be broken into of quartz crystal, pure white quartz filled entirely or in part with the metal. Break into such a pocket and the mercury pours out, often to the amount of several ounces, and even pounds—over six pounds of the pure metal has been saved from a single such pocket.

The mines are on a hill at an altitude of near three thousand feet, or eight or nine hundred feet above the furnaces. The ores are packed down on the backs of mules. How profitable the mines will prove remains to be seen by experience. The principal mines of this region are within an extent of about six or seven miles, and on the line of the leads many are prospecting, digging tunnels, etc., the majority of which must bring only disappointment and loss. Yet some will, in all probability, make money. More mines of the same metal have been found a few miles distant, which we did not visit.

We commenced the season's work a year ago, intending to work up the Coast Range to the Geysers, then to go into winter quarters. Here we were, within less than four miles (six or seven by the trail) of the spot, late in the season. Although it was Sunday and looked like rain, this four miles must be made and the season's work finished. The next day we might not do it, if the rains set in, as they bid fair to; so, Sunday, November 10, we started on our way. Mr. Wattles and another man from the mines accompanied us.

First we went up a very steep slope through a forest, then through chaparral to the summit of the ridge. We rode along the crest several miles, commanding one of the most sublime views for wild scenery since leaving the southern country. We rose to the altitude of about 3,500 feet. Around on all sides was a wild and rough country. We were higher than the country south, but on the north mountains rose higher than we were. The air was very clear, although the day was cloudy, the sun appearing only at times.

To the west of us rose Sulphur Mountain, a sharp conical peak 3,500 to 3,700 feet high, sharp, regular, and covered with dark green chaparral, like a great green mat. At times clouds curled over its summit, at others they rolled away and the sharp cone stood out against the cloudy sky. To the southwest lay the beautiful valley of the Russian River, some villages scattered in it, mere specks in the distance; beyond it were the rough ridges between it and the sea, against which heavy masses of cold fog rolled in from the Pacific.

To the south we could see the broken country lying between the Napa and the Russian River valleys, the Santa Rosa Valley, the black peak of Tamalpais by Tomales Bay, and the Bay of San Francisco. To the southeast was Napa Valley with rugged ridges east of it; and at the head of the valley stood the grand St. Helena, towering over a thousand feet above us, its rugged sides scarred by the elements, sometimes clear and distinct, at others with heavy manes of clouds drifting across its rocky brow for two thousand feet down its sides.

To the north was the deep, almost inaccessible canyon of Pluton (Pluto's) River, more than a thousand feet beneath us, and towering beyond it stood Mount Cobb, which, though not a thousand feet higher than our position, was enveloped in masses of cloud at times, although occasionally its tall pines stood out marvelously clear. Then we could see the valley of Clear Lake and the rugged mountains around it.

The scene was not merely beautiful, it was truly sublime. But we turned from the ridge, down the steep sides of Pluto's Canyon, and soon lost all this extensive view. The hill was so steep that we walked, leading our mules. On descending the slope, we saw the pillar of steam rising, several miles distant, and when more than a mile, we could see the Geyser Canyon very distinctly and hear the roaring, rushing, hissing steam.

We were soon on the spot. The principal springs or geysers are in a little side canyon that opens into Pluton Canyon.

Here let me say, by way of introduction, that the geysers are not *geysers* at all, in the sense in which that word is used in Iceland—they are merely hot springs. Their appearance has been greatly exaggerated, hence many visitors come away disappointed. They were first seen by white men some nine or ten years ago, and such very extraordinary descriptions were given, that it was supposed that the whole world would flock to see the curiosity. All the facts were magnified, and fancy supplied the entire features of some of their wonders. But a company preëmpted a claim of 160 acres, embracing the principal springs and the surrounding grounds, built quite a fine hotel on a most picturesque spot, and at an enormous expense made a wagon road to them, leading over mountains over three thousand feet high. But the road was such a hard one, the charges at the hotel so extortionate, and the stories of the wonderful geysers so much magnified, that in this land of "sights" they fell into bad repute and the whole affair proved a great pecuniary loss. The hotel is kept up during the summer, but the wagon road is no longer practicable for wagons and is merely used as a trail for riding on horseback or on mules.

The springs cover an extent of a number of acres, but the principal ones are in a very narrow canyon with very steep sides. They break out on the bottom and along the sides up to the height of 150 or 200 feet, and on a little flat nearby. There are hundreds of springs—of boiling water—boiling, hissing, roaring. The whole ground is scorched and seared, strewn with slag and cinders, or with sulphur and various salts that have either come up in the steam or have been crystallized from the waters.

Passing over the flat we saw several of these—many in fact —here a boiling spring, there a hole in the ground from which steam issues, sometimes as quietly as from the spout of a teakettle simmering over the fire, but at others rushing out as if it came from the escape pipe of some huge engine. The ground is so hot as to be painful to the feet through thick boots, and so abounds in sulphuric acid and acid salts as to quickly destroy thin leather—it even chars and blackens the fragments of wood that get into it.

Near some of the springs a treacherous crust covers a soft, sticky, viscous, scalding mud; one may easily break in, and several accidents more or less serious have thus occurred. Quite recently a miner was so badly scalded as to be crippled, probably for life. Sulphur often issues with the steam and condenses in the most beautiful crystallizations on the cooler surface. Specimens of sulphur frostwork are of the most exquisite beauty, but too frail to be removed. We crossed this table and descended into the canyon above the geysers and followed it down. I found some flowers out in the canyon above, in the warm steamy air, of a species that elsewhere is entirely out of flower.

One can descend into the canyon and follow it down with safety, a feat that seems utterly impossible before the trial. Here is the grand part of the spectacle. Here are the most copious streams, the largest and loudest steam-jets, the most energetic forces, and the most terrific looking places. Standing part way down the bank at the upper end of the active part, where the canyon curves so that all its most active parts are seen at a glance, the scene is truly impressive. It seems an enormous, seething, steaming cauldron. Steam or hot water issuing from hundreds of vents, the white and ashy appearance of the banks, the smell of sulphur and hot steam in our faces, combined to produce an entirely novel effect.

We descended and followed down the canyon, threading our way on the secure spots. Hot water or steam issued on all sides—under us, by our side, over us, around us. Sometimes the whole party was enveloped in a cloud of vapor so that we could not see each other, at other times this was blown away by the winds. Once the sun came out from between the clouds and shone through this steamy air down on us, lurid, yet indistinct. In one place a rocky pool of black rock several feet in diameter, filled with thick, black water—black from sulphuret of iron, black as ink—was in the most violent agitation. It is the most peculiar feature of all the geysers and is well called the Witches' Cauldron. The water, black and mysterious, boils so violently that it spouts up two or three feet from the surface, inclosed in this rocky wall.

A considerable stream of hot water issues from this canyon, and a short distance below are sulphur banks where hundreds, or even thousands, of tons of sulphur could be cheaply obtained. A curious fact is that a low order of plant, like *confervae* or "frog spawn" grows in this hot water, most copiously in water of 150° F., and even on the margins of springs of a temperature of 200° F., and over surfaces exposed to the hot steam. As the springs are at an altitude of 1,600 or 1,700 feet, the water boils at a temperature of about 200° F., so these plants literally grow in boiling water! I have obtained specimens, but owing to their character, they were very unsatisfactorily preserved.

We returned to the house, where our friends had ordered dinner, but they were very tardy in getting it. We had been a long time without news, and a man brought a recent slip, an "extra" of telegraphic news, from Petaluma, telling of the bombardment and taking of Charleston (which has since proved untrue), which called forth three hearty cheers. After a tedious wait dinner was announced, and after we had eaten we went out to saddle our mules.

Suddenly the sky had become darker and more threatening. We were soon off, but on reaching the ridge summit we were enveloped in a thick, driving, drizzly fog that rolled in from the sea, soon wetting us to the skin, and making our teeth chatter. The wind, cold and raw, swept over the ridge with terrific violence—in places it almost stopped our mules—there was no rain, but much wet. Yet we rode along light-hearted, for we had reached the goal of our summer's hope, we were on the back trail, the rain had deferred its coming a month later than usual—for us.

Dripping with wet we plodded our way through the thick clouds and fog. Luckily our guides were well acquainted with the trail, for in a fog one is easily lost in these wilds and it would be terrible to be caught out on such a night. But we got back safely, and had scarcely got secure in the cabin when the rain fell in torrents. We got supper, then built a big fire and dried as well as we could before it. But *how* it rained! It came through the roof in many places, but we were tolerably secure. We listened to stories of the mountains, got partly dry, picked out the driest places to spread our blankets, and were soon asleep.

I awakened several times, each time to hear the wind howling through the pines, the rain pattering on the thin roof and sides of our cabin, and dripping on the floor. But it stopped before dawn, and the morn was tolerably fair. We packed our saddlebags with specimens, and two boxes besides, which were packed on my mule, so all the mules were loaded. We mounted the hill for our return, visited the Pioneer Mine, were shown the trail to take over the mountains, bade adieu to our kind but rough host, and started on our way.

The feature of our return was the grand view we had of the rocky peak of St. Helena. We found the best view from a ridge over 2,000 feet high, where we could take in the whole mountain from base to top at a glance—its 4,500 feet of crags and rocks furrowed in canyons.

We got back before night and made our preparations for departure. The rainy season had set in, we must hurry back to San Francisco. In our absence Hoffmann had attempted the ascent of Mount St. Helena. He got within 400 feet of the top, and reckoned it about 4,500 feet high or a little less.[8]

Our preparations for departure were premature, for it began to rain in the night very hard, and it rained nearly all the next day. We counseled about going on in the rain, for it might not stop for a week, but decided to wait one day for it to stop—longer we could not for want of provisions for men and mules. But in the afternoon the rain ceased.

Wednesday, November 13, was a fair day, but not clear. We were up betimes and pushed on. The amount of rain fallen was large, the roads muddy and slippery, the foliage once more bright and clean. The storm had been snow back on the mountains. We pushed on to St. Helena that day, where Professor Whitney and Hoffmann left us, going on by stage.

Thursday, November 14, we had some rain, but not much, and we made a long march, stopping at Suscol once more. We had landed on the shores of California just a year ago this day and had intended to celebrate it, but our only celebration was sundry glasses of lager and punch in the evening, which was dark and rainy. It rained most of the night, and at morn it had not ceased. So we took our breakfast at the neighboring "hotel"—there are no "taverns" here—loaded up in the rain and pushed on. Luckily I had an old overcoat, one left by Hoffmann, which protected me much. It continued to rain much of the way to Benicia, where we arrived at 2 P.M. It was uncomfortable enough—roads muddy and slippery, the deep dust of two weeks ago slippery mud now, a drizzling rain in our faces as we plodded on our slow way.

At Benicia we put all our things in the wagon, harness and all, and I sent Averill and Peter on to Clayton at the foot of Mount Diablo, with the mules, where they are to be kept during the winter. That night Mike and I took the steamer for San Francisco, bringing the wagon along. It lacked only a week of a year that we had been actually in the field.

Then came getting ready for office work, clearing up, etc. And now we are fairly at work—and irksome enough it is, I assure you. To sit down in the office, write, compare maps, make calculations, and plot sections, is harder work than mountain climbing. I begin to long for the field again—were it not so wet, for it has rained much since.

A few items in summing up. I have been the entire year in the Coast Range between San Bernardino and Clear Lake. The Coast Range is not a single chain, nor even parallel chains, but a system, radiating, like the spokes of a wheel, from a point near the great mass of San Bernardino.[9] I find on looking at my notes, that since taking the field I have traveled 2,650 miles on my mule; 1,027 on foot, and 1,153 by other means—total in the field, 4,830 miles. During the same time my letters, official and otherwise, have amounted to just 1,800 pages (note size), of which 604 (including this letter) were my "Home Journal," and about 400 were my letters to Whitney reporting progress, etc. In addition to this, over a thousand pages of "Field Notes" of about the same size—some writing to do amid the inconveniences of camp life, surely! These figures will show that I have not been entirely idle, as the data are strictly accurate.

Last Saturday my friends, Blake and Pumpelly, sailed for Japan. They go at the invitation of the Japanese Government, to aid in developing the resources of the mines in Japan. Blake stuck hard to have me go with them, but my place looks too secure here, and that looks precarious. Had I been free I would have rejoiced at the chance.

On the other hand, some most influential men of the state want me to go as commissioner to London next spring, in charge of the California collections at the World's Fair. There is much scrambling and wirepulling for the office of commissioner, but I could get it if I would put in. But the Survey here is a surer thing for me. I have proved myself capable of successfully managing a party, carrying it through times of discomfort, and even hardship, accomplishing much labor, and doing it economically. Whitney feels that I cannot be spared here. I have assurance that should our appropriation next year be cut down I will be the last man discharged—that is, should it be too small to keep a larger party in the field, I will have my place if no one else stays but Whitney. All this is very gratifying, for the actions of legislatures in such matters are very uncertain. We will, of course, get *some* appropriation, it remains to be seen *how much*. The present tardiness in paying us up is most unfortunate. It is a very great inconvenience. Even the Governor has not been paid any salary since May last. We hope for better things soon.

I have a pleasant room, hired by the month, and eat my meals at restaurants. I may take board by the week soon—am not yet decided.

Thankful to an overruling Providence who has thus far guarded me and kept me in times of danger—this is Thanksgiving night—I have just returned from a dinner at Professor Whitney's.

NOTES

1. Serranus Clinton Hastings (1814–93) was a native of Jefferson County, N. Y. He lived for a time in Indiana, then in Iowa, where he became Chief Justice of the Supreme Court after serving a term in Congress. He came to California in 1849 and settled at Benicia. He was immediately appointed Chief Justice of the Supreme Court of California (1849–51). In 1851 he was elected Attorney-General and served two years. Turning to business and the practice of law, he made a large fortune from lands and was reputed to be worth nearly a million dollars in 1862. He founded the Hastings College of Law in 1878. His wife, Azalea Brodt, of Iowa, died in 1876. They had four sons and four daughters (*History of the Bench and Bar of California,* ed. Oscar T. Shuck [Los Angeles, 1901]).

2. The name Sebastopol was retained until about 1867, when it gave way to Yountville. There is a Sebastopol in Sonoma County, with which this place should not be confused.

3. George Yount was one of the outstanding American pioneers in California in the period preceding the gold discovery. He was born in Burke County, North Carolina, 1794, whence the family moved to Missouri. He enlisted in the War of 1812 and then, and subsequently, engaged in Indian fighting. After several years of farm life in Missouri, during which he married and had children, he set out for the west and engaged in trapping. He was with the Patties on the Gila in 1826, and came to California with Wolfskill in 1831. After hunting sea otter near Santa Barbara, he came to San Francisco Bay and settled for a time at Sonoma, where he became acquainted with M. G. Vallejo. He obtained a grant of land in Napa Val-

ley, the Caymus Rancho, which was confirmed in 1836. To obtain land it was necessary to be baptized a Roman Catholic, and it was thus that he obtained the middle name Concepción. He died at the Caymus Rancho in 1865 and was buried at Yountville with Masonic honors ("The Chronicles of George C. Yount," ed. Charles L. Camp in the *California Historical Society Quarterly*, April, 1923, II:1; *George C. Yount and his Chronicles of the West*, edited by Charles L. Camp, Denver, 1966.

4. Joseph Henry (1797–1878) of Princeton and of the Smithsonian Institution. The name was never adopted and the mountain is now known as St. John Mountain (2,370 feet).

5. Calistoga.

6. The maps referred to were probably editions of Colton's *Map of California*, published from 1855 to 1860.

7. *Torreya californica*.

8. The United States Coast and Geodetic Survey gives the height as 4,338 feet. The first known ascent was in 1841, when a party from the Russian settlement at Fort Ross visited the mountain and gave it its name. For several years there was on the summit a metal plate bearing the names of Wossnessenski and Tchernich, with the date 1841. Dr. W. A. Wossnessenski was a naturalist traveling under auspices of the Russian Government; E. L. Tchernich was proprietor of a ranch at Bodega Bay. In one account it is said that Princess Helena de Gagarine accompanied the party to the summit and that she christened the mountain for her aunt, Helena, the Czarina. It is somewhat of a speculation as to whether the name should be Mount Helena, as named for the Czarina, or Mount St. Helena, as named for her patron saint (Honoria Tuomey and Luisa Vallejo Emparán, *History of the Mission, Presidio, and Pueblo of Sonoma* [1923]; *History of Napa and Lake Counties, California* [San Francisco, 1881]; Whitney Survey, *Geology*, I, 86–87).

9. This conception is erroneous; more extensive knowledge later disclosed a different relationship.

BOOK III
1862

CHAPTER I

THE RAINY SEASON

Floods—Sacramento under Water—The Money Question—A Muddy Journey to San Jose—Results of the Floods—The Chinese.

San Francisco.
Sunday, January 19, 1862.

THE rains continue, and since I last wrote the floods have been far worse than before. Sacramento and many other towns and cities have again been overflowed, and after the waters had abated somewhat they are again up. That doomed city is in all probability again under water today.

The amount of rain that has fallen is unprecedented in the history of the state. In this city accurate observations have been kept since July, 1853. For the years since, ending with July 1 each year, the amount of rain is known. In New York state—central New York—the average amount is under thirty-eight inches, often not over thirty-three inches, sometimes as low as twenty-eight inches. This includes the melted snow. In this city it has been for the eight years closing last July, 21¾ inches, the lowest amount 19¾ inches, the highest 23¾. Yet this year, since November 6, when the first shower came, to January 18, it is *thirty-two and three-quarters inches* and it is still raining! But this is not all. Generally twice, sometimes three times, as much falls in the mining districts on the slopes of the Sierra. This year at Sonora, in Tuolumne County, between November 11, 1861, and January 14, 1862, seventy-two inches (*six feet*) of water has fallen, and in numbers of places over five feet! And that in a period of two months. As much rain as falls in Ithaca in *two years* has fallen in some places in this state in *two months*.[1]

The great central valley of the state is under water—the Sacramento and San Joaquin valleys—a region 250 to 300 miles long and an average of at least twenty miles wide, a district of five thousand or six thousand square miles, or probably three to three and a half millions of acres! Although much of it is not cultivated, yet a part of it is the garden of the state. Thousands of farms are entirely under water—cattle starving and drowning.

Benevolent societies are active, boats have been sent up, and thousands are fleeing to this city. There have been some of the most stupendous charities I have ever seen. An example will suffice. A week ago today news came down by steamer of a worse condition at Sacramento than was anticipated. The news came at nine o'clock at night. Men went to work, and before daylight tons of provisions were ready—eleven thousand pounds of ham alone were cooked. Before night two steamers, with over thirty tons of cooked and prepared provisions, twenty-two tons of clothing, several thousand dollars in money, and boats with crews, etc., were under way for the devastated city.

You can imagine the effect it must have on the finances and prosperity of the state. The end is not yet. Many men must fail, times must be hard, state finances disordered. I shall not be surprised to see our Survey cut off entirely, although I hardly expect it. It will be cut down, doubtless, and some of the party dismissed. I see no help, and on whom the blow will fall remains to be seen. I think my chance is good, if the thing goes on at all, but I feel blue at times.

I finished my geological report on Tuesday, it is 250 pages on large foolscap, besides maps, sketches, etc. I have my botanical and agricultural work yet to do.

San Francisco.
Friday, January 31.

WE have had very bad weather since the above was written, but it has cleared up. In this city 37 inches of water has fallen, and at Sonora, in Tuolumne, 102 inches, or 8½ feet, at the last dates. These last floods have extended over this whole coast. At Los Angeles it rained incessantly for twenty-eight days—immense damage was done—one whole village destroyed. It is supposed that over one-fourth of all the taxable property of the state has been destroyed. The legislature has left the capital and has come here, that city being under water. This will give us a better chance for our appropriation, but still the prospect looks blue. There is no probability that we will get enough to carry on work with our full corps.

Wednesday, January 29, was the Chinese New Year, and such a time as they have had! I will bet that over ten tons of firecrackers have been burned. Their festivities last three days, closing tonight. This is their great day of the year. They claim that their great dynasty began 17,500 and some odd years ago Wednesday—a pedigree that beats even that of the "first families of Virginia."

All the roads in the middle of the state are impassable, so all mails are cut off. We have had no "Overland" for some weeks, so I can report no new arrivals. The telegraph also does not work clear through, but news has been coming for the last two days. In the Sacramento Valley for some distance *the tops of the poles are under water!*

San Francisco.
February 9.

I WROTE you by the last steamer and also sent a paper. I have sent a paper by each steamer for some time and will send another by this. A mail now occasionally gets in, but many letters and papers must have been lost. For papers and printed matter the "Overland" is a total failure.

Since I last wrote the weather has been good and the waters in the great valleys have been receding, but there is much water still. I have heard many additional items of the flood. Judge

Field, of Sacramento City, said a few days ago that his house was on the highest land in the city and that the *mud was two feet deep in his parlors* after the water went down. Imagine the discomforts arising from such a condition of things.

An old acquaintance, a *buccaro*, came down from a ranch that was overflowed. The floor of their one-story house was *six weeks under water* before the house went to pieces. The "lake" was at that point sixty miles wide, from the mountains on one side to the hills on the other. This was in the Sacramento Valley. *Steamers* ran back over the ranches fourteen miles from the river, carrying stock, etc., to the hills.

Nearly every house and farm over this immense region is gone. There was such a body of water—250 to 300 miles long and 20 to 60 miles wide, the water ice cold and muddy—that the winds made high waves which beat the farm homes in pieces. America has never before seen such desolation by flood as this has been, and seldom has the Old World seen the like. But the spirits of the people are rising, and it will make them more careful in the future. The experience was needed. Had this flood been delayed for ten years the disaster would have been more than doubled.

The telegraph is now in working order, and we had news this morning—up to 5 P.M. last night from St. Louis—surely quick work. But the roads will long be impassable over large portions of the state.

San Francisco.
Monday, February 10.

An assistant in the zoölogical department was stung by a stingaree so badly in his foot that he has been very lame the last six weeks and came to the city a short time ago. As he will be nearly helpless for some time yet he is going East by this steamer and I will send this by him to be mailed in New York.

We can get no information as to whether our money will be on hand the first of March, as has been promised, or not.

Rates have risen here to three per cent a month lately, which shows how hard the money market is. It has been but one and a half per cent up to the last month.

San Francisco.
March 9.

As the "money question" just now engrosses more of my thoughts than any other subject, I may be pardoned if I tell you something about it, for "out of the abundance of the heart the mouth speaketh."

The act which created the Survey appropriated $20,000 to begin it, and the next legislature added $15,000 to continue it. That would have carried us on until about May of this year. The treasury was then in good condition, and all looked well. Ten thousand dollars was handed over at the start, and $10,000 more came on during the first year of our work. But scoundrels interfered with the treasury. A transfer of $250,000 from one fund to another deranged state finances so that we could realize none of our last appropriation, but as it was necessary to carry on the work and the money was promised in the fall, Professor Whitney borrowed money to go on with the work. Fall came, but none of the promised funds. We must wait until March, then till May. None of the state officers had received their salaries during the latter part of the last year, but before they left office, the Comptroller and the Treasurer hit on a plan to relieve their own cases and that of their fellows. Taxes are paid in twice a year, spring and fall. They called in, in advance, over $250,000 in the fall, not due until spring, and thus paid off all *their* back salaries and the claims of certain friends. A part of this money belonged to the Interest Fund, supposed to be kept inviolate for payment of the interest on the state debts; about $110,000 or $115,000 of this was used, and of this, $86,000 is still out, to be paid out of the first moneys due the treasury.

The new set of officers came in, but couldn't be paid. The Assembly seized on $60,000 of a select fund, the Swamp Land

Fund, and are now trying for $100,000 more, and will doubtless take it. If they do, the Lord only knows when we will be paid. With this new move, I went to Sacramento three days ago and had a long and confidential talk with the Treasurer, who is an old friend of mine here, and found out all this and much more, the upshot of which is, that we will probably have to wait until next December for our pay! The state now owes me $1,400, a thousand of which was solemnly and surely promised before this time. Professor Whitney is still worse off, for he has borrowed several thousand dollars to carry on the work. You need not wonder that I feel blue—disgusted, indignant, and mad. I had hoped for better things, that the rule of scoundrelism in the state was over—it is *not* over—but "still lives." I have no fears of not getting it at some time, it is provided for by special appropriation, and, of the last $15,000, $10,000 are audited and stand first on the list of claims from the General Fund. There is no talk of repudiation—only, we can't get our money when it is due.

Meanwhile, the most influential state officers are all favorably disposed toward our work, and see its immense importance to the material interests of the state, but this doesn't pay us for past labors. Yet I have hopes that something will be done to relieve the state, and with its relief, I hope for our relief.

The floods have still more deranged finances and make some action imperative. The actual loss of taxable property will amount to probably ten or fifteen millions, some believe twice that, but I think not even the latter sum. The Treasurer says that the next tax list will cut down the taxable property about one-third of the whole amount, or probably about fifty million dollars, as each man will get as much taken off on his property as is possible. I suppose the actual loss in all kinds of property, personal and real, will rank anywhere between fifty and a hundred million dollars—surely a calamity of no common magnitude!

On Saturday, March 1, Ashburner left for the East on business of the Survey, to do some chemical work there. He wanted to visit the New Almaden Mines before he sailed, so Saturday, February 22, I started with him to make the proposed visit. It was a rainy, wet, dull day, but the city was gay with flags, for it was Washington's Birthday. The shipping in the harbor was even gayer than the city itself.

We took steamer to Alviso, at the head of the Bay of San Francisco, then stage for seven miles to San Jose. The roads were *awful*. We loaded up, six stages full, in the rain, and had gone scarcely a hundred rods when the wheels sank to their axles and the horses nearly to their bellies in the mud, when we unloaded. Then the usual strife on such an occasion. Horses get down, driver swears, passengers get in the mud, put shoulders to the wheels and extricate the vehicle. We walk a ways, then get in, ride two miles, then get out and walk two more in the deepest, stickiest, worst mud you ever saw, the rain pouring. I hardly knew which grew the heaviest, my muddy boots or my wet overcoat. Then we ride again, then walk again, and finally ride into town, having made the seven miles in four hours' hard work. The pretty village was muddy, cheerless, and dull beyond telling, but I called on Mr. and Mrs. Hamilton and had a pleasant time.

We intended to take stage to New Almaden the next day, but finding the roads so muddy, and the rain continuing, we feared being shut up there, so resolved to come back on Monday. Sunday I heard Hamilton preach, called on a friend and lunched, then took dinner and spent the evening with Mr. Hamilton.

The next day we started back and had the usual amount of walking in the mud, but it did not rain. We got within two and a half miles of the boat, where we found a stream had broken over its banks and had made a new bed and had cut up the road so that it was impossible to get across with the stages. After much delay a part of the passengers got across on the backs of horses, some getting down, and all thoroughly wet. Ashburner went on, I went back with our carpetbags. We had gone back a mile or two in the mud, when a man overtook us on horseback saying a boat had come up to take the ladies and baggage across. So the stages turned back. But on arriving at the break we found the boat gone, and after another delay, we again started for town, where we arrived in due time, having consumed over seven hours in that muddy operation. Hamilton chanced to meet me at the stage office, and insisted on my going to his house, which I did, and spent two days there. I came back on Wednesday, when the water had fallen so that stages could cross.

San Jose has not suffered with floods, but much by too wet weather. Roads are impassable, and nearly a million pounds of quicksilver has accumulated on hand, the roads being too muddy to get it to Alviso. It was still very wet. Apricot trees were in blossom, and the hills began to look green. The foothills had suffered with drought the last few years and the artesian wells in the valley had begun to fail. They will work well enough now, I think, for a few years at least.

Ashburner got off on Saturday and much we miss him, I assure you. He was a jolly good fellow, good at a joke, good-natured, philosophical, and with a great fund of humor and of anecdote.

Last week we got word of a worse condition of the treasury than we had anticipated, so I started immediately for Sacramento to see the Treasurer. Some of the results I have given in the first part of this letter, but of the trip some items may be of interest.

I left here at 4 P.M. on Thursday, March 6, by steamer. Night came on before we reached the mouth of the Sacramento River, but it was a glorious afternoon and the views of the mountains were lovely before sunset. I went to my berth early, but some gamblers were playing within a few feet of me until near morning, and it was but poor sleep that I got. Early in the morning I went to a hotel in Sacramento and got my breakfast and brushed up for business. That dispatched, I had some time to look at the city. Such a desolate scene I hope never to see again. Most of the city is still under water, and has been for three months. A part is out of the water, that is, the streets are above water, but every low place is full—cellars and yards are full, houses and walls wet, everything uncomfortable. Over much of the city boats are still the only means of getting about. No description that I can write will give you any adequate conception of the discomfort and wretchedness this must give rise to. I took a boat and two boys, and we rowed about for an hour or two. Houses, stores, stables, everything, were surrounded by water. Yards were ponds enclosed by dilapidated, muddy, slimy fences; household furniture, chairs, tables, sofas, the fragments of houses, were floating in the muddy waters or lodged in nooks and corners—I saw three sofas floating in different yards. The basements of the better class of houses were half full of water, and through the windows one could see chairs, tables, bedsteads, etc., afloat. Through the windows of a schoolhouse I saw the benches and desks afloat.

It is with the poorer classes that this is the worst. Many of the one-story houses are entirely uninhabitable; others, where the floors are above the water are, at best, most wretched places in which to live. The new Capitol is far out in the water —the Governor's house stands as in a lake—churches, public buildings, private buildings, everything, are wet or in the water. Not a road leading from the city is passable, business is at a dead standstill, everything looks forlorn and wretched. Many houses have partially toppled over; some have been carried from their foundations, several streets (now avenues of water) are blocked up with houses that have floated in them, dead animals lie about here and there—a dreadful picture. I don't think the city will ever rise from the shock, I don't see how it can. Yet it has a brighter side. No people can so stand calamity as this people. They are used to it. Everyone is familiar with the history of fortunes quickly made and as quickly lost. It seems here more than elsewhere the natural order of things. I might say, indeed, that the recklessness of the state blunts the keener feelings and takes the edge from this calamity.

It was a rainy, dull day. I left the city at 2 P.M. Friday, and as I came down the river saw the wide plain still overflowed, over farms and ranches—houses here and there in the waste of waters or perched on some little knoll now an island.

San Francisco.
March 16.

WE have had severe storms in the mountains, and for near two weeks the telegraph was stopped, but on Thursday news again began to come, and on Friday the word was that Manassas was occupied by Federal troops. The city was wilder with excitement than I have seen it before over war news. The legislature adjourned. The streets were filled. From the top of our building a hundred flags could be counted floating in the stiff breeze. Hurrahs were heard in the streets, and as night came on a hundred guns were fired on the plaza, the bands in the theaters played patriotic tunes, the streets were crowded with people—out in front of every bulletin board were black masses of men straining their necks and eyes to see if anything new was posted. Saturday brought confirmations of the news and the excitement did not go down, flags were flying, boys and Chinese exploded firecrackers, and columns in the papers teemed with telegraphic news that had accumulated along the way during the break. I trust that the way is opened now for a speedy close of this unfortunate war.

We have had more heavy rains since I last wrote, and when we can get out again, if we get out at all, I don't know. It is time for the rains to cease, but they don't.

I have long deferred writing about the Chinese in California, and will now devote a few words to them, as it may be a matter of some interest. What the "Nigger Question" is at home, the "Mongolian Question" is here.

As you all know there is a great Chinese immigration to this state. But they come not as other immigrants, to settle; they come as other Californians come, to make money and return. Every Chinaman expects to return, and even if he dies here, his body is returned for burial in the "Celestial" land. Dead Chinamen form an important item of freight on every ship leaving this port for China. It is estimated that there are at least fifty thousand Chinese in the state, and many authorities rate the number still higher; this would make about one-sixteenth to one-eighteenth of the whole population. They all land here, and from three to five thousand live in this city. Whole streets are devoted to them. But most of them are at work over the country—most of the placer mining in the state is in their hands. They come here a "peculiar people," and stay so: they very seldom learn the language, and they adopt none of the customs of the country. They come with all the faults and vices of a heathen people, and these are retained here.

I wish you could walk with me through the Chinese part of the city. All the trades and professions go on as at home; they have their aristocracy and their masses, their big men and their poor ones, their fine stores and their poor groceries, but all is in true Chinese style. Although the houses are American, they are fitted up Chinese and furnished Chinese. Signs hang out in Chinese, and the cragged Chinese characters are written on everything. In the better class of shop they have also an English sign. These are some I copied: Sam Loe, Ning Lung Co., Wau Kee, Wing Yung, Yang Wo Poo, Tsun Kee, Kip Sing, Hee Sun, Wau Hup, Hang Kip & Co., Hong Yun, Chung Shung & Co., Hing Soong & Co., Lun Wo, Tong Yue, Tin Hop. In this evening's paper I read that a murder was committed this morning in the city—Ah Choe killed Ah On. There is, at times, a Chinese theater, and there has been, and may possibly be yet, a temple for their worship.

They keep up their old customs and dress here, as well as their language. Men wear their long hair braided into a "tail," which is either done up around the head or hangs down to the heels behind. These are held in the greatest veneration; a man losing his "tail" falls into disgrace upon returning to his own country, and it is made a crime now under the laws of this state for anyone to cut off a Chinaman's "tail."

The morals of this class are anything but pure. All the vices of heathendom are practiced. The women are nearly all the lowest prostitutes, and there seems no way of improving them. They are outside of all the ordinary means. There is a mission station, with a chapel, but it fails to reach the masses. I was there a few Sundays ago, but understood but little. There were about forty or fifty Chinamen, the missionary, and his wife. The service was entirely in Chinese, all the audience, except myself, Chinese. In front of me sat a rich Chinese merchant, his finger nails over an inch long, looking like the huge talons of some bird of prey; but, with him, emblems that he did no hard work. For the Chinese wear long finger nails, as Americans wear certain styles of dress, to show that they are above work.

NOTES

1. A table of the monthly rainfall in San Francisco for the five years, 1860 to 1864, will be useful for reference here and throughout Professor Brewer's Journal. The figures are from Alexander McAdie's *The Clouds and Fogs of San Francisco* (1912). Rainfall varies greatly between different parts of the state, but the San Francisco figures may be taken as a general index of conditions throughout central California at least.

Month	1860	1861	1862	1863	1864
January	1.64	2.47	24.36	3.63	1.83
February	1.60	3.72	7.53	3.19	0.00
March	3.99	4.08	2.20	2.06	1.52
April	3.14	0.51	0.73	1.61	1.57
May	2.86	1.00	0.74	0.23	0.78
June	0.09	0.08	0.05	0.00	0.00
6 Months	*13.32*	*11.86*	*35.61*	*10.72*	*5.70*
July	0.21	0.00	0.00	0.00	0.00
August	0.00	0.00	0.00	0.00	0.21
September	0.00	0.02	0.00	0.03	0.01
October	0.91	0.00	0.52	0.00	0.13
November	0.58	4.10	0.15	2.55	6.68
December	6.16	9.54	2.35	1.80	8.91
6 Months	*7.86*	*13.66*	*3.02*	*4.38*	*15.94*
Annual	*21.18*	*25.52*	*38.63*	*15.10*	*21.64*
Seasonal	*22.27*	*19.72*	*49.27*	*13.74*	*10.08*

Annual rainfall is from January to December; seasonal rainfall is from July to June. The seasonal rainfall for the twelve months ending June 30, 1862, is the heaviest on record; the lightest is 7.42 in 1850–51. The average seasonal rainfall over a period of many years is approximately 22 inches.

CHAPTER II

TAMALPAIS AND DIABLO

Mount Tamalpais—Sausalito—Marin County—Spanish Grants—Patriotic Outbursts—Sonoma County—Stockton—Lunatic Asylum—In Camp Again—With Starr King on Mount Diablo—A Vast Panorama—Alone on the East Peak—Mighty Oaks—The Story of Mr. Marsh.

In Camp, near Martinez.
April 27, 1862.

For three weeks I have been too much occupied to write journal so am more than a month behind, but I am once more in camp and hope to have time now to bring it up. We went into camp four days ago, Wednesday, April 23, for the summer.

We are at present at work on a map of the portion of the state lying about San Francisco Bay—north to Petaluma and Napa, south to the quicksilver mines of New Almaden, east to beyond Mount Diablo, and west to the ocean—a very small portion of the state, but an important region, embracing some 4,500 square miles (about 3,000 square miles of land) in the heart of the state. We did much work on it last year and want to finish it and publish it along with other work this year. The region lying north of the Golden Gate and the bay could be best looked up on foot, and Hoffmann and I did it after spring opened, while the rest were at work in the office.

Wednesday, March 26, Hoffmann and I started, with compasses, barometer, etc.—an unpleasant day and rainy evening. We went up to San Rafael, about twenty miles north of San Francisco. The following day we struck back into the hills a few miles and stopped all night at a ranch-house. The man was absent, but his wife, a Boston lady, who showed much intelligence and even refinement, was at home and hospitably entertained us. She was one of the prettiest women I have seen in this state.

The whole region between the bay and the sea is thrown up into rough and very steep ridges, 1,000 to 1,600 feet high, culminating in a steep, sharp, rocky peak about four or five miles southwest of San Rafael, over 2,600 feet high, called Tamalpais. On this was once a Coast Survey station, an important point.[1]

On March 28 we were up early and were off to climb this peak. A trail led through the chaparral on the north side. We reached the summit of the ridge, got bearings from one peak, and started along the crest of the ridge to the sharp rocky crest or peak. The wind was high and cold, fog closed in, and then snow, enveloping everything. We were in a bad fix—cold, no landmark could be seen—to be caught thus and have to stay all night would be terrible, to get off in a fog would be impossible. We waited behind some rocks for half an hour, when it stopped snowing and the fog grew less dense; we caught glimpses of the peak and started for it.

The last ascent was very steep. We climbed up the rocks, and just as we reached the highest crag the fog began to clear away. Then came glimpses of the beautiful landscape through the fog. It was most grand, more like some views in the Alps than anything I have seen before—those glimpses of the landscape beneath through foggy curtains. But now the fog and clouds rolled away and we had a glorious view indeed—the ocean on the west, the bay around, the green hills beneath with lovely valleys between them.

We got our observations, remained two hours on the summit, then started for Sausalito,[2] eight or ten miles in a direct line south. It was a hard walk, over hills and across canyons—our distance was doubled. It was long after dark before we found Sausalito, where we stopped at an Irish hotel. We ate a hearty supper, then sat in the kitchen and warmed and talked. Hogarth never sketched such a scene as that. The kitchen, with furniture scattered about, driftwood in the corner, salt fish hanging to the ceiling and walls, lanterns, old ship furniture, fishing and boating apparatus, a Spanish saddle and riata—but I can't enumerate all. Well, we stayed there all night and for several hours the next morning, then took a small boat for San Francisco, along with a load of calves and pigs piled in the bottom.

Sausalito is a place of half a dozen houses, once "destined" to be a great town; $150,000 lost there—city laid out, corner lots sold at enormous prices, "water fronts" still higher—for a big city was bound to grow up there, and then these lots would be worth money. The old California story—everybody *bought land* to rise in value, but no one *built*, no city grew there. Half a dozen huts and shanties mark the place, and "corner lots" and "water fronts" are alike valueless. This was on the same ranch with "Lime Point," where $400,000 was asked of Uncle Sam for a spot of land worth $100 at the highest figure, to build a fort on, but never bought.

April 4 we started on our second trip. We went up by steamer to near Petaluma,[3] forty-five miles north of San Francisco, then struck back over the hills toward San Rafael, climbing all the principal ridges, getting the geological data and notes for a correct topography. We were in the hills three days, and hard days they were. From the summits of these steep hills we got some most magnificent views, while many of the valleys are perfect gems of beauty, nestled among these rugged hills. The finest grazing district I have yet seen in the state is among these hills, lying near the sea, moistened by fogs in the summer when the rest of the state is so dry. But the curse of Marin County is the Spanish grant system. The whole county is covered with Spanish grants, and held in large ranches, so settlers cannot come in and settle up in smaller farms. The county is owned by not over thirty men, if we except men who have small portions near the bay or near some villages. As a consequence, there is but one schoolhouse, one post office, etc., in the whole county, although so near San Francisco.

From the County Surveyor's map I find that there are 330,000 acres of land in the county, of which less than 2,000 acres were ever public lands—in fact, only 1,500 acres besides some little islands. The rest of the county is covered by twenty-three "grants" of which seventeen are over 8,000 acres each! I will give the size of a few of these grants, or ranches, which I copied from the map named—13,500 acres, 21,100, 15,800, 56,600, 48,100, 21,600, 19,500, 13,500, 23,000, 12,200, and so on, but the rest are under 10,000 acres. No wonder that the country does not fill up.

We returned to San Francisco April 8 and remained until April 14. In the meantime, at ten o'clock Saturday night, April 12, came the news of the battle at Pittsburg Landing, supposed to be a much greater victory than it really was.[4] Such an excitement in the streets. People were out, at midnight a salute of cannon was fired, and the streets were as noisy all night long as if it had been Fourth of July. A great concert was in progress, the oratorio *Creation,* when the dispatch was read on the stage. The music stopped, people cheered, wild with excitement, *Yankee-Doodle* and *The Star-Spangled Banner* were sung and played—probably the first time those pieces were ever introduced in that sacred oratorio.

Sunday was a day of excitement and rejoicing. From Telegraph Hill, an eminence on one side of the city, someone counted 356 flags floating in the stiff breeze, besides the multitudes of small flags, too inconspicuous in size to be noticed. T. Starr King delivered a patriotic sermon that night, the anniversary of the fall of Fort Sumter, which, although probably hardly appropriate for Sunday, was nevertheless a most brilliant and eloquent performance. The crowded church could scarcely be restrained from bursting out in enthusiasm during some passages.

Monday, April 14, Hoffmann and I started off for another trip. We went up to Petaluma, spent two days there, then footed it across the ridges to Sonoma, then across to Napa—a most interesting but laborious trip.

A valley extends north from Petaluma to the Russian River—a mere low range of hills, scarcely perceptible, separating the Petaluma and the Santa Rosa valleys. They are of exceeding beauty at this time of the year, and both Petaluma and Santa Rosa are thriving villages surrounded by rich farming lands. From the high hills between Petaluma and Sonoma villages, a most beautiful view can be had of all these valleys, the bay, Mount Diablo, and the region south. We enjoyed these views much.

Sonoma Valley, north of San Pablo Bay, is surrounded by high hills on three sides, and is a plain running up a number of miles, as if San Pablo Bay once occupied the ground, as it probably did in earlier times. Sheltered from the winds by high ridges on the west and north, the climate is milder than most other places on the bay, and is now obtaining some considerable renown for its vineyards and its wines.

From Sonoma we kept on east toward Napa, but did not go quite to the latter place, but returned to Sonoma, crossed the hills by stage to Petaluma Creek and returned by steamer to San Francisco. All the places I have mentioned are on your map, north of the bay.

But in speaking of these trips I forgot quite an item. I had been invited to lecture in Stockton on the evening of April 10, and went up the night before, arriving that morning. I was well received, hospitably entertained, and had a good time generally—got tall puffs in the three daily papers for my lecture.

The State Lunatic Asylum is there, and as the trustees were to have a meeting, I was invited to go up, which I did, and spent several hours visiting the institution while they were transacting their business. There are more insane in this state, by far, in proportion to the whole population, than in any other state in the Union. I need not dilate on the reasons. High mental excitement, desperate characters, disappointed hopes of miners, the unnatural mode of life incident to mining, separation of families, and the indiscretions and infidelity to the marriage vows incident to these separations—these and other reasons have produced this frightful result.

Four hundred and thirty unfortunates are crowded into this institution, only intended to accommodate two hundred and fifty. The proportion of men is more than twice as great as women—there were over three hundred men there. Near two hundred were out in a great yard, surrounded by high brick walls, in the warm, bright, spring sun—some lolling on benches, some talking, swinging on ropes in a sort of gymnasium, "speechifying," singing, praying, etc. Some were sullen and silent, some talkative, some bombastic, some modest.

One spoke to me in German and was delighted to hear me answer in the same tongue. He announced that he was the Emperor of Austria, but was illegally deprived of his liberty here, that Queen Victoria wished him to marry one of her daughters—all of which he told me over and over and besought earnestly my assistance in helping him obtain his rights. He followed me closely for half an hour, closely imitated my actions in everything I did, and before I left the building I heard he had formally applied to the keeper of his ward for permission to accompany me to England.

We went up to the tower and enjoyed a perfectly magnificent prospect. The great plain around Stockton is some forty or fifty miles wide from east to west, and to both the north and south stretches to the horizon, literally as level as the sea and seeming as boundless. In the west and southwest lies the rugged Mount Diablo Range, to the northwest lie the ranges north of Napa, while along the eastern horizon the snowy "Sierras" (Sierra Nevada) stretch away for a hundred miles, their pure white snows glistening in the clear sun.

That evening I lectured—had a good house for so small a

place—and all seemed well pleased. After the lecture I was invited with a few others to the house of the mayor of the city, where an oyster and champagne supper awaited us. Both were freely partaken of and it was two o'clock before we left.

Next day they pressed me to receive fifty dollars for my lecture. I can receive nothing, of course, for any outside services while in my present position. I refused all except traveling expenses. They insisted and I refused—at last they paid my traveling expenses, gave me a very neat meerschaum camp pipe (worth ten dollars) and fifteen dollars toward the botanical wants of the Survey. This was evidence enough that my "effort" was appreciated. I came back April 11, delighted with my trip, and already my "meerschaum begins to color."

Tuesday last, April 22, I sent on the party to go into camp here, near Martinez, and I followed the next day. Our party consists of: Averill, who goes as mule driver, clerk, etc.; Hoffmann, topographer; Schmidt, our new cook, who promises well; Gabb,[5] our paleontologist, young, grassy green, but decidedly smart and well posted in his department—he will develop well with the hard knocks of camp; Rémond,[6] a young Frenchman, who will be with us for about two weeks. We commenced by drinking a bottle of champagne presented by a young lady of San Francisco.

Well, for three days we have been hard at work here and expect Professor Whitney up tomorrow to join us. The hot sun begins to tell on the faces, necks, and hands of the party. Rémond's nose looks like a strawberry, red and fiery. Hoffmann's is like a well-developed tomato, while Gabb's nose is today more like the prize beet at an agricultural fair. Skins are red, faces burned, necks more scorched, but I think all will come out right in a little while.

Camp 70, near Mount Diablo.
May 18.

Professor Whitney joined us, and on Wednesday, April 30,

we moved about nine miles up the valley, south. The next day Hoffmann and I visited a ridge about two thousand feet high, about six miles from camp, quite a hard day's tramp. Heavy clouds wreathed the whole summit of Mount Diablo, but we had a fine view of the green hills near and around us. A shower caught us on our return and wet us, but how unlike the rains of the city. The smell of the rain on the fresh soil and green grass was decidedly refreshing.

That night the wind was high, and for three days we had intermittent but heavy rains. We stuck to our tents, but on Friday, May 2, Professor Whitney left for San Francisco. It rained too hard to cook outdoors, so that day and Saturday we got our meals at a tavern near. On Sunday, May 4, we got a fire lit and dinner half cooked when a heavy shower put it out. We went to the tavern again. In the afternoon there was less rain and I sent Averill to Pacheco, seven miles, for letters.

The rain was a godsend to the farmers. The soil had begun to bake and crack so that the growing grain could not get on farther. Everything has "greened up" marvelously, and this region, so brown, dry, dusty, and parched when we visited it last fall, is now green and lovely, as only California can be in the spring. Flowers in the greatest profusion and richest colors adorn hills and valleys and the scattered trees are of the liveliest and richest green.

Tuesday, May 6, the camp came on here, about ten or twelve miles. I waited to observe barometer to get the height of the camp above the sea, then footed it across here, a part of the way across hills, the rest across the Pacheco Valley, a plain of many thousand acres, several miles wide, sloping gently from Mount Diablo northwest to the Straits of Carquinez.

The plain is covered for miles with intervals of scattered oaks; not a forest, but scattered trees of the California white oak (*Quercus hindsii*), the most magnificent of trees, often four to five feet in diameter, branching low. They are worthless for timber, but grand, yes magnificent, as ornamental trees, their great spreading branches often forming a head a hundred feet in diameter. Across this great park the trail ran.

On arriving I found the camp pitched in one of the loveliest localities, a pure rippling stream for water, plenty of wood, fine oak trees around, in a sheltered valley, but with the grand old mountain rising just behind us.

Professor Whitney had met us, and a party was here from San Francisco to visit us and climb the mountain. A little town, consisting of a tavern, store, etc., is rapidly growing up scarce twenty rods from camp, where a "hotel" accommodated them. The party consisted of Rev. T. Starr King, the most eloquent divine and, at the same time, one of the best fellows in the state, Mrs. Whitney, Mr. and Mrs. Tompkins (a lawyer whose wedding I had attended last winter), Mr. Blake[7] (a relative of Mrs. Whitney and an intimate friend), and a Mr. Cleaveland (one of the officers in the United States Mint)—a better party could not have been selected.

That evening, a lovely moonlight May evening, we lighted a great camp fire, and our visitors enjoyed it, so new to them, ever so charming to us. The moon lit up the dim outline of the mountain behind, while our fire lit up the group around it. We "talked of the morrow," spun yarns, told stories, and the old oaks echoed with laughter.

Wednesday, May 7, dawned and all bid fair. We were off in due season. I doubt if there are half a dozen days in the year so favorable—everything was *just right*, neither too hot nor too cold, a gentle breeze, the atmosphere of matchless purity and transparency.

Five of our party, Professor Whitney, Averill, Gabb, Rémond, and I, accompanied our visitors. They rode mules or horses; we (save Averill, who was to see to the ladies) went on foot. First, up a wild rocky canyon, the air sweet with the perfume of the abundant flowers, the sides rocky and picturesque, the sky above of the intensest blue; then, up a steep slope to the height of 2,200 feet, where we halted by a spring, rested, filled our canteens, and then went onward.

The summit was reached, and we spent two and a half hours there. The view was one never to be forgotten. It had nothing of grandeur in it, save the almost unlimited extent of the field of view. The air was clear to the horizon on every side, and although the mountain is only 3,890 feet high, from the peculiar figure of the country probably but few views in North America are more extensive—certainly nothing in Europe.

To the west, thirty miles, lies San Francisco; we see out the Golden Gate, and a great expanse of the blue Pacific stretches beyond. The bay, with its fantastic outline, is all in sight, and the ridges beyond to the west and northwest. Mount St. Helena, fifty or sixty miles, is almost lost in the mountains that surround it, but the snows of Mount Ripley (northeast of Clear Lake), near a hundred miles, seem but a few miles off. South and southwest the view is less extensive, extending only fifty or sixty miles south, and to Mount Bache, seventy or eighty miles southwest.

The great features of the view lie to the east of the meridian passing through the peak. First, the great central valley of California, as level as the sea, stretches to the horizon both on the north and to the southeast. It lies beneath us in all its great expanse for near or quite *three hundred miles of its length!* But there is nothing cheering in it—all things seem blended soon in the great, vast expanse. Multitudes of streams and bayous wind and ramify through the hundreds of square miles—yes, I should say *thousands* of square miles—about the mouths of the San Joaquin and Sacramento rivers, and then away up both of these rivers in opposite directions, until nothing can be seen but the straight line on the horizon. On the north are the Marysville Buttes, rising like black masses from the plain, over a hundred miles distant; while still beyond, rising in sharp clear outline against the sky, stand the snow-covered Lassen's Buttes, *over two hundred miles in air line distant from us*—the longest distance I have ever seen.

Rising from this great plain, and forming the horizon for three hundred miles in extent, possibly more, were the snowy crests of the Sierra Nevada. What a grand sight! The peaks of that mighty chain glittering in the purest white under the bright sun, their icy crests seeming a fitting helmet for their black and furrowed sides! There stood in the northeast Pyramid Peak (near Lake Bigler), 125 miles distant, and Castle Peak (near Lake Mono), 160 miles distant, and hundreds of other peaks without names but vieing with the Alps themselves in height and sublimity—all marshaled before us in that grand panorama! I had carried up a barometer, but I could scarcely observe it, so enchanting and enrapturing was the scene.

Figures are dull, I admit, yet in no other way can we convey accurate ideas. I made an estimate from the map, based on the distances to known peaks, and found that the extent of land and sea embraced between the extreme limits of vision amounted to eighty thousand square miles, and that forty thousand square miles, or more, were spread out in tolerably plain view—over 300 miles from north to south, and 260 to 280 miles from east to west, between the extreme points.

We got our observations, ate our lunch, and lounged on the rocks for two and a half hours, and then were loath to leave. We made the descent easily and without mishap or accident—a horse falling once, a girth becoming loose and a lady tumbling off at another time, were the only incidents. The shadows were deep in the canyon as we passed down it, but we were back at sunset. Our friends were tired, some of them nearly used up. With us, the day was not a hard one.[8]

The party sat rather silent by the cheerful camp fire that evening, some through fatigue, others from thought, but all seemed happy. Tompkins was the most fatigued, in fact, nearly used up. Next day he said to me, half confidentially, "Last night I felt humiliated, much so. I had ridden most of the way up the mountain and back again and was so nearly used up that I could have gone but little farther. You went up on foot, carried a barometer, picked out trails for us, and did the work, and came back apparently as fresh as when you started." I told him he must enjoy the luxury of camp life, sleep on gravel stones, put off his "boiled shirt," and come down to bacon and beans, to really enjoy life and health.

Thursday, May 8, was a cool blustery day and most of our company left for San Francisco. Mrs. Whitney and Mr. Blake left the last of the week, dining with us the last day. We were up the mountain in the "nick o' time," for it has not been perfectly clear since, and three days after our first ascent the mountain was white with snow, and snow remained on it for three days. We had very cold weather, especially nights.

We continued our work in this region. Although Mount Diablo is the initial point of all surveys for this part of the state, yet, strangely enough, its topography had not been mapped. Nor was the geology at all understood, although it is a most important spot as furnishing a key to many formations of the state, and of great pecuniary interest from the fact that the only coal mines worked to any advantage in the state are on the north side of the mountain and within five miles of the summit.

Last week we spent in exploring and examining. We climbed the mountain twice, once passing over it to the south and west sides, another time crossing the crest to the east peak, lower by three hundred feet than the main peak, but rocky and almost inaccessible. Its views were the more picturesque. Both days the Sierras were veiled in clouds and the distant view shut out, but all within forty or fifty miles, an area of ten thousand or twelve thousand square miles, was in perfectly plain view.

I cannot detail each day's work, and were I to do it you would find much sameness in the pictures, for all would be of the same or similar scenes. Yesterday Professor Whitney left for San Francisco to try and raise some money. Our state legislature has lately adjourned, and a distinguished state senator, Mr. Banks, from San Francisco, came up here two days ago. Yesterday he came to camp, took dinner with us at 5 P.M. and spent the evening. Several others came in, and we had a jolly time.

Today has been a very quiet day. We pitched the tent in a new place this morning, for the grass was worn out in the old site and we needed a "new carpet." The boys went down to town, but a few rods distant, where a new brewery has just been started, bought a barrel of lager, brought it up in a wheelbarrow, and it reposes under a tree. Now they are out by the camp fire—it is evening—and have drawn a pail of it. I look out and see them lying around, each with his pipe and basin of beer; the bright blaze lights up the scene, and makes it one fit for a painter.

Livermore Pass.
June 1.

WE remained at work in the Mount Diablo region until the twenty-eighth of May. I resolved to make another measurement of the mountain for the sake of seeing where some errors lay in our previous measurements, and also to determine the altitude of the northeast peak, which we have named Mount King.[9] So that morning I started, with Averill and two barometers, leaving Hoffmann at camp to observe station barometer. A citizen went with us to find the trail, for the people of Clayton have resolved to spend a hundred dollars in making the trail passable for "travelers," to attract visitors.

The morn was gloriously clear, but before arriving at the summit cold winds and clouds came on, which hung over the summit at intervals during the day. There are two peaks, one about three hundred feet below the other, and between them a gap. Leaving Averill on the main peak, I descended into the rocky gap over eight hundred feet below. Through this gap the wind roared with a violence almost terrific at times—the temperature only 44° to 46°—and at intervals the clouds rushed through like a torrent. I spent an hour and a half, observing every fifteen minutes, in that peculiarly grand spot—not grand because of the view, but because of the surrounding rocks and the clouds which rushed among them, dissolving again after leaving the mountain, driven by the fierce wind that was concentrated by coming up a canyon that had its head in this gap.

Then I took my way along the crest to Mount King. It is but half a mile from the gap, but it is a crest of naked rocks, over and among which one has to pick his way, not always without danger. I spent two and a half hours in observations. The wind was not so high, and the clouds enveloped me only once, but they hung much of the time over the main peak where I had left Averill and our companion.

I shall never forget those hours. The solitude—total silence save the howl of the winds in the gap below—the cragged rocks around me, and the enchanting scene spread out below—the great San Joaquin plain seeming more like a sea than ever—cloud shadows chasing each other over the green hills at the base of the mountain—and the clear blue sky in the southeast beyond the reach of the clouds that formed over our summit. The distant Sierras were in view, but not with that distinctness with which we saw them before.

I got back to the other peak, joined my shivering companions, and we were back before sunset, tired enough. But we were to leave the next morning and a month would elapse before I could mail letters again, so after making calculations, I wrote until midnight.

We have made five measurements, but have discarded two that were made under the poorest circumstances, saving only the best. Let me tell you how they agree: height of Mount Diablo above mean tide—(a) 3,879.6, (b) 3,876.0, (c) 3,874.1: mean 3,876.4 feet.[10] These three measurements varied less than five and a half feet, from the lowest to the highest, although the distance from the summit to the tidewater (at Martinez) must be over fifteen miles. We have made a most accurate observer of Averill. These measurements are the result of eighty-two barometrical observations, and not less than three hundred have been taken on all our work in this vicinity.

Wednesday, May 28, we came on, but neither letters nor money had arrived. So I left Gabb at Clayton. We went on to Marsh's, about fourteen miles east. The road led down a canyon torn by last winter's floods. The road was considered passable, but one not used to our life would pronounce it utterly impassable for any wagon, yet we brought our wagon through with only one slight break, making the fourteen miles in four and a half hours.

Before the canyon emerges into the San Joaquin plain it winds into a flat of perhaps two or three hundred acres surrounded by low rolling hills and covered with oaks scattered here and there, like a park. And *such* oaks! How I wish you could see them—nearly worthless for timber, but surely the most magnificent trees one could desire to see. I measured the circumference of about thirty, near camp, that were over fifteen feet around three and a half or four feet from the ground —eighteen, nineteen, or twenty feet are not uncommon—with wide branching heads over a hundred feet across—one was seven feet in diameter with a head a hundred and thirty feet across.

Well, Gabb did not arrive with letters and funds until Friday night. In the meantime we examined the region about there and got acquainted with Mr. Marsh. I promised you something of this man's history last fall, and I will now give the *facts* gleaned from several sources—some of the main ones from him himself, the rest from reliable authorities.

Doctor Marsh was born in Massachusetts, graduated at Harvard University, spent some time in Canada, Wisconsin, Illinois, and other western territories at an early day, and finally, in the employ of the American Fur Company, came here to California in 1836, just after the confiscation of the missions by the Mexican Government, and bought a large ranch on the east side of Mount Diablo. Here he lived, and later married a Boston lady who came to this state. He had various adventures, was robbed, had all the experiences incident to the life of a wealthy *ranchero* here in early times, counted his cattle by the tens of thousands, and lived like a patriarch.[11]

Sometime in the summer of 1856 he was a little unwell, and, while he was reclining on a bed in his sitting room, a seedy, tired, hungry looking man entered, late in the evening, and wished to stay all night. The man was an American, and the Doctor, suspicious of his countrymen, especially of such looking men, refused to keep him.

"But, Sir, I am tired—have lost my way—I saw a light here and came in."

"Have you any money?"

"No."

"A man has no business to be traveling here afoot and alone, without money."

"I can't help that. I am tired—my feet are sore. The great Joaquin plain lies beyond, as I hear, without houses—I don't ask for money, not even food, nor a bed—let me lie before the fire."

At last the wealthy doctor gave his consent and began various inquiries.

"Where are you from?"

"San Francisco—just out from the States."

"What State?"

"Illinois."

"What is your name?"

"Marsh."

"Marsh?"

"Yes, Marsh."

"Marsh—from Illinois? Where were you born?"

"Wisconsin."

"What county, and when?"

In the meantime, the doctor, becoming interested, rose up from the bed and sat upon the edge. A few more questions were asked in quick succession, much to the astonishment of the traveler.

"Pull off your boot," says the doctor. "Now your stocking."

It was done, disclosing two toes grown together to their ends.

"Man—you can stay, for you are my son. I supposed you died a boy!"

The doctor, while in Canada, had taken a French mistress, who had gone with him to the western states. By her he had two children, a boy and a girl. While he was absent he heard of her death and that of the children. The mother and daughter did die, but not the boy, who was brought up by a Kentucky farmer in Illinois.

He married, lived a poor man, and finally came to California in 1856 to seek his fortune, as one account says, or to seek his father, of whom he had accidentally heard—the latter is probable. However, he arrived, poor, friendless, landed in San Francisco, started in the search, lost his way, saw a light, went to it, and found it to be that of his father's house.

The old man kept him for a time, merely as a hand on his ranch—I think about four to six months—when the old man was murdered by some Spaniards.

Young Marsh threw in his claim for his share in the ranch, that he was a legitimate son, etc. After much legal investigation, a Roman priest from Montreal testified to the fact that Marsh had called this woman his wife. Other witnesses were found who declared that Marsh had introduced her as his wife. They had lived in a region where magistrates were few and far between, and marriages were not always solemnized in that way to be legal. Well, he got the ranch, sharing the property with a half-sister by the second wife, a girl now ten years old.

Here are the facts for you to build a romance upon—the long separation, the strange meeting, the murder of the old man by which this before-neglected man (I was about to say outcast) so soon became rich, coming here at just the right time.

He has a beautiful ranch—the old man claimed eleven leagues, but only three leagues, or over 13,300 acres, were confirmed—with the finest ranch-house I have yet seen in the state.

But here, for the sake of romance, let the facts stop—do not let me go on and tell how the present occupant is a boor—how old chairs with rawhide bottoms occupy rooms with marble mantles, how the fine mansion is surrounded with a most miserable fence, with hogs in the yard, some of the windows broken, and things slovenly in general. Such is the true story.

Well, yesterday morning we left that camp. The morning was clear and warm, and as we came out on the plain, in the hot sun, a beautiful mirage flitted before us—the illusion was perfect. It seemed as if a marshy lake lay along ahead of us, a few miles off. Often the trees were reflected as perfectly as if it were really water, but on approaching, it would keep away from us. About noon the wind rose, however, and with it the mirage left the plain.

We passed on beyond the trees and came into camp here, a camp as unlike the last as day is unlike night—among foothills utterly treeless—before us stretching the plain, also treeless, not a bush even greeting the eye for many miles. From a nearby hill yesterday we could look over an area of at least two hundred square miles and not see a tree as far as the river, where, ten miles off, there is a fringe of timber along the stream. The hills are already brown. Over them swept a fierce wind and it took all hands to get the tent pitched. All night long the wind howled, the tent shook and rattled, nor has it ceased today, although tonight it is less.

NOTES

1. The height of the east peak is 2,586 feet; that of the west peak is 2,604 feet (U.S.G.S.). The Coast Survey station was on the west peak. The early Coast Survey reports call it Table Mountain. Tamalpais is a local Indian name—*Tam-mal* (bay country) and *pi-is* (mountain); but see Erwin G. Gudde, *California Place Names*, 1960, p. 313.

2. Brewer spells it *Saucelito*, an erroneous form current for a short time. *Sausalito* is Spanish for "little willow grove" (Nellie Van de Grift Sanchez, *Spanish and Indian Place Names of California* [2d ed., San Francisco, 1922]).

3. Rudesill's Landing.

4. The dispatches called it "the bloodiest battle of the age." It was reported that fifty thousand were killed.

5. William More Gabb (1839–78).

6. Auguste Rémond died in June, 1867.

7. Gorham Blake.

8. Starr King's glowing account of this day will be found in Charles W. Wendte, *Thomas Starr King: Patriot and Preacher* (1921), p. 126.

9. This name for the east peak has not persisted. Starr King's name has, however, been established on a dome above Yosemite Valley.

10. In the Whitney Survey, *Geology* (1865), I, 10, the height is given as 3,856 feet. The United States Coast and Geodetic Survey figure is 3,849.

11. The episode of Dr. Marsh and his son is corroborated from other sources. John Marsh (1799–1856) was born in Danvers, Mass., and was a member of the class of 1823 at Harvard. He lived for a while at Fort Snelling, Minnesota, and at Prairie du Chien, Wisconsin, but not in Canada. Later he went to Missouri and to Santa Fè. In 1836 he moved to Los Angeles and was the first man to practice medicine there. In 1837 he came to San Francisco and for a year explored in the northern part of California. In 1838 he purchased the ranch near Mount Diablo, where he became, with Sutter, Yount, and a few others, one of the outposts of civilization. George D. Lyman, *John Marsh, Pioneer*, New York, 1930.

CHAPTER III

THE DIABLO RANGE SOUTH

Corral Hollow Revisited—Sheep-Herders—Mount Oso—Cañada Del Puerto—Impenetrable Canyons —Hot Days on the Plain—Orestimba Canyon—A Rodeo—Pacheco Pass—The Road to San Juan.

Orestimba Canyon, Camp 76.
Sunday, June 15, 1862.

Two most laborious weeks have passed since I have written anything—weeks of care, toil, and anxiety—to write all the incidents of these two weeks that have crowded upon my mind for writing would take more time than I can devote and would perhaps tire you.

I think I told you that I was instructed to work up the San Joaquin plain, at the foot of the Diablo Range, examining that mountain chain as far as is possible from the plain, up to Pacheco's Pass, then cross the San Joaquin River about the last week in June, and join Professor Whitney on the east side.

We made ample preparation for the trip. We would be a month away from all mail facilities and all stores save fresh meat, so we profited by previous experiences and were well provisioned and prepared. So soon as we left Mount Diablo, and the party was fully in my charge, I got things going like clockwork, inaugurating a more strict discipline than we had last summer, and giving each man his place and work. Thus far everything has moved on capitally.

At Zimmermann's we heard that all the ferries across the San Joaquin were still impassable, up to Firebaugh, eighty miles up the river, and that to get to Stockton was practically impossible. Two men had recently arrived from there on horseback, but had to ride in the water four miles, often swimming their animals.

Tuesday, June 3, we raised that camp, and went on to Camp 72, at the mouth of Corral Hollow. The morning was peculiarly clear. The plain looked like the sea, or rather a great bay, the snowy Sierras seemed scarcely ten miles distant, instead of sixty to a hundred as they really were. We passed out upon the plain, and passed the fifteen miles across it by a very obscure trail, the mirage flitting about us most beautifully.

All work at the Corral Hollow coal mines had ceased since last fall and only a few men hung around there. A torrent, like a river, had swept down the canyon last winter, destroying the road, so we camped at the mouth of the canyon, where there was water, six miles below where we camped last fall—a solitary spot.

We had done most of the work last fall, so but little was now to be done. Nevertheless, on Wednesday, June 4, Hoffmann and I climbed a high hill several miles south for bearings for topography, etc., while Gabb explored the canyon for its geology.

As we have worked south it has grown hotter—today was over 90°, but on the hills, if we could get in the shade of a tree, a most delicious breeze fanned us. I have never experienced any more delicious air than we have these clear days, provided one can be in the shade and rest from exertion. I have long ceased to wonder that the natives are so universally lazy. Any man is justified in being lazy in such a climate. But the evening was still more delicious, cool—you would say *warm* at home—a delicious breeze, no dew. The moon was bright, and we lay in the open air and sang songs of home until the rocky sides of the canyon echoed again. I assure you that no song is oftener sung than "Sweet Home."

Thursday, June 5, we rode up the canyon and again visited the mines and prospectings, all now deserted for a time. My previous opinion of the worthlessness of the mines was confirmed strongly. Over $100,000 has been spent in that desolate place—and more will probably be sunk there!

We have been sleeping in the open air since this month began, and it seems like old times again. Friday, June 6, we were up early and out upon the plain, passed up a few miles and turned into Lone Tree Canyon, Camp 73. Here another canyon opens into the plain, a lone tree stands near its mouth and gives the canyon a name. We went up the canyon about one and a half miles to where another tree stands, a poor apology for a cottonwood, but still a *tree*. Here we camped. A man lives here, and has a flock of sheep, and feeds them over the hills. He seemed a very intelligent man, and lived the best of all of these sheep herders that we have seen, although living entirely alone. He was glad to see us, for we were the first human beings he had seen for three weeks, and as many more weeks may pass before he sees any others. What a lonely life must be his! Summer and winter he must be here—his visitors few and far between. Sunday and week day are alike to him. Up at dawn, he gets his breakfast, and drives his sheep out in the early cool air of morning twilight. He carries a little bag in which is his noon meal. He watches his sheep among the hills the entire day, and at night brings them into the fold, or "corral," beside which he sleeps, to keep away coyotes, wolves, and bears. Such is the monotonous life of hundreds of sheep herders in California.

Saturday, June 7, I intended to climb Mount Oso, but our mules got away and so much time was wasted that it could not be done. The feed is poor and water has been poor and alkaline since leaving the region of Mount Diablo, so that both horses and men are affected by it. The mules are so disgusted with affairs that they utterly refuse to stay unless a part of them are tied up. I don't blame the poor animals. But the water is much better and more abundant in this region than usual, from the heavy rains of last winter. Although this canyon heads back many miles among high hills, yet, this man said, during the five years he has been here he has never before seen water in it, summer or winter. Now there is a small stream.

Sunday, June 8, was a quiet Sunday in camp—no one seen save ourselves, and even our mules were absent, for we had picketed them a mile and a half up the canyon where the feed was better.

Monday, June 9, I was up early and, with Hoffmann, started for Mount Oso. Averill was left in camp to observe station barometer, and Gabb was sent back to the last camp with another barometer to get the difference of altitude for obtaining the height of the mountain. We crossed a table-land, furrowed by canyons, to Hospital Canyon, then up that. It is a wild canyon, which breaks through from behind Mount Oso. Here we met another solitary herder, who anxiously inquired the news from the outer world. The news we gave was fresh to him, although we have heard none for over two weeks.

We rode up the canyon two or three miles, often with high walls of rocks on both sides. Occasionally it widened out into little valleys, surrounded by steep hills, where scanty grasses, now dry, support the sheep. A few cottonwoods grow along the creek, and in them hundreds of cranes have built their nests—great awkward birds, with their maltese-colored plumage, long slim necks, and longer slimmer legs.[1] At the sight of us they would fly out in flocks, squawking and screaming, but some, more tame, would stand their ground and look down on us with dignity. Great numbers of other species of birds also congregate in these canyons. They are marvelously tame, for they have never learned to fear *man*—he is too rarely seen.

We tied up our mules and climbed the ridge. It was steep and long, but the summit was gained. We found the mountain to be 3,400 feet high. The view was magnificent. Back of the treeless hills that lie along the San Joaquin plain, there rises a labyrinth of ridges, furrowed and separated by deep canyons. These ridges rise 3,200 to 4,000 feet high, with scattered trees over them, sometimes, but not often, with some chaparral.

This region is twenty-five to thirty miles wide and extends far to the southeast—I know not how far, but perhaps two hundred miles. It is almost a *terra-incognita*. No map represents it, no explorers touch it; a few hunters know something of it, and all unite in giving it a hard name. Two different ones, one a companion of old Grizzly Adams, have described it to us as "a hell of a country," and so far as our observations go they were not far from correct. We got into the margin of it on the west last summer, from the San Jose Valley, and were now peeping into it from the east.

We found Mount Oso rightly named—Bear Mountain, of the Spaniards—for the whole summit had been dug over by bears for roots. Many tracks were seen, and trees scratched and broken, but we saw no bears, much to the surprise of the herder in the canyon below, for he had never been up the canyon without seeing some, but we kept on the ridges and thus avoided them. Only one rattlesnake was seen, making six seen this trip so far. We were back safely at sundown, and, tired enough, were in our blankets early.

I have been interrupted while writing this page by a great rumpus outside the tent—Schmidt calls for a light—something in his trousers leg—the lights are carried out—his pants carefully rolled up, for scorpions, centipedes, and tarantulas are all abundant—but only a huge mole cricket is found under his drawers—a common insect, but harmless—a perfectly huge, wingless cricket, as thick as a man's little finger. We laugh over the affair—the rest go to bed, and I return to writing.

Tuesday, June 10, we came on about fifteen miles to Camp 74, in the Cañada del Puerto, or, to translate, "Door Canyon."[2] It was a tedious ride along the plain, for since we left Corral Hollow we have had no road. We take our way across the trackless plain, sometimes sandy, at others hard, gravelly soil where we can trot a little, but oftener a clay soil, now dry and cracked by the heat, so that the mules must pick their way slowly. The cracks are from two to four inches wide, running in all directions. You cannot imagine how tedious it is thus plodding along, two or three miles off from the foothills to avoid the gulches that come down from them. Sometimes we come to one and must follow along it a mile or more to find a place to cross, then keep on our weary way.

Once we came upon a herd of thirty or forty antelope, a kind of small deer. We came up to within two hundred yards, or less, of them, when they galloped leisurely away. They are most beautiful and graceful animals, with slender legs, large ears, and erect heads. They are as fleet as the wind when really alarmed, but these only ran a short distance and then stopped to graze again.[3]

The Puerto Canyon breaks through the ridges, which are made by strata of rocks tilted up at a high angle. The outside ridge is of hard rock, and the canyon comes through by a very narrow "door," which gives name to the valley behind, which widens out. There is a fine stream now—soon it will dry, however—with cottonwoods growing along it. We camped in a beautiful spot just within the "gate."

Here we found another sheep ranch—a cabin, and two men keeping four thousand sheep. One of the men was going to Grayson's Ferry, and would carry a letter to the express, so I sat up until midnight writing to Whitney. One of these men was a fine looking man—had been here five or six years—but he lived decidedly slovenly. His cabin was dirty beyond description, his scanty utensils dirtier, his bed a dirty sheepskin, and his blanket merely another dirty sheepskin. He says that in the summer, when they take the sheep back farther into the hills, months will pass without their seeing any other persons. Last summer, for three months, they only saw three men, hunters probably, who had come through the mountains.

Wednesday, June 11, was spent in examining the hills back for three or four miles. We looked at the total lunar eclipse until the moon was entirely obscured, then turned in. I watched it still as the shadow slowly passed off, but sleep accidentally cut short the observations at an early hour, and the bright morning sun was the next thing I saw. Strange enough, a slight shower had passed late in the night, a rare thing for this time of the year. I found my blankets and "pillow" (figurative) wet, but the rain had not disturbed me.

Thursday, June 12, as it was desirable to get back into the mountains, and, this being a very favorable spot to try it, I resolved on a two days' excursion to see what was beyond the outer ranges of hills and to get bearings to connect with back points from some of the higher ridges within, for marking localities on our maps. So Hoffmann, Averill, and I started on mules up the canyon for a ridge supposed to be fourteen miles in a direct line from camp.

We traveled hard all day and found on plotting our maps on returning that we had made less than ten miles in a direct line from camp. For six miles the creek breaks through very steep ridges from 1,000 to 2,500 feet high, all made up of strata of sandstone tilted up at an angle of fifty to seventy degrees. The western side of each hill was made by the broken edges of the strata, so they were very steep. The numerous rocky precipices made a picturesque but desolate landscape. The side canyons were generally dry, but sometimes had small streams or pools of bitter, alkaline water, and the rocks often looked as if covered with frost, from the alkali that exudes from them. I scraped off about a pound of alkali, in one place, to take for analysis. It is nearly white as snow and is a mixture apparently of carbonate of soda, salt, gypsum, and Epsom salts, of which the first named ingredient is the most abundant.

Sometimes the canyon is a mere gorge between high rocks, at others it widens out into pretty little valleys with grass (now dry as tinder) and scattered trees. We passed these sandstone ridges and got into the metamorphic region beyond, where the strata have all been roasted and changed by volcanic agencies and tossed about in grand confusion. Here we found decidedly a "hard road to travel," among and over rocks, now up, now down—places were passed which you members of civilized countries would pronounce absolutely impassable. Indeed, Averill (who unfortunately does not hold out well in places of danger or difficulty) once wanted to turn back, but I shut him up quickly and kept on, and the rest followed. We found it impossible to reach the point we started for, so came to a halt, near night, at a convenient spot.

From the exceeding abundance of grizzly tracks, it was but natural to suppose that we might be visited in the night, so we slept "conveniently" near an easy tree to climb and built a bright fire. But I bet Hoffmann a keg of beer, to be drunk at the first place where it could be got, that we would neither *hear* nor *see* a bear in the night. We built a bright camp fire, and in our scanty blankets lay down beside it, our saddles for pillows, the clear sky above, and these rugged mountains about. You cannot appreciate the peculiar pleasure of sleeping thus, in such solitudes. The stillness seems almost deathlike, but I do enjoy it. Tired enough, we slept soundly. I only awoke once, to replenish the fire, for it was cold.

But our sound sleep won for me the beer, for we found large bear tracks within a hundred feet, or less, of us in the morning —he had passed during the night. It was light moon, when bears love most to roam, but all hunters unite in saying that it is the rarest thing in the world for a grizzly to seriously disturb a sleeping man. I have never heard of a man being thus attacked. They often come up and smell the man, but if he lies perfectly quiet he will not be molested. The difficulty is, to lie quiet while an animal more ferocious than the lion and stronger than the strongest ox is thus examining you. But our friend that night took no such liberties. He apparently passed down the canyon, stopped and turned around when near us, then passed on.

We were up at dawn the next day, climbed a hill and got bearings sufficient for our purpose, then returned to camp, where we arrived just at night. At camp, we met three men lately from San Francisco, who gave us a "treat" of news—the evacuation of Corinth. They were on their way up the valley, on a vacation trip—one was a professor from Oakland College.

Saturday, June 14, we were up and off early, raised camp, passed out of the canyon, and again struck up the trackless plain. You cannot imagine how tedious it is to ride on this plain. The soil and herbage is dry and brown, few green things cheer the eye, no trees (save in the distance) vary that great expanse. Tens of thousands of cattle are feeding, but they are but specks save when they cluster in great herds near the water—often for miles we see nothing living but ourselves, except birds or insects, reptiles, and ground squirrels. We can only ride at a walk, and in the clear air long distances seem so short that we appear to make no progress. We ride on many a weary league, while a mountain ahead that seemed scarcely a dozen miles distant, at most, becomes no nearer. We camp, and then travel another day toward it, but approach it not. The Sierra is ever in sight, the brilliant summits seem ever in the same position. Peaks a hundred and fifty or more miles distant change their looks but little by our changing place forty or fifty miles.

It is hot, too, on this plain, and every day the mirage flits before us—seeming cool waters ahead that are never reached. Although this has become so familiar a thing that we look at it as a matter of course, yet I never cease to wonder at it. Daily observation has made me familiar with it. Science has explained its mystery, but its beauty, its poetry, remains ever the same to me.

We passed up about fifteen miles, then turned up the Orestimba Canyon and camped about three miles from its mouth. Here we were on a cattle ranch, away from the infernal sheep. We struck a good place to camp. It was a lovely Saturday night. We had wood, and built a bright camp fire, the first we had had since striking the San Joaquin plain—except that one back in the mountains. But the feed was poor, so, early in the morning of Sunday, June 15, we raised camp and moved up the canyon about a mile to where the feed was better and camped in a most charming spot—a little plain with high hills each side, feed tolerable (though dry as hay), a clear stream of pretty good water, and a great sycamore for shade. We were in camp again and all in order before ten o'clock.

It was a lovely and quiet Sunday. It followed a hard week's work, and was spent quietly as could be—no one but ourselves to molest. The day was intensely hot, we could see the Sierra with marvelous distinctness out of the mouth of the canyon, which expands to over a mile wide, and the night which followed was yet clearer than the day—the sky of the intensest blue, and the stars as numerous as on a clear winter's night at home.

San Francisco.
July 6.

As we came on farther and farther, the air grew day by day hotter and drier, and all the evidences of a drier climate increased. All the herbage on the hills was dry enough to burn, and the plain brown and dry as hay. At Orestimba the ranch runs to the river, six or eight miles distant, and often for several summers in succession all the cattle must be driven there for water. Until this past winter and spring it has been five or six years since there was water in the canyon where we were now camped!

Monday, June 15, we climbed some hills about 2,200 feet high, a few miles from camp. It was intensely hot. I know not how hot, but over 90° all day in camp where there was wind, and vastly hotter in some of the canyons we had to cross. It was 81° long after sunset. The Sierra seemed but a few miles away, and the night was intensely clear.

Tuesday, June 17, was still hotter. Hoffmann and I rode a few miles up the canyon, then a laborious walk up a ridge

2,400 feet high, to get bearings. Whew, how hot it was! On the top, where there was a breeze, it was 89°. How hot it was in some of the still canyons I have no idea, but in one, not the hottest, I got in the "cool shade" of a tree, where there was some breeze and found it 105°. It seemed positively parching. But on sinking into the deep, narrow canyon we came to a stream, found a deep hole, and took a most delicious bath. We saw a deer, also a panther—were quite near the latter, but he made off. This is the panther which is known here as the California lion.[4] I have heard much of them, but had no idea that they were such formidable looking animals. This fellow could have carried off a man easily. Bear signs were numerous.

A hot night followed this hot day—few slept well. Mosquitoes, numerous, huge in size, and ferocious in disposition, haunted us, but they molest me less than the other men.

I had not spent a day in camp, save Sundays, since leaving Diablo, so Wednesday, June 18, I resolved to stay in camp, write, wash, and do odd jobs. It was still hotter—above 100° most of the day, and 87° after eight o'clock in the evening. With a fresh breeze it stood at 102° for hours, and so dry that heavy woolen clothes after washing were perfectly dry in less than two hours. It may have been sooner. I looked in two hours after leaving them out and found them dry.

Well, I washed my clothes, a job I positively hate—I would rather climb a three-thousand-foot mountain—and to make matters more aggravating, just as I was in the midst of it, along came two women, one young and quite pretty, who were assisting as *vaqueros*. A rodeo took place near camp, and several thousand head of cattle were assembled, wild almost as deer. Of course it takes many *vaqueros* to manage them, all mounted, and with lassos. A rodeo is a great event on a ranch, and these women, the wife and daughter of the ranchero came out to assist in getting in the cattle. Well mounted, they managed their horses superbly, and just as I was up to my elbows in soapsuds, along they came, with a herd of several hundred cattle, back from the hills. I straightened my aching back, drew a long breath, and must have blushed (if a man can blush when tanned the color of smoked bacon) and reflected on the doctrine of Woman's rights—I, a stout man, washing my shirt, and those ladies practicing the art of *vaqueros*.

The cattle were "marshaled," in a close body on the plain near, with hills on either side, surrounded by *vaqueros* on horseback. The rancheros from adjoining ranches were on hand to select such cattle as belonged to them and get them out. A rodeo is always a spirited scene. The incessant "looing" of these three thousand cattle, the riding, the lassoing, etc., form a scene that an eastern farmer cannot well conceive from description. This is considered a fine stock ranch—several thousand head of cattle and hundreds of horses are kept.

Just after sunset a fine deer, a buck, came down to within three hundred feet of camp and leisurely looked at us. We had no gun loaded, and he trotted away unmolested, much to our chagrin.

Thursday, June 19, we raised camp and came on to Camp 77 at Ranch San Luis de Gonzaga, at the eastern end of Pacheco's Pass, twenty-seven miles—twenty-four of which were along the plain—with a temperature all day of 95° to 102°. Before entering the pass, we struck a road—a mere trail, yet it was a road, and what a luxury! We trotted our mules gaily along it. It is a novelty, for we saw our last road three weeks ago!

We camped in a very windy place. A strong current of air pours over Pacheco Pass the entire summer—from the west, cool, to supply in part the demand of the hot San Joaquin plain—and we were camped directly in this channel.

We passed a herd of antelope today on the plain.

Friday, June 20, we climbed a peak nine or ten miles south of camp, a conspicuous point, from which we had a prospect of a large extent of very rough country and a very great extent of plain beyond—a long and laborious day's work.

In crossing some of the ridges that lay in the channel of this great wind current over the depression here in the mountain chain, the wind was so intense that our mules at times were blown out of the trail several feet—they could scarcely stand. The oak trees often lay along the ground, in the direction of the wind. They were not uprooted, but had grown in that way in that perpetual wind.

Saturday, June 21, we climbed another peak, about eight miles northwest—nothing of especial interest. Sunday was a quiet day in a windy camp. I wrote for some time in the tent, but at last the wind grew too high for that. We took a pleasant swim in a stream near, the first swim for the season. That night was the windiest we have had this season. We had to pile saddles, stones, anything heavy, on our blankets while we crawled in them, yet the wind would blow through them.

Monday, June 23, I visited some hills alone, and sent Gabb and Hoffmann on a longer excursion back, with the mules. The event of the day was their meeting, in a narrow ravine, a large she-grizzly with a cub. Now this is the worst kind of a customer to meet, and as they came upon her very suddenly, matters did not look well. She faced them at first, scarcely thirty feet distant, then slowly retreated. They took the hint, and both parties escaped unharmed; the two bears leisurely climbing the steep bank of the ravine on one side, the geologists climbing, less leisurely by far, the steep bank on the other side.

We have ascertained that the San Joaquin River is entirely impassable for its whole length. My instructions were to work up to this point, cross the river wherever we could, and meet Professor Whitney in the Sierra. He was, however, to write us at San Juan, and I was to send over the pass to that place for letters and funds. It was forty-five miles distant. But my program must be changed. I resolved to cross the pass with the party and work north, perhaps to Martinez again, 180 miles, and thence ship to Stockton by steamer, unless the plan should be again modified by future contingencies.

Tuesday, June 24, we raised camp and started over the Pacheco Pass, and stopped in the pass after a ride of eighteen miles. First our way was up steep ridges to the height of about 1,500 feet (1,200 above camp)—a heavy hill for our wagon, although the road was tolerably good. This pass is a toll road, so it has been put in repair for about twenty miles. It took $1,800 to make thirteen miles of it passable after the rains of last winter. The Overland route formerly ran through this pass.

After reaching the summit we descended through narrow, wild, and exceedingly picturesque canyons, in places so narrow on the bottom that our wagon could scarcely pass, in other places along a shelf worked in the steep sides. We camped in the pass six miles from the west side, near a little "hotel" established when the Overland ran.

Two days were spent in examining this region—one day in climbing Pacheco's Peak, a very conspicuous peak from all sides of this point, the other in climbing a peak about eight miles north. We were thus enabled to work up the general topography entirely across this chain. Everything shows that we are now in a moister climate, getting fresh breezes from the sea. It is cooler, and trees are more frequent and green.

I ought to have said that the San Luis Ranch which we had just left contains nearly fifty thousand acres, and that it has the feed of as much more territory. They keep an immense number of sheep—some eighteen thousand belong to the ranch, but twice that number are brought in from elsewhere and rent pasture. The ranch is about eighteen miles long and six or eight wide. One valley, near where we camped, a plain nearly surrounded by hills, has as much as fifteen thousand or twenty thousand acres in it. On getting farther west the feed becomes better as we get away from the sheep.

Friday, June 27, we went on to San Juan, twenty-seven miles, first down the canyon for six miles, then across the plain over twenty miles. As soon as we left the toll road we struck the plain and most of the way had a good road, but where the road belongs to the public the bad places had received just as little work as would possibly make them passable. We had to cross four bad gulches, or arroyos—decidedly bad places—from ten to twenty-five feet deep, with very steep sides. In one instance, where we crossed the San Benito, the bank is perpendicular for over twenty feet, and down this the narrowest possible road was cut—steep, sidling, and narrow. We got down safely by tying a long rope to the wagon, passing it over the top, and three of us holding on from the top of the bank, to keep the wagon from capsizing. The leaders were taken off and the wagon was got down by two mules. The San Benito is now a small creek, but last winter it must have been a river twenty or twenty-five feet deep and from a quarter to a third of a mile wide.

The plain of San Juan is vastly greener and more luxuriant now than it was last year at this time, owing to the wet winter. We camped on the old spot where we were camped a year ago.[5]

NOTES

1. Great blue heron (*Ardea herodias hyperonca*).
2. Or, "Canyon of the Pass."
3. *Antilocarpa americana.*
4. *Felis oregonsis californica.*
5. The letter following this is missing. Very likely it was lost with the burning of the SS. *Golden Gate,* mentioned in the next chapter. From Brewer's notebooks, however, we can trace the movements of the party. They went on to San Jose, climbed Mount Hamilton again (on July 2), moved on to Walnut Creek, and reached Camp 82, on the Suisun road, two miles from Benicia, on July 18. In the meantime Brewer visited San Francisco, San Jose, and New Almaden.

CHAPTER IV

UP THE SACRAMENTO RIVER

Benicia—Sickness at Rag Canyon—Steamer to Red Bluff—Cottonwood Creek—Indian Types and Customs—Heat and Fever—Branches of the Sacramento—Darlingtonia—Views of Mount Shasta.

Camp 82, near Benicia.
Sunday, July 20, 1862.

I AM glad enough to be here, although our camp is not in a pleasant place, yet it is preferable to the city. The crowds of the city make me feel sad and lonely. I feel restless and long for the quiet of camp life—quiet, yet active—rich in that excitement that arises from the contemplation and study of nature, but quiet in all that relates to strife with the busy, bustling world.

We are camped in a little valley among the treeless hills that lie on this side of the Strait of Carquinez. On crossing from Martinez a great change comes over the landscape. There are open groves, beautiful trees, and cool shade; here, not a tree to break the wind or to invite rest in its shade. Yet the land is fertile, and rich fields of grain lie on every side. Although Sunday, the clatter of a reaping machine is borne on the breeze from a neighboring field. I heard it before I was up in the morning. It sounds strange to hear it, the sound mingled with the tones of the bells of the churches and convent from the neighboring town.

Benicia is a very dull place—scarcely any business, although once the rival of San Francisco and the capital of the state. A large city was "laid out," and thousands were prospectively rich. But, alas, the abundant "lots" and "water fronts" were held at such high prices—everyone must get rich rapidly by the rise in the value of his "city property"—that no one could buy. All speculated and none built—the same old Californian story—so that capital hesitated about coming, trade kept away and sought cheaper quarters, and industry, the natural foe of speculation, stood aloof. The "city" remained in its "lots," and now Benicia—the "City of Benicia"—is merely a little, dull, miserable town of not over five hundred inhabitants, and were it not for its United States Arsenal and the shops of the Panama steamers, where they make their repairs, there would be *nothing* here.

Another curious history is just now developing here, a true Californian episode, and one of those which more than any other retard the progress of the state. The entire point of land, some fifty thousand acres, has been held under an old Spanish grant. Under this title it has all been sold and converted into farms, now valuable. Two villages, the "cities" of Benicia and Vallejo, have grown up upon it. At last—at this late date, when men who bought property in good faith and paid its full value have been living in undisputed possession for periods of from five to fourteen years—the United States court rejects the claim, and immediately the whole is considered "unoccupied public lands." What a field for the squatter, the sovereign squatter, that highest type of the American citizen!

Instantly the whole is "taken up" by squatters. Within sight of our camp are numbers of these "actual residences." The squatter goes in some man's field, even his orchard, and, usually in the night, erects his "residence," a board shanty worth at most five to ten dollars. He spends one night on it, which is sufficient for him to swear that he has erected a residence and resided there. He thus lays claim to 160 acres of the "government land." In this way men have actually squatted on lots in the city of Benicia occupied by brick stores. Of course, the population is intensely excited. One man tore down the shanties erected on his farm. The squatters in turn tore down his fences and burnt his fine barn. The rest of the residents now are doing nothing—grain is ripe, the fields dry, but they dare do nothing, for the free and independent squatter who would take possession of a man's farm is none too good to burn his house, if molested. Courts now will quarrel, lawyers make money, some men get money, and honest industry suffer.

The story of these Spanish grants is a long one, and a black one. Our central government has much to answer for. This case ought to have been settled long ago—there is neither reason nor justice in such delays. If the claims are poor now, they could not have been good fourteen years ago. Political corruption has other sins to answer for as well as the rebellion at home.

This case here is by no means an isolated one—every year tells of numbers of such. *The* great drawback on the settlement of this state has been, is now, and will be for years to come the insecurity of land titles, and for this southern politicians more than northern are the cause, but the explanation is too long to write now.

Camp 83, Rockville.
July 25.

TUESDAY, July 22, we came on here, about seventeen miles. We are camped in the field of a Secessionist, but our Stripes and Stars float from our tent. This is about twenty miles northeast of Benicia and five miles from Suisun. We have been here now three days and will leave tomorrow. It is very hot, myriads of fleas annoy us and pests of ants invaded our beds one night. North of the marsh that skirts the bay for some miles is a very rich agricultural region, teeming with grain.

Camp 84, Rag Canyon.
August 6.

SATURDAY, July 26, we left Rockville and struck north up the Suisun Creek—a lovely ride—beautiful valley, fine oaks, rugged hills on both sides. We crossed the divide and struck into

Rag Canyon,[1] a canyon that enters the Puta Creek.[2] For the last ten or twelve miles of our ride we passed but one house, that inhabited by a man and two squaws. We camped in a pretty spot under two fine large oaks, in the narrow valley. A little water here came to the surface in the bed of the brook. We proposed to spend Sunday and Monday there.

"Man proposes, but God disposes," says the old proverb, and thus it has been in our case. Here I have been sick nearly ever since the last date of writing—eleven days—sick, very sick a part of the time, but now am recovering, and we hope to be off in two days more.

San Francisco.
Sunday, August 10.

THURSDAY, August 7, I raised camp and returned to within five miles of Suisun, and the next day went to Suisun, took steamer and arrived here the same night. I am much better, but, considering it unsafe to ride so soon, I sent the party on up the Sacramento Valley, where I will join them in about a week, at Red Bluff, the head of navigation on the Sacramento.

On my arrival here I found the whole city in excitement and mourning over the loss of the *Golden Gate*, a terrible calamity. Everyone has lost friends and acquaintances by that accident. I sent some papers telling about it.[3]

Red Bluff, on the Upper Sacramento.
Sunday, August 17.

BUSY, although weak and out of sorts, I got ready and at 4 P.M. of Tuesday, August 12, left San Francisco by steamer for Sacramento. It was a most lovely afternoon—the beautiful bay was crossed, the sun set, gilding in the most golden colors the bare hills, now either brown or a rich straw color. Mount Diablo stood up, a most majestic object, until shut out by the shades of evening. We were in the "sloughs," as the many mouths of the Sacramento and San Joaquin rivers are called, when the moon rose from the plain as from the sea. The illusion was heightened by its blood-red color and distorted shape as it rose from the low horizon. The river was low, although still eight feet above the low water of previous years. The bed has been so raised, however, that we stuck on a bar for seven hours, and only arrived at Sacramento at eight o'clock in the morning.

Wednesday, August 13, I took a short stroll through that city before going on board the steamer bound for Red Bluff.[4] Everywhere one sees the effects of the flood in that unfortunate city, and, indeed, the water was still over a part of it. The morning was intensely hot, in strong contrast with the San Francisco weather.

Although the channel of the Sacramento is insufficient to carry off all the water of wet winters, yet it is rapidly filling up, each year increasing the difficulty. Previous to 1848 the river was noted for the purity of its waters, flowing from the mountains as clear as crystal; but, since the discovery of gold, the "washings" render it as muddy and turbid as is the Ohio at spring flood—in fact it is perfectly "riley," discoloring even the waters of the great bay into which it empties. A man pointed out to me a spot at the mouth of the American River at Sacramento where, in 1849, he had sounded the river and found it fifty feet deep. He had seen a schooner sink there, so that only a little of her masts stuck out a short distance. Now there is a luxuriant growth of young willows on the mud bank that occupies the spot. Last winter's floods alone are supposed to have raised the bed of the river at Sacramento six or seven feet at least—that is, in spots, so as to raise the water that much.

Red Bluff is at the head of navigation on the river—three hundred miles above Sacramento by the river, but only half that distance by land. Stern-wheel steamers, drawing but eighteen or nineteen inches of water run up. Our boat was the *Gem*, and we towed a barge with two hundred tons of freight, quite an impediment to rapid progress. We got off at 11 A.M. I had plenty of books along, and although it was very hot, 90° to 96° every day, yet I enjoyed that trip much.

Thursday, August 14, we kept on our slow and winding way, often on bars and shoals that took long to get over. A wide plain borders the river on each side. We caught distant views of the mountains, but generally we saw only the river and its banks, which were more or less covered with trees—willows, cottonwoods, oaks, and sycamores—with wild grapevines trailing from them. Some of the views were pretty indeed. When it got dark, we tied up, it being impossible to run in the night, owing to snags, bars, and rapids.

On Friday morning, August 15, as I stood on the deck of the steamer, who should come down to the shore to see us pass but Averill. The party were camped on the river bank, and by chance I saw them. It was a relief, for it showed that they were safe thus far.

Our progress this day was slower than before. Many bars and rapids in the crooked river were but slowly surmounted. During the day Lassen's Buttes stood out in clear outlines in the east—two majestic sharp peaks, their sides streaked with snow. Before night we saw Mount Shasta rising above the horizon, clear in outline, although its snowy crest was 150 miles distant in air line.

We tied up that night at Tehama, a little village on the bank. A circus was the excitement of the time and I attended. Such an audience! At least two-thirds were Digger Indians, who enjoyed the riding much, but were decidedly undemonstrative as to the rest. After it, there was the usual excitement of "side shows"—the bearded woman, the stone eater, etc.—the agent of each yelling for custom. There were gambling tables in saloons, where *monte* was in progress—the usual music, women, liquor, piles of gold and silver on the tables, etc. It was decidedly a scene to be remembered.

Saturday we again went on and arrived here in the afternoon. My party had arrived just before me and had encamped near town. This is a stirring little town of a few hundred inhabitants—saloons, taverns ("hotels"), and corrals being the chief features, for here pack trains and teams start for the whole northern country, Oregon, etc. But, oh, how hot it is! I am now writing at eleven o'clock at night, and it is 94° in my room—it has been 100° to 102° most of the day. I went to church this morning—an audience of about twenty-five only—in the schoolhouse.

Here the low hills close in on the river, and here begins a most interesting country to visit. I went out to camp a little while this afternoon, but I shall stay at the hotel until we leave here, then take to camp again.

Camp 94, North Fork of Cottonwood Creek.
Sunday, August 24.

TUESDAY, August 19, the camp went on, but I rode with a gentleman to visit the Tuscan Springs, some eight miles east, in a wild region of low, desolate, forbidding hills. These springs have some repute—a house, bathhouses, etc., have been erected—but they are not quite a Saratoga or an Avon in either surroundings or accommodations, for they are hot, dry, dreary, and desolate. The waters are varied in the different springs, but all are sulphury, stinking, with salt, alkali, iron, etc. They have much repute for the cure of certain diseases. Not the least interesting feature was the evolution of gas along with the water. It is so abundant, and burns so well, that it is collected, purified, and burned to heat the baths. The whole region around shows strong marks of volcanic action.

Wednesday, August 20, I got up at 4 A.M. to take the stage, but had to wait three hours before it came along, heavily loaded with mail matter and boot full. Only two men could ride inside, the rest on top. A hot ride of forty miles brought us into Shasta about two o'clock—over hills and through gulches, very dusty. The last four miles was through a mining region.

The boys had arrived ahead of me and had camped near town. We spent two days there, camped in a dusty, dirty, hot corral, but the best place about. On getting in camp I felt vastly better. I am now recruiting rapidly, notwithstanding the heat and dust, but have not near my desired strength yet. I weigh but a hundred and fifty—twenty pounds less than last summer, and fourteen less than I weighed a month ago.

Yesterday, August 23, we came here, where we will stop about three days, then return to Shasta and wait for Professor Whitney. The whole aspect of the scenery is changed on arriving in this region. Across the head of the Sacramento Valley stretches a great table-land, perhaps 1,200 feet above the sea, and north of this a chain of mountains stretches across and unites the Sierra Nevada with the Coast Range. This table-land stretches south from the foot of these mountains. Passing through, it would not be noticed that it is a table, so cut and furrowed is it with canyons and gulches, but from any considerable height it seems like an immense plain, if one is high enough so as not to notice its furrowed ravines.

From all the more elevated points, Mount Shasta is a most sublime object. Distant about sixty to seventy miles in an air line, so clear does it seem in its sharp outline that to an eye unaccustomed to our scenery here it would seem scarcely ten miles off. Its whole top is white with snow, save where cragged rocks peep through. The snow line is a well-defined, perfectly level line all around.

Later.

ALL the way from Shasta here is a placer region—a high table-land, furrowed into innumerable canyons and gulches. The soil is often a hundred feet thick—a very compact, red, cement gravel. In this is the gold, especially in the gulches or ravines. Here in an early day miners "pitched in"—many made their "pile" and left, others died. Little mining towns sprang up, but as the richest placers were worked out they became deserted and now look dilapidated enough.

We came here yesterday from Shasta, passing through Middletown (which is on the wrong road on the map), Briggsville (now nearly deserted), Horsetown, Piety Hill—all mining places.

The scenery here is as unlike as can be anything that we have passed through before. It is a dry, hilly country, with high mountains along the north, the soil very dry and covered with scattered trees and bushes. There are gardens, etc., in the valleys, but generally the land is barren from drought. The whole region is scarred by miners, who have skimmed over the surface and left the region more desolate than before.

Many are mining here still, and many years will elapse before all the placers here will be exhausted. Water is supplied during the summer by ditches, dug for miles in length, by which the mountain streams are carried over the lower hills and the water used for washing the dirt. When these ditches cross gullies, the water is carried in a wooden trough, or, as we in the East would say, a "race," but here they are universally called "flumes." We passed one of these flumes yesterday that ran across the valley for a distance of over five thousand feet, most of the way over fifty feet high and in places over ninety feet—it was a magnificent work, known as the Eagle Creek Ditch.

Shasta.
Sunday, August 31.

Our special object in visiting the region of our last camp, seventeen miles distant, was to see the development of Cretaceous rocks, and trace their relations to the gold-bearing slates. We wished also to collect fossils, in which that locality is peculiarly rich.

Monday, August 25, with Mr. Hubbard, the man who lived where we camped, we rode three or four miles to a gulch, where, he said, fossils were abundant. The gulch was a ravine excavated into the soft shales, in places carrying a small stream of water; in other places the water sank and the gulch was dry. We descended into it, rode some distance, but saw no fossils.

A little later we found lots of fossils. We heard that a man named Wheelock, living about a mile from our camp, had a very large and fine fossil, an *ammonite*, so we called to see it, and get it if possible. I found him a brother-in-law of my old friend, H. A. Ward, of Rochester, my most intimate friend in Paris, whom I also knew in Munich. He was as glad to see me as if I had been an old acquaintance. I called there in the evening and ate peaches and nectarines to my heart's content.

Both Mr. Hubbard and Mr. Wheelock had small orchards—fruit gardens we would call them—watered or irrigated from a mining ditch. The luxuriance of the fruit trees must be actually seen to be appreciated. These oases in the general dry, desert landscape, are cheering and beautiful. Trees of two years' growth are as large as ours would be of four or five, and loaded with fruit. Cherry trees of only four years' growth were as large as the most thrifty ones of ten or twelve years' growth in New York. The peaches were fine, both as to size and quality, and the nectarines positively delicious. Apples were good, but hardly reaching those of central and western New York, *I* think.

Quite a number of Indians, "Diggers," were about—they often stopped near camp and stared wonderingly at us. Sometimes there would be a group of five or six in a trail that ran within a rod of our tent—the men, with their bows and arrows and long hair; the women, with their faces horribly tattooed and their heavy, thick, stiff, and coarse black hair cut off square, just even with their eyes in front but hanging down over the sides of the face and back of the head to the neck. Sometimes the women had burdens, a bundle or basket on the back carried by a strap across the forehead; sometimes they came with children, which were often entirely naked. Such groups would stop, just at evening while we were talking and smoking, and stand within a rod of us for some time, looking intently at us, at everything around us, then pass on with very few words among themselves. Sometimes, during the day, two or three squaws would come along, sit down in the hot sun within three or four rods of the tent, say nothing, but listlessly watch us for half an hour together.

These Indians are peaceable and nearly harmless when in no larger numbers than they are here, notwithstanding the unnumbered wrongs they have endured from the mining population of whites. As I hear of these wrongs, of individual cases, related from time to time, I do not know which feeling is the stronger awakened, that of commiseration for the poor Indians, or of indignation against the barbarous whites. There are now "Indian troubles" at various places in the upper part of the state—white men are murdered, etc., troops are out—and as yet I have not heard a single intelligent white man express any opinion but that the whites were vastly more to blame than the Indians.

But they are a low, very low, brutal-looking race. A number lived on the bank of Cottonwood Creek, about a half or three-quarters of a mile from our camp. Still nearer was their burial place, on the table-land at the top of the bluff bank of the creek, which is here sixty to eighty feet high. It is a pen, fenced with rails, perhaps two or three rods in diameter, under some oaks. The dead are buried in a sitting posture, the knees drawn up to the chin, blankets wrapped around. Clean, coarse sand from the stream is afterward carried up from the creek, in a tin pan, and piled over the grave, so that when finished it is a conical pile of sand, perhaps two or three feet high. This little inclosure was filled with such sand piles. There had recently been a fire just outside, where I believe the goods of the deceased were burned—of this I am not sure.

On the night previous to our arrival, a man of note had been buried, and the yells, screams, and noise over the grave were heard over the whole neighborhood for a mile or two. A few days afterward we passed the grave in the morning, and as we rode along, three women were there "mourning." One was on the grave dancing—a sort of uncouth hopping motion—the other two were sitting on the ground near, and all were howling and uttering words in a peculiar tone, anything but plaintive or mournful. They paid no attention to our presence as we rode along. I could not but think of some of my lady friends at home, great sticklers for fashionable mourning, and wish that they might see these Digger women, who are probably just as rigorous in their fashions—only these take it over the grave, in the solitude, among the trees, instead of flourishing in *crêpe* and black in crowded parties, where the depth of their grief might be "seen of men" and its precise depth judged of by the intensity of the black or the amount of the *crêpe*.

These Indians are very dark, black as our darkest mulattoes, and not as intelligent looking as the negro. Often as we were exploring the gulches for fossils we would see them following and watching us from the banks, keeping nearly concealed, with peculiar Indian secretiveness. Once, Rémond and I came across a fine place to swim, in Cottonwood Creek. We stripped and had a bath, delicious in that hot weather. As we came out we discovered two squaws, young women, quietly regarding us from the bushes against the bank, but a few rods off.

We secured a fine lot of fossils, which we packed up the night before leaving, and Thursday, August 28, came back here. The weather had been so hot that we were up at dawn, and off in the cool of the morning, and were here by two o'clock, but the last three hours of our ride were intensely hot, and the dust almost insupportable.

Camp 98, Strawberry Valley.
Base of Mount Shasta.
Sunday, September 14.

AFTER our long siege in the intense heat, three months with the thermometer nearly all the time above 90°, and often for days or even a week together with it above 100°, that hot week was "the straw that broke the camel's back" with most of my little party. Hoffmann, Schmidt, and Rémond were taken down with fever, and Averill also was at least nearly down. He did not get down entirely, but became unfit for duty. Sunday, August 31, the three were miserable enough. It was very hot. They lay out under a horse shed during the day, at night they went into the tent.

Monday, September 1, was another hot day. Our position in camp, dirty and hot, was uncomfortable enough, I assure you. Near us was the county hospital; and the steward, or rather manager, a graduate of Union College in New York, had visited Averill, who was also graduated there. He saw our plight, and most generously tendered us the use of a room at his house, near the hospital. We accepted it and moved our men that morning. After righting things at camp, I went there also and spent the night with them, as in fact I did for several nights, sleeping on the floor.

Tuesday, September 2, Professor Whitney arrived, much to our relief. We were anxious to get out of that hot place and up into the mountains as soon as possible for the men. Saturday, September 6, Averill and I were up at the first tints of dawn and had our wagon loaded before sunrise. While waiting for breakfast, which we got at a neighboring house, we had a most glorious sunrise, the sun rising directly beyond the high, snow-covered Lassen's Butte. The peak was gloriously illuminated. We drove into town, where the rest had gone the previous evening, and started. Schmidt and Hoffmann were barely able to ride, Rémond had to be left behind. So we left a barometer with him, that he might keep up a series of observations to aid us in our determinations of the height of Mount Shasta.

We started east, crossed the Sacramento, struck northeast, passed through Buckeye, and camped at Bass's Ranch, seventeen miles from Shasta. The Sacramento is a clear, swift, cold stream, in a narrow valley, almost a gorge where we crossed it. Then we rode over a table-land, much gullied into gulches, covered with sparse, scattered trees, such as are indicative of an intensely hot and dry soil and climate. Scattering gold diggings occurred the whole way, but all were dry now—they can only be worked in the wet months. We arrived at Bass's Ranch, in a very pretty but dry valley, at noon (thermometer 100°), where we held up for the day. Heat, watching, anxiety, and perhaps also indiscretion in eating had almost made me sick again.

Sunday, September 7, we were all up at dawn and started before seven. We had hoped to be at Mount Shasta at full moon, but the detentions had so delayed us that we must travel Sunday and were yet three days behind time. Five miles over a hilly road brought us to Pit River. Here let me say that the upper Sacramento, above where it turns from the east to the south, is here known altogether as the Pit River, while the name Sacramento is retained for the branch that runs nearly straight south from the west base of Mount Shasta.

We crossed the ferry, crossed some hills to McCloud's Fork, a swift stream of pure, cold water, green as the Niagara and cold from the snows of the mountains beyond. We followed up that a few miles, then crossed the ridges to the Sacramento Fork, where we camped for the night. Such a hilly road—all up and down—now winding along a mere shelf hundreds of feet above the river, then descending into ravines. The country between these forks is dry, the hills mostly covered with bushes and scattering trees, but after this day we had very different scenery, for we were in a wilderness of mountains and continually rising. We were glad enough to get into camp, for all were tired. We had had a hard day's drive, although we had come but twenty-one miles.

We had seen a number of Indians, and at the ferry where we camped that night there were a number more. We heard that many had recently died. There were some graves on a knoll near camp, and a number of squaws kept up an incessant howling, moaning, screeching, and thumping on something until dark. Their noise was positively hideous, but then this is their way of showing respect for the dead. They ceased when it got dark, but commenced again soon after dawn.

Monday, September 8, we were up again at dawn. We crossed the Sacramento Fork by ferry, and all day followed up that stream, making twenty-one miles. It was certainly, together with the next day's ride, the most picturesque road I have traveled in this state—in fact, I think that I ever traveled. Sometimes down to the level of the river—then crossing ridges, sinking into ravines—sometimes a narrow way where two wagons cannot pass for half a mile at a stretch, the steep mountain on one side and the swift stream hundreds of feet below on the other. None of your magnificent roads, such as one sees in Switzerland, where at such places a parapet guards from all danger; but rough, sidling, the outer wheel uncomfortably near the soft shelving edge—bridges, without rail, made by laying poles or split timber on the beams, spanning deep ravines, where the mules went over trembling with fear. The road is pretty well engineered. The fifty miles that we passed over, rough as it is, cost, we were told, $40,000, and our tolls up and back were $25.50.

The valley ran nearly straight toward Mount Shasta, and at times we got most glorious views of that peak. Its snow-covered head rose magnificently far above everything else—with what wonder and awe we regarded it, the goal of our trip! The many stories we heard of the terrors of ascending it—many declaring that no man ever had succeeded in reaching the highest summit, although many had nearly succeeded—were fiction, as we shall see farther on.

When within a mile of our camping place for the night I came on a patch of a very curious and rare plant, the *darlingtonia*, a sort of pitcher plant, as yet found in no other locality, the wonder and admiration of botanists.[5] It is needless to say that I stopped and filled my box while the rest went on—my mule expressing great indignation thereat, hardly being restrained by the rope with which I tied her to a tree, making the woods hideous with her braying and awakening the echoes of the rocks with her remonstrances. Professor Whitney rode my mule, so I had old Blanco again, my mule of last year.[6]

We had a beautiful camp that night among the pines and firs at Sim Southern's. He entertained us with some most marvelous stories of his attempted ascent of Mount Shasta—marvelous indeed to hear, but received with some allowance, and more so after we had been on the ground. In fact, popular testimony was that with him "truth is stranger than fiction."

Tuesday, September 9, we continued on our way. In a few miles we passed the Castle Rocks (Devil's Castle of the map), most picturesque objects to behold. A granite ridge rises very abruptly from the valley, its crest worn into the most fantastic forms—pinnacles, minarets, battlements, domes, and peaks. Some of these rise perhaps three thousand feet above the valley, and the chain of Castle Mountains is much higher beyond. We were in sight of them a long time and each turn of the road disclosed a new view of them. In crossing a spur from this chain that runs down to the river we had the most magnificent view of Mount Shasta that we have yet had. It appeared up the valley, the foreground of mountains opening to show it, the great cone rising high, its upper six thousand feet streaked with glistening snow, its outlines sharply cut against the intensely blue sky, its sides steep beyond belief.

Next we came to the Soda Springs. These are close by the river, here merely a large mill stream in size, its waters green and cold, and traces everywhere of what a torrent it must be during the winter rains. The waters of the spring are highly charged with carbonic acid—so are called "soda" springs, for they sparkle like soda water—and hold iron in solution. They have a considerable reputation for curative powers.

Here we left the immediate side of the stream and struck up an inclined table-land, rising a thousand feet more in the next nine miles to Strawberry Valley Ranch, where we camped. This is the base of the mountain.

On the last two days' ride we had met much lava. It seemed to have run over the country after it had its present general features but not the present details. The streams have, in many places, cut through the bed of lava into the softer slates beneath. These slates were for the most part very hard, for they had all been baked and altered by heat. The last nine miles from the Soda Springs was entirely over lava. Much of the last two days had been through fine forests—pine, fir, cedar, and spruce, with various other trees. Many of the cone-bearing trees were large and grand beyond anything the eastern states know of. Trees six or eight feet in diameter and 200 to 250 feet high were not rare.

A sort of wide valley runs through on the west of the mountain, in which both the Sacramento and Shasta rivers may be said to rise, that is, there is no ridge lying between these two rivers at this point. As this valley is but three thousand feet in elevation, it presents from this side a vast portion of the entire height of the peak.

NOTES

1. The name is probably derived from an early settler in the vicinity, named Wragg.

2. The official spelling is now Putah, suggesting an Indian origin. Brewer and the Whitney Survey publications use the original form, Puta, derived from the Spanish *Cañada de las Putas*.

3. The SS. *Golden Gate* was burned at sea near Manzanillo, July 27, 1862, with a loss of 198 passengers and crew, and property valued at $2,000,000.

4. Brewer writes this name Red Bluffs, but the singular is found in the Whitney Survey reports and is now the established form.

5. *Darlingtonia californica* Torrey, California pitcher plant, was first collected by Brackenridge of the Wilkes expedition in 1841; the flowers were first obtained by Dr. G. W. Hulse, U.S.A., in the fifties (W. L. Jepson).

6. In a letter to Professor Brush written a few weeks later Brewer says of this mule: "Peace to her memory—after having ridden her nearly two thousand miles, first and last, I sold her yesterday to a packer, and this morning I saw her leaving in the train, packed, braying furiously for the

rest of 'our party.' She evidently regretted leaving the ranks of Science. Botany was her favorite pursuit, she had a keen eye for plants and has made the *graminae* her special study along various sections of the Pacific Slope for the last thirty years, possibly longer." This letter was published in the *California Historical Society Quarterly*, Vol. VII, No. 2 (June, 1928).

CHAPTER V

MOUNT SHASTA

Shasta Rumors—Strawberry Valley—A Timberline Camp—An Early Start—Steep Snow Slopes—The Red Bluffs—Sulphurous Springs—The Highest Pinnacle—A Clouded View—Swift Descent—A Squally Night—Comparisons—Barometrical Observations—Indian Episodes—Bass's Ranch.

>Camp 98, Strawberry Valley.
>Base of Mount Shasta.
>Sunday, September 14, 1862.

From what has been written from time to time, you have seen that the ascent of Mount Shasta was an item of "Great Expectations"—it seemed indeed the grand goal of this trip. How we had barometers made a year and a half ago to measure it, how our summer's work was planned to bring us here in August but deferred so late by sickness of the party, how we had come prepared and had collected information on the subject from all possible sources, you know already.

Here let me say that Mount Shasta is the highest point in the state,[1] that it has long been an object of admiration and wonder, that it has been ascended by a number of persons, and yet absolutely *nothing* was known of it except its existence—of its geology and structure not a word can be found anywhere; of its height, matters were nearly as vague.

Lieutenant Williamson and Colonel Frémont *guessed* that it was seventeen thousand feet high, and hence it went thus into all the maps and authorities. But many doubted that it was so high, and last year a man got on top with a barometer and gave its height as less than fourteen thousand feet—but he had a poor instrument and, moreover, had no good facilities for measuring the mountain.[2]

We inquired of everybody who might possibly know anything about the matter, about getting up. The stories we heard would fill an amusing volume. Mr. A tells us that the ascent is easy, that all creation can be seen from there, that he has been up and never will forget the day. We inquire on certain particulars, find that he doesn't know on which side he went up, the appearance of the top, nothing. We doubt if he has ever seen it. Mr. B says he got *nearly* to the top, but that no living man has ever reached the summit—that long before reaching the top a man's breath gives out, his nose bleeds, his head aches, etc.—and that no man could possibly make the ascent of the last cone. In fact, many told us that it was an impossibility to reach the highest summit—some on account of ice, some because of sulphurous vapors, some because of its steepness, etc.

On our road here the stories became more divergent. One man told us that it was a common trip, that more than five hundred persons had made it the present year. We were elated. But that evening we camped at Mr. Sim Southern's (long be his name remembered), only twenty-two miles from here, and he told us that the first eight miles were easy, a good road, he had *drawn a ton of hay* up to the camping ground once! Whew, how easy! (The camping ground is only reached by a trail, in places steep as the roof of a house for a thousand feet together, and in others through thick chaparral.) He said that he had *nearly* reached the top, but an impassable glacier had stopped every person from going farther. Such were the stories, which would fill a volume—a few grains of truth, and an abundance of pure fiction—so, as I write facts, I will pass them.[3]

We reached this place the evening of Tuesday, September 9. Here are two houses, one a "hotel," the other a ranch house, but also a sort of public house, as hay is sold to travelers for their teams.

At this Strawberry Ranch we are camped, the most lovely camp we have had.[4] It is on a stream of clear, cold water, an open forest back of us, of splendid trees—pines 150 to 200 feet high, and cedars almost as large. In front, in unobstructed view, lies "The Butte," as Mount Shasta is generally called here. We are a little over three thousand feet altitude, while the mountain rises in one grand slope over eleven thousand feet more at one view—the base, up to eight thousand feet, clothed with pines, cedars, and firs, all above that a desolate waste of rock and snow. The mountain is a great irregular cone, with steeper sides by far than any other large mountain I have ever seen. How we gazed on it that night from camp as the moon rose. I even got up in the night to see it by the better illumination of the moon after midnight.

The air is cool and delicious, in most pleasing contrast with the hot parching air we have felt for the last three months—all are inspired with it.

Wednesday, September 10, we spent in camp, making preparations. Apparatus was tested, provisions got ready, guide found—not a professional guide, but a Mr. Frame, who was up once, plain, unostentatious; he says it is practicable, that he has been on the top, that he will show us the way, but we must furnish our own *muscle*. Two other men came up from the Soda Springs, eight miles distant, to make the ascent with us—Perry and Campbell, both capital men.

Thursday, September 11, we started for the "camping ground." This is at the upper limit of the timber, just where the snow begins this year. We are the only persons who have been up this year, and it was doubtful before we started if it could be ascended—the quantity of snow is greater, and the snow line is lower.

We start, all on foot, our three friends with their baggage packed on two horses, we with ours on three mules. Two of the mules, Jim and Nell, carry the blankets and provisions, while Kate carries a very miscellaneous collection of bags, containing instruments, canteens of water, botanical box, etc. Hoffmann, Professor Whitney, and I each carry a barometer, Averill a tripod. We walk and lead the mules—first across a meadow, swampy—Nell sinks to her belly in the bog, is released and is repacked, but she looks disgusted enough.

We soon take to the woods and follow a trail directly toward the mountain—the first four miles up a very gentle slope, among trees that must be seen to be at all appreciated—pines, firs, and cedars, all of species peculiar to the region west of the Rocky Mountains. Cedars over 100 feet high and 4 to 6 feet in diameter, firs 200 feet high, sugar pines often 200 and even 250 feet high, possibly even more. Fire had been through the woods and hundreds of trees had fallen, some this year, but more in past years. I had the curiosity to measure one prostrate tree. It was about 7 feet diameter at the base and lay along the ground for 225 feet, and was then burned off where it was still 9 inches in diameter. The burnt tip must have been 40 feet longer. These gigantic trees, straight as arrows, formed a magnificent forest.

The last four miles was up a very steep slope, nearly a thousand feet per mile—part of the way through the pines, part through thick chaparral of manzanita. Blocks and ledges of lava are the only rocks seen, and dry, dusty soil—no springs, no streams, until we reach the camping ground, where there is a small stream that runs from the snow but sinks soon into the porous lava that forms the mass of the mountain. Mount Shasta is an extinct volcano, and this will explain what I have to tell of its appearance.

Camp 99 was here, at the upper edge of the timber, streaks of snow coming down below this level. It is about 7,400 feet altitude. Here the trees are still numerous, although scarcely forming a "forest." They are of only one species, a grand fir, *picea nobilis*,[5] many of the trees over four feet in diameter—one near camp is six feet—yet they cease entirely but a few hundred feet higher. Above this timber, for at least six thousand feet, the cone rises naked, bleak, and desolate. Here we camped at 2 P.M. and spent the rest of the afternoon about there. All are in our blankets right early, although we have a big camp fire, for we must be off early.

In bed early, to be sure, but scarcely to sleep; all are too much excited—the mules bray, the dogs that came along bark, the wind howls, it is cold, yet sleep is a duty. I awake—surely it is time, but my watch tells me it is only eleven o'clock. An hour later the thing is repeated.

The next thing I hear is Professor Whitney arousing the party—it is after two o'clock. All are soon astir; breakfast is soon got by the bright light of the moon. Hoffmann keeps his blankets; he does not feel strong enough after his recent sickness to make the attempt, and great is his disappointment. He stays to observe barometer.

It is nearly half-past three when we at last get fairly under way. Before us, in silent majesty, lies the immense mountain mass. How cold its snows look in the bright moonlight! All are anxious, for there is the uncertainty as to whether we can succeed.

For half a mile or more the route lies over loose blocks of lava, or dry, sandy, ashy soil; then we strike a strip of snow lying in a gulch, and up this we follow. But what a path! Such a grade! Hour after hour we toil on it—the bright moon gives way to gray dawn, and then the sun comes up and gilds the summit ahead and throws its dark shadows into the valleys below—and yet we are on that slope. Both at the right hand and the left are sharp ridges. The lava first wreathed into curious forms when it flowed down there, then, in later times, weathered into fantastic shapes—walls, battlements, pinnacles shooting up hundreds of feet, more forms than can be described.

The ascent continues to grow steeper as we approach the red bluffs, a wall of yellow or red lava and ashes. For two thousand feet or more below it the average slope is not less than forty degrees. At last we mount this wall. It is eight o'clock, and we are now thirteen thousand feet up. That snowy slope of five thousand feet perpendicular has not carried us on at most over two miles horizontal distance!

The last thousand feet has been hard—the air is so light that one is very short winded, must stop often, and resting does not appear to restore the strength. One pants for more air, but the air refuses to strengthen him. It is now very cold; we stop on the sunny side of this rocky wall to rest and hang up a barometer to see how high we are. We enjoy a scene that bursts on our view from the east—a great gorge filled with snow, perhaps hundreds of feet thick, cracked and melted into fantastic shapes.

We now follow up the ridge.[6] Cold as it has been, the cold wind becomes colder, and Professor Whitney has his fingers frostbitten. We toil on almost in silence, for no one has breath to spare for talk. Our three friends are ahead—two carry canteens of water, the other a canteen of cold coffee. Averill carries a bag containing our lunch, thermometers, etc. Professor Whitney and I generally bring up the rear, for each of us carries a barometer—and had each a baby it would not require more constant vigilance to protect it from injury.

After leaving the red bluffs we had about three and a half hours' climbing to reach the summit—a part of the way up a steep slope over hard lava, strewn here and there with loose rocks—in fact, it was the same bed that formed the red bluffs. We were often on the snow. We wore colored goggles to protect the eyes when the sun shone.

The day was unfavorable—the first cloudy day of the fall. At times we would be enveloped in cloud, shutting out all distant views; then again the cloud would blow away and disclose glimpses of the landscape two miles beneath us. The snow, especially near the summit, has melted into a very curious form. I never saw anything like it in Switzerland. Imagine the snow sliced, or gashed into slices, from one to four feet deep, the slices running from east to west—not perpendicular, but leaning to the south, or down hill—the slices melted so that they present their sharp edges to the sun. Thus the whole surface was melted, and it made the worst possible going, especially with the barometer to carry, as we had to step on these edges. They were frozen when we went up, so they would generally bear us, but on our return they were thawed, and we broke into these clefts so that the going was worse than ever.

As I said, this is an extinct volcano. About three hundred feet below the highest summit there is a hollow, with very steep ridges rising on two sides, probably once a crater, and in this is a *boiling* spring, the boiling water and steam charged strongly with sulphurous gases. These frequently make people sick who breathe them. There is much sulphur mixed with the soil and we collected fine specimens.

Then came the last hard tug, and at about noon we reached the highest point. This is a mere pinnacle of lava, shooting up into the air—difficult of access, and only reached with some daring. One has but a small hold in climbing on it; I would never trust myself to it on a windy day. It is accessible only by a narrow ridge, while a fall from any one of the other three sides would precipitate one many hundreds of feet below on the rocks.[7]

It was entirely impossible to hang barometers there, so I rigged up a support twenty feet lower and hung both barometers to it. I had to use my coat to shade them from the sun, so I sat and shivered in the cold, for the thermometer was 26° F.

We stayed there an hour and a half. It was curious to note the effect of the thin air and fatigue on the men. All were more or less drowsy and sleepy, all complained of headaches, eyes were bloodshot and red. My lips and fingernails were of a deep blue, so were those of Campbell. But no one bled at the nose, as is common. We lunched, and as some began to get sick, the rest started, leaving Professor Whitney and me with the barometers. The clouds grew thicker before we got through. Averill,

Campbell, and Frame all became very sick and vomited severely, from the effect of the rarified air. The barometer stood at only about 17½ inches.

We got only occasional glimpses of the landscape beneath, but enough to show how magnificent it must be in the clearer weather earlier in the season. In the west is a perfect wilderness of mountains extending all the way to the Pacific, chain beyond chain, many with snow on them, all now dim, however, for the valleys are filled with smoke and the tops are more or less obscured by clouds. To the northwest lies the great valley of the Klamath River. In the north are the Siskiyou Mountains, but we saw only glimpses of them. We had no view of the east.

To the south there are mountains for about fifty miles, then the great Sacramento Valley. The latter was entirely filled with smoke and haze, the surface level as the sea, and rising above it was the sharp Lassen's Butte, remarkably distinct although near a hundred miles distant. It rises like an island of black rock and white snow from this sea of fog, a grand object.

The descent was much more rapid than the ascent. We reached the red bluffs in one hour, a distance that required three and a half hours in the morning to ascend. The fog grew very dense, and so cold that our beards were white as snow, mustaches frozen, and faces blue—the way not plain, and the guide ahead. But when we reached the red bluffs all was then safe, so far as the way was concerned. Then the rest went on, for Professor Whitney and I could not travel half so fast down those steep slopes as the rest, owing to the barometers. Some of them got into camp an hour and a half before we did.

Such a descent—sliding, sometimes on our feet, sometimes on our "bases," down the soft snow, which the sun had now thawed. The latter mode of descent, although rapid, was soon rendered uncomfortable to me by the giving way of the main seam of my pants, and the consequent introduction of large quantities of snow.

More snow fell also during the last half of the descent, not a storm but rather like a quiet squall. Since noon the clouds had been forming over the sky, which was now nearly covered, at a height of ten thousand or eleven thousand feet. When we got below this line, we had a peculiar scene, but the effect was grand. The clouds, like a curtain, cut off all the mountain above us and also the tops of the mountains in the west, but below, the air was smoky, and through this we could see the streaks of sunshine here and there. The effect was very peculiar and striking. Then we saw a heavy shower over the mountains in the northwest.

We got back to camp before dark, found Campbell and Frame asleep in their blankets, pretty well used up. We were all of us tired enough. A hearty supper, a good smoke, and we were soon in our blankets.

It grew very cold, the clouds grew thicker, it grew colder, began to freeze vigorously, and the wind grew high. It snowed quite briskly, quite a fierce squall, decidedly an unpleasant night to be sleeping out. If some of you wish to realize what it was, choose some squally night when the thermometer sinks to 20°, when the snow comes with driving wind, take your blankets and go out on some bleak hill, lie down on the ground and try to sleep and enjoy yourself.

I believe that I fared the best that night. I got my blankets in the shelter of a bushy tree that partly broke the force of the wind, and I slept with my thick clothes on, yet the snow did blow into my blankets—and ugh! how cold it was—but I covered up closer, hauled my head under my shawl, and slept soundly and sweetly, save perhaps an hour spent in shivering.

Everything froze up tight that night, ice was an inch thick on the stream near, and in the morning the ground was white with snow in places. But the sky cleared up, and the mountain stood out again in clear outline against the bluest sky.

Mount Shasta is about 14,500 feet high—the precise height we cannot yet give. It is higher than any mountain in Switzerland and only 1,200 feet less than Mont Blanc.[8] Yet it lacks the grandeur of the Swiss Alps, and it is entirely destitute of many of the elements of beauty that they possess. In this dry climate, where there is no rain during the summer—although immense quantities of snow fall on it in the winter—no *glaciers* form.[9] Much of it is so steep that the snow blows or slides off, and at this time in the year the snow lies only in patches and streaks over the upper seven thousand feet.

In a rainy country these bare patches would be more or less clothed with alpine plants, and streams of water would come down the sides everywhere. Not so here. The waters of the melting snows are drunk up and absorbed by the porous lava rock of the mountain, so its watercourses are dry. The soil is a mixture of decomposed lava and volcanic ashes and would undoubtedly possess great fertility with water, but as it is, it is barren.

Above the timber the mountain is naked, and all above nine thousand or ten thousand feet is a scene of unmixed and unrelieved desolation—rock, snow, and dry soil, that is all. No plant cheers the eye, no insect or bird appears. This barren scene succeeds immediately the failure of the upper zone of timber. In Switzerland, or the Tyrol, or the eastern United States, above the timber there would still be vegetation—first, green pastures; while up among the eternal snows, where a rock was bare of ice or snow, some alpine plants would be warmed into life. The Alps are grand in their beauty, Mount Shasta is sublime in its desolation.

Geologically it is as barren as it is botanically. It is a great cone of lava, nothing else. Not, like Etna, made up of an almost infinite number of small eruptions, but it seems as if it had been formed in a comparatively short period, by a few gigantic eruptions. It appears to belong to a series of volcanoes that formed a line—like the Central American volcanoes of the present day—that had their time of greatest activity during the Tertiary period. This line extended nearly north and south—Lassen's Butte is another of them, in the south—but the chain ran north to an unknown distance into Oregon.

Battle Creek.
Sunday, October 5.

ON the morning of Saturday, September 13, we were at our mountain camp at the snow line on Mount Shasta—a clear, lovely morning. Hoffmann was immediately sent back to Strawberry with a barometer, and all the rest also left in the morning. I alone remained, with a barometer and with my botanical box. I first collected that full of plants near camp, and then for four hours observed barometer quarter-hourly.

How I enjoyed those hours of solitude, so far from men, such a picturesque spot! Near me the grand forests, behind me the lovely valleys below, before me the grand old peak, its outlines so beautifully cut against the intensely blue sky. I gazed on it for hours, as I lay there, not with the awe that I did two days ago, but with even more admiration.

This day closes my thirty-fourth year, the morrow is my birthday. Six years ago yesterday I was on the Great St. Bernard, in Savoy—how unlike that view from this! My mind wanders to the Swiss Alps and the views I saw there. And then it wanders home and to loved ones there, and then to battlefields and scenes of carnage and blood and sorrow in the East, and to hospitals where men are enduring the keenest of physical sufferings—but all is quiet here, so quiet that no wonder thought and imagination wander.

At 3 P.M. I started to return—eight weary miles to walk, with barometer, and with botanical box and bag heavy with specimens. It was after dark when I got back.

Our barometrical observations between the Strawberry Valley camp and the snow line did not calculate up satisfactorily. So, on Monday, September 15, with Schmidt and one pack-mule, botanical box, and barometer, I returned. The day was fine, and we had again the lovely view and grand forest. We botanized and got the desired observations. Our appetites were good, and we had a fine chunk of venison, as well as bacon, etc., and fared well. We stayed there all night. I observed again until the afternoon of Tuesday, then returned. Averill had been out and caught a mess of fine trout.

Wednesday, September 17, we were up early, ate a hearty breakfast of venison and delicious trout, raised our camp, and were off, returning by the same road that we came, instead of going by way of Yreka as we had intended. We got but $17\frac{1}{2}$ miles that day, and camped at the Sweet Briar Ranch that night.

Thursday, September 18, we were off early, as we intended to get to the Sacramento ferry that night. During that day, as on others, we saw many Indians. They are "lower" than you have any idea of—sometimes nearly naked—men with merely the "breechcloth," women with scarcely more, although generally the latter (theoretically) wear a part, at least, of a skirt of civilized style.

Our noon lunch was where we lunched on the way up, and I had an amusing adventure buying some potatoes from a squaw. She could talk no English, while my knowledge of "Digger" was equally poor. Here was a white man, who lived with two squaws—or rather they lived with him—their faces horribly tattooed, but they wore dresses. Near, along the river, was an Indian camp, and as we went by an old woman passed us in "deep mourning." She wore but very little clothing, her form bent, skin wrinkled, and not only her face but her entire body was painted black, with patches of pitch on it in places to make it still more hideous. She threw her skinny arms about, screeched some, and showed her grief in the Digger style. The camp was such as might be expected. I will not attempt to picture it, but some of the least orthodox of our party questioned if such beings had much of a soul, or at least, thought the orthodox rule of having them all lost was uncharitable to say the least. I do not wonder that missionaries have met with no more success among this miserable tribe and race.

We arrived at Dogtown (one house) a little too early to camp, and we thought that with diligence we might reach a house about six miles farther, where hay might be obtained for our animals, so we pushed on.

I have already described the roads, often mere narrow dugways, where two teams cannot pass. On such places we generally ride some distance ahead of the wagons to look out for meeting teams. But in one place the wagon had nearly caught up with us, when we were suddenly face to face with a wagon. Both stopped, and we parleyed and palavered. It was nearly sunset. He was loaded with four thousand pounds of freight. After a careful examination we found that we could not pass; it was impractical to draw either wagon back. The track was perhaps four hundred feet above the river, and, in passing, the outer wagon could not fail starting and finally bringing up in the water below.

At last we unloaded our wagon and set it carefully so far over the edge that his could pass, which it did, clearing ours only three inches. We loaded up, but as we could not reach the desired haven, we tied up in a cheerless place for the night. We had a little barley for the mules, but no hay, so we tied them to the bushes, brought water half a mile, got supper long after dark, provoked, ill humored, and uncomfortable.

But one must be bad off indeed if a joke will not pass, and even here Averill found time to "sell" me most beautifully. There was a cleared field near, and Averill found that it belonged to a man named Campbell. As I was hurrying for a pail of water, he wanted me to "go to the house and see if Mrs. Campbell would consent to allow our mules to be picketed in the field." "But where is the house?" "Right over there—follow that path." I did—found only a camp of Indians, half a dozen, some half naked. I inquire for "Mrs. Campbell," no answer—

again inquire—a grunt from a half-naked man. I can't make myself understood—follow with more questions—a squaw in some blankets jabbers, but I can't understand a word. I hear Averill laugh, and soon the "sell" appears—"Mrs. Campbell" is one of the squaws of the camp. I afterward learn that "Mrs. Campbell" is in fact two squaws, and in the morning I see several youthful Campbells, of semi-Digger type.

Friday, September 19, we made a long heavy drive of twenty-three or twenty-four miles, and camped that night at Bass's Ranch, near Pit River. We saw many Indians again that day. I had got far ahead of the party and stopped to let my mule drink at a stream. A party of Indians were gathering acorns near. A man was up a tree cutting off the limbs with a hatchet; the rest below, gathering or eating the acorns. They ate them raw, like hogs, although they are very bitter and astringent.

One young squaw especially attracted my attention from her costume which was the neatest Indian costume I have seen— a fillet of many colors around her forehead, pretty buckskin moccasins neatly embroidered in colors, and around her hips a girdle perhaps a foot wide, with fringe around its lower edge perhaps six inches long, neatly woven or braided, of several bright colors. Her body above the hips, and legs below the short skirt (which was perhaps but eighteen or twenty inches wide) were bare. Her limbs were better formed than those of any other squaw I have seen, and, in fact, her appearance was rather pleasing—that is, she did not excite the feelings of deep disgust that the others did. But she sat down as I passed her, and commenced eating acorns, reminding me of a baboon eating nuts.

These Indians gather large quantities of acorns for winter food. We saw them also catching and drying salmon. The squaws carry enormously heavy loads in a conical basket, which is wide above and comes to a point below, held by a band across the forehead, the point of the basket resting against the rump. One of the Indians came up to me and talked some time, but the only words I could understand were "Klamath," "Shasta" (he pronounced it *Tschasta*), and "tobacco." I gave him some, and he looked very unthankful and sour because there was not more of it.

We stopped for an hour at noon. An Indian came along whose only dress was a piece of deerskin hung by a string over one shoulder so that it covered one hip, coming partly around in front and behind, reaching from the hip nearly to the knee. It was rather too scant for civilized eyes.

The next two days we spent near Bass's Ranch, examining some interesting limestone ridges near, cut through by trap dykes, and in places affording good marble. The next day we went back to Shasta. Hoffmann was so used up by his partial ascent of Mount Shasta that he was unfit for further duty and he had to take to his bed immediately on arriving at Shasta. Schmidt, too, had got worse, and had to take to the bed again —a bad prospect. Rémond had entirely recovered. His quiet life after we left had cured him. For fifteen days he had made hourly observations on the barometer, and this had so confined him to the house that he had had no relapse from overexertion.

NOTES

1. The altitude of Mount Shasta was found by the United States Coast and Geodetic Survey to be 14,162 feet, as the result of very accurate measurements completed in 1904. With this figure it ranks sixth in California, being exceeded by Whitney, Williamson, North Palisade, and Russell, in the Sierra Nevada, and White Mountain, in the Inyo Range.

2. In a note contributed to *The American Journal of Science and Arts* (2d series, Vol. XXXVI, No. 106 [July, 1863], 123), Whitney says: "A careful and elaborate series of barometrical observations by the State Geological Corps of California, made in September, 1862, has fixed the elevation of Mt. Shasta at 14,440 feet. Previous to this the height of Shasta had been variously estimated at from 13,905 to 18,000 feet. The number 13,905 was the result of a barometrical observation made by Mr. W. S. Moses, August 20, 1861; 18,000 feet was the height as estimated by the Pacific Rail Road expedition, under Lieut. Williamson [Lieut. H. L. Abbot in *Pacific Railroad Reports*, Vol. VI, Pt. I, 36]; Frémont's estimate was 15,000 feet [*Geographical Memoir Upon Upper California* (1848), p. 25], which is much nearer the truth than Williamson's. It is a very curious fact,

that the height of Mt. Shasta, as given by the author of Colton's Atlas and author of the article on California in the New American Cyclopaedia is 14,390 feet, which is a very close approximation. Where these figures were obtained, I have been unable to ascertain. [Footnote: "Wilkes says 'it is said to be 14,350 feet; but Lieut. Emmons thinks it is not so high.'"] It is pretty certain that they were not the result of any actual measurement, as it is known that Mr. Moses was the first person to ascend the mountain with a barometer." An account of the ascent of Mr. Moses and party, from the journal of Richard G. Stanwood, one of the members, is published in the *California Historical Society Quarterly*, Vol. VI, No. 1 (March, 1927).

3. The first ascent appears to have been made in 1854. Between that date and 1862 a considerable number of people had been to the summit. Several ladies accomplished the feat in 1856 (San Francisco *Bulletin*, September 23, 1856).

4. This spot was for a long time known as Sisson's, or Sisson, a name unhappily abandoned recently for the pretentious and confusing Mount Shasta City.

5. The tree described here is not now called by this name; it is a subalpine variety of red fir called *Abies magnifica shastensis* Lemmon. Lower on the mountain the forest is (or was in Brewer's day, before it was burned) largely white fir (*Abies concolor*), douglas fir (*Pseudotsuga taxifolia*), incense cedar (*Libocedrus decurrens*), yellow pine (*Pinus ponderosa*), and sugar pine (*Pinus lambertiana*). The timberline tree is white bark pine (*Pinus albicaulis*). For descriptions of the natural history of Mount Shasta see an article by Ansel F. Hall in *Sierra Club Bulletin*, Vol. XII, No. 3 (1926), and C. Hart Merriam, "Results of a Biological Survey of Mount Shasta, California," *North American Fauna*, No. 16 (Division of Biological Survey, United States Department of Agriculture, Washington, 1899).

6. The route described is, throughout, the one followed by a majority of climbers today.

7. Recalling his experiences on Mount Shasta in addressing the members of the Appalachian Mountain Club at its tenth anniversary, in Boston, March 5, 1886, Professor Brewer said: "When we got to the top we found people had been there before us. There was a liberal distribution of 'California conglomerate,' a mixture of tin cans and broken bottles, a newspaper, a Methodist hymn-book, a pack of cards, an empty bottle, and various other evidence of a bygone civilization" (*Appalachia*, Vol. IV, No. 4 [December, 1886], 368).

8. Mont Blanc is 15,781 feet, more than 1,600 feet higher than Shasta. Several peaks in Switzerland, including the Matterhorn (14,780), Monte Rosa (15,217), Dent Blanche (14,318), and Weisshorn (14,804), exceed it.

9. There are, nevertheless, glaciers of considerable magnitude. The Whitney party ascended by a route which avoided them, excepting at the top, where they were covered with snow. Clarence King first reported the Shasta glaciers, which he discovered in September, 1870 (*Atlantic Monthly* [March, 1871]; *American Journal of Science* [3d series, 1871], I, 157; see also King, *Mountaineering in the Sierra Nevada* [1872], chap. xi). The subject is discussed at length by Israel C. Russell in *Glaciers of North America* (1897).

CHAPTER VI

WEST AND EAST OF THE SACRAMENTO

Weaverville—Hydraulic Mining—Mount Balley—A Street Fight—Another "Balley"—Volcanic Remains—The Sacramento Valley—Deer Creek—Chico—Pence's Ranch—Table Mountains—Tunnel Mining—Disbanding at Marysville.

Oroville.
October 15, 1862.

AT Shasta City the party broke up for a time owing to sickness of several members. Tuesday, September 23, Professor Whitney left before daylight for San Francisco. Schmidt and Hoffmann were better, but both unable to ride.

After breakfast, leaving orders for the rest, Rémond and I started on mules for Weaverville, forty miles distant among the mountains west of Shasta. Over a mountain six miles to Whiskey, a little mining place on Clear Creek—once clear, but foul enough from mining now—then up that eight more to Towers—merely a public house, and a very pleasant one at that. This is on the great Yreka road, many heavy teams are met, and the road is dusty almost beyond endurance. The day is hot. At Towers the Weaverville road branches off. This is a toll road. It is only twenty-six or twenty-eight miles long, but this spring $38,000 was spent on it—this tells something of the country it must pass through. It was a fine road and well engineered.

We went up Crystal Creek a few miles, then over Trinity Mountains at an elevation of over three thousand feet, and then went down a ravine into the valley of Trinity River—a very picturesque road, with magnificent fir, spruce, pine, and oak trees. Some of the scenery is decidedly fine. We stopped all night at Buckhorn Station, a tavern and stage station—cold enough, heavy frost and some ice. I have only a linen coat along—decidedly too thin.

September 24 we were up early and off—six miles to the Trinity River, which we crossed by a ferry (our tolls for this trip were $5.50 on our two mules), then over Brown's Mountain, 2,800 to 3,000 feet high, then into the valley of Weaver Creek, and by a little past noon were in Weaverville, always called simply Weaver.

We had heard of fossiliferous strata of Cretaceous rocks there, and as we were striving to solve the difficult problem of the geological age of the Sierra Nevada and of the gold-bearing slates, this was supposed to be a most important point to visit. We spent that afternoon in making inquiries, and found that all our information had been wrong. We found in a drug store a single fossil, said to have come from diggings six miles distant; so the next morning, September 25, we started for the place, Douglas City, on the Trinity River, six miles south. We found no fossils there, but very extensive "hydraulic" diggings.

The river here makes a curve. A stratum of soil twenty or thirty feet thick forms a flat at the curve of the river, of limited extent. The "bed rock" beneath this is of metamorphic slates, much twisted, contorted in every shape by former volcanic convulsions, and much of it very hard. The soil above is *very* hard, like rock itself, made up of loose rounded bowlders, cemented by a firm red clay into a mass as hard as ordinary sandstone. In this the gold is found.

Deep ditches are cut, not only through this, but deep down into the hard bed rock beneath, often twenty or more feet into the latter, and running out to the river. In these are the "sluices"—merely long troughs for conveying the water. The bottoms of these sluices are made of blocks sawed from the ends of partially squared timber, so that the end of the grain is presented to the surface, sometimes of a double row, thus—

[NOTE: Professor Brewer's diary contained several diagrams done in brownish ink with a very fine pen. It was not practicable to reproduce these photographically nor to trace them, hence the reduced diagrams shown herein are very accurate and careful redrawings by Miss Lois North, but are not photographic copies.]

sometimes, however, of but a single row of blocks. These do not lie perfectly square and level, so, as the water flows swiftly over them, they cause a ripple, like water flowing swiftly over the stony bed of a stream. The bottom of the box or trough, below these blocks, is perfectly tight, and quicksilver is poured in and collects in all the holes between the blocks.

Ditches, from miles back in the mountains, bring the water up against the hillside, far above the surface of the flat, and a flume, or "raceway," built on high stilts, over seventy or one hundred feet high, brings the water directly over the "claim." A very stout hose, often six inches in diameter, conducts the water down from this high head, and has at its end a nozzle like that of a fire engine, only larger. Now, this stream of water, heavy and issuing with enormous force from the great pressure of so high a head of water, is made to play against this bank of hard earth, which melts away before it like sand, and all flows into the sluices—mud, bowlders, gold. The mud is carried off in the stream of thick, muddy water; the bowlders, if not too large, roll down with the swift current; the heavier gold falls in the crevices and is dissolved in the quicksilver, as sugar or salt would be in water. In some mines these sluices are miles long, and are charged with quicksilver by the thousands of pounds. This washing down banks by such a stream of water under pressure is "hydraulic mining." After a certain time the sluices are "cleaned up," that is, the blocks are removed, the quicksilver, amalgamated with the gold, is taken out, the former being then driven off by heat—"retorted"—and the gold left.

From this flat near Douglas City over a million dollars has already been taken, and it looks as if as much more was yet to be got. Owing to the want of water at present the sluices were dry, and men were only preparing for water when the rains should commence.

The amount of soil removed in hydraulic mining must be seen to be at all appreciated. Single claims will estimate it by the *millions of tons*, the "tailings" (refuse from the sluices) fill valleys, while the mud not only muddies the Sacramento River for more than four hundred miles of its course, but is slowly and surely filling up the Bay of San Francisco. In the Sierra, the soil from hundreds of acres together has already been sluiced off from the rock, which it formerly covered even to the depth of 150 feet! I have seen none of the *heavy* mining as yet, although I have seen works and effects that one would imagine it would take centuries to produce, instead of the dozen years that have elapsed since the work began.

Well, this we found, but we found no fossils, and no formation at all possible to contain them.

A high mountain rises to the northwest of Weaver, called here Mount Balley. This we resolved to climb the next day. A citizen who wished to make the ascent resolved to accompany us. So, on September 26, we made the trip. The mountain is 7,647 feet high, and at the altitude of over 6,300 feet there is a small lake, a very uncommon thing in this part of the state. Ice is got out of this in the winter, and is packed down on mules in summer, selling at six to eight cents per pound, so a pack trail leads up to that point. The slopes of the mountain to this height are covered in some places with chaparral and in others with forests, mostly of the same species of trees that we found at Mount Shasta. Some of the spruce trees were especially grand. I measured some that were over twenty-five feet in circumference and must have been over two hundred feet high.

Above the lake it is very steep, and our friend puffed so hard that we went on ahead, he following up more slowly. The summit is a very narrow ridge of granite. We had ascended it from the south side, while the north side is so steep that bowlders will go bounding down several hundreds of feet if once started. The view was very picturesque.

Trinity County is a very rough one—too rough to survey, so there is no "public land," that is, surveyed land, in the entire county. In the middle of this we were perched at an altitude of over 1,300 feet higher than Mount Washington or Mount Mitchell, the highest land east of the Mississippi River. The view in every direction was extensive, and the roughest region I have yet seen, all broken into mountains from five thousand to nine thousand feet high, many of them streaked with patches of snow. Immediately to the north the mountains were not only high but very rugged and broken, the canyons thousands of feet deep and the sides *very* steep and rocky. We enjoyed that trip greatly, and were back in time for a late supper.

I met in Weaver a Mr. and Mrs. McClure, who had been fellow passengers on the steamer when we went out, and I had a pleasant time with them.

Before leaving Weaver I must speak of the town. It is a purely *mining* town—no other interest there except such as bears upon that—so it is like California in bygone times. There are a few valley ranches, very small, in the county; they raise something, but everything else must come from the outside. Hay is generally $50 per ton, and the livery-stable man told me that he had paid $160 per ton. Our mules cost us $1.75 each day to keep—other things proportionately dear. Freights from Red Bluff are about four cents per pound, so while all kinds of goods are dearer than outside, they are dearer in proportion not to their *value* but to their weight. In winter freights are often eight, ten, or even twelve cents. It is curious to note the effect on prices in a place where the increased price of a pound of silk is the same as the increased price of a pound of iron—where you buy your stockings and shirts, as well as sugar and tea, by the pound—at least really if not nominally.

Sluices run through the town, but the town has no limits, no corporation. The houses grew more and more scattered up and down the valley—some are "washed out"—one church has been sluiced around until only enough land remains for it to stand upon.

There are multitudes of Chinese—the men miners, the women "frail," very frail, industrious in their calling. There is a Chinese temple there, with its idols and fixtures, decidedly a curious concern; for in this land of religious liberty the Chinese, of course, introduce all their heathen rites of worship.

There are twenty-eight saloons and liquor holes in the place, and gambling and fighting are favorite pastimes. After the third fight had come off in the streets Rémond remarked to me, "I teenk dat de mineeng customs are petter preserved in dees plaze dan in any town I yet see in dees state." He was quite right.

One of these fights deserves more than a passing notice. A brawl occurred in a groggery between an Irish bartender and a tough, plucky little teamster. They adjourned to the street to fight it out, when a constable stopped it, then went his way. When he was out of sight they adjourned to a neighboring corral to have the fight out quietly. A crowd came in to see the sport. I was in there letting my mules drink, so I got them tied up, perched myself on the fence, and calmly witnessed the preparation for the contest. Teamster pulled off shirt, took an extra hitch in his belt, pulled up his pants above his boots, and announced that he was ready, asking Paddy whether he preferred a fair stand-up fight, or rough and tumble. Paddy preferred the "fair stand-up," "peeled" also his shirt, announced that he was ready, and they "sailed in." No one interfered—they pommeled each other and Paddy got knocked down. Teamster says, "Come up here, my boy," and waits for him to get up. Paddy is up and at it again—a little more pommeling—they clinch, hit, strike, squirm, and writhe. Both get down, part of the time one on top, part of the time the other, spectators standing calmly around giving them plenty of room, when the constable makes his appearance and stops the "sport" by separating the men and arresting both. Paddy, who has the worst of it, seems satisfied, but Teamster jocosely requests constable to absent himself just fifteen minutes, then come back and he may arrest and be damned. The fight, however, stops, the people mostly disperse, but the belligerents are not taken into custody. I come down from my lofty position on the fence and finish with my mules. The constable gone, Teamster proposes that now is a good time to quietly settle it, but Paddy, who has one eye closed and the other terribly black, objects and threatens to knife him. Teamster coolly offers to settle it right there with knives, if Paddy prefers that way. Men, however, here interfere and Paddy is got away.

Another street fight had less of note in it. That night I heard a noise in the street, inquired in the morning what it was, found there had been a fight in front of a gambling saloon across the street, and that a miner had been stabbed in five places, probably fatally.

We had often heard that "Weaver is a great place for amusement," "a lively little place"—we found it so.

The morning of September 27 we left—a chilly morning. We returned over the same road we had come on, and stopped all night at Towers. To the south of Towers and west of Shasta rises a high, conspicuous mountain, 6,357 feet high, also called Balley—an Indian word meaning "high mountain." September 28 we climbed it. Two men accompanied us. As we had to ascend from the base, and there was some chaparral, the view was more extensive, but less picturesque, than from the "Balley" near Weaver. We started to return, but our two companions went down by another way to hunt. While we were on the top we saw two rainstorms coming down the mountains from the far north. We lost our way on the way back, when in the labyrinth of foothills at the base, and were caught in a cold rain. We waited until it ceased, then found our way back at last, having walked six miles extra because of our mistake, making it decidedly a heavy day's work. That night and the next morning were very cold.

September 29 we went back to Shasta. Here we were rejoined by Schmidt who had been to Red Bluff. He had chills and fever every other day. Hoffmann had gone down the valley, too badly used up for hard work for the season.

Marysville.
October 20.

OCTOBER 2 we left Shasta City. First, six miles southwest to the river, over the table-land before described—a table furrowed entirely into hills and gulches—then crossed the river, then rose to another table. This was unlike the first; it is in reality the head of the great Sacramento Valley plain, which at its northern end rises into level tables, perhaps six hundred to eight hundred feet above the sea, often for many miles without streams or deep gulches, bearing no gold, covered at this place with scattered oaks and pines, the soil dry, but barren because of its dryness only. The surface is as hard as a paved road, the trees and shrubs have a dry aspect, although they are mostly evergreen, and all the herbage is long since dried up and gone.

We camped on Cow Creek, about twelve miles from old Fort Reading. We had heard of coal mines discovered up this creek, about fifteen miles from our camp, so on October 3 Rémond and I started for them, intending to be gone two days. Our ride was a gentle ascent, sometimes passing over tables sloping to the west and elevated perhaps a hundred feet above the surrounding country. These were remarkable, being made up of strata of volcanic ashes, sometimes mixed with bowlders of lava. This had covered the entire region, but up about fourteen miles from camp we found where the stream had cut through these strata into the sandstone beneath, which is rich in fossils —shells of many species were as thickly imbedded in the rocks as if the sea had but lately left these shores.

It was a solitary region, with houses only at intervals of several miles. We did not find the coal mines, did not find the men who were to show us where they were, could find no place where we could get shelter or feed for our animals, so returned the same night. It rained in the night, and drove us into the tent.

San Francisco.
October 26.

SAFELY back here again I will go on with my journal, but first, a few more words on the great interior valley of the state, that of the San Joaquin and the Sacramento.

This great feature is a vast valley, often thirty to forty miles wide, a perfect plain enclosed by high mountains on both sides, its only opening, the Straits of Carquinez, being less than a mile wide. One can start on this plain, near Shasta, and travel southeast *four hundred and fifty miles* in a nearly straight line, without crossing any hill of any considerable height—that is, if a road is run near the rivers.

The extreme north end rises in a table-land a few hundred feet high, but the valley does not taper to a point—it is cut off nearly square, where it is at least thirty miles wide, by the mountains that extend across the north part of the state. But the eastern edge is modified by a range of hills that stretches east from Lassen's Peak, down into the valley, way to the river, near Red Bluff, so that the upper end of the plain spreads out above it, something like a letter T. Now these hills mentioned are mostly of lava and need more than a passing notice, for they impart features to the landscape so unlike anything else that I must make these preliminary explanations so that you may understand what I will have to write for some time to come.

Lassen's Peak, and in fact, that whole part of that chain, like Mount Shasta, is a gigantic extinct volcano, perhaps about twelve thousand feet high—a volcano not only much higher, but vastly greater in every respect of magnitude and effect than Etna. It is flanked by a considerable number of smaller cones, old volcanoes, from one thousand feet high, up to that of the main peak itself, many of these cones being much higher and greater than Mount Vesuvius.

Here, in a former age of the world, was a scene of volcanic activity vastly surpassing anything existing now on the earth. The materials from these volcanoes not only formed the mountains themselves and covered the foothills, but also came down on the plain for more than a hundred miles. Sometimes volcanic ashes covered the whole region many feet thick, then sheets of molten lava would flow over it, hardening into the hardest rock, then ashes and lava again. Thus were formed beds of enormous thickness, regularly stratified, descending with a gentle slope toward the Sacramento River, and even crossing it in one place near Red Bluff.

But all volcanic action ceased ages ago, and the snows and rains falling on the high lands about Lassen's Peak formed streams which radiate from it. They have worn deep canyons, channels, in this lava, often a thousand feet deep, but generally less. Between these are table-lands, sometimes strewn with loose bowlders of lava, at others showing a surface of nearly naked lava with only enough soil to support, here and there, low cragged shrubs and a few herbs during the wet parts of the year.

Saturday, October 4, we went on north to near the mouth of Battle Creek. It rained a little, and we camped at a miserable little tavern, where we took our meals. A part of our way led over a table, perfectly treeless, and for six miles apparently as level as a floor. Bear Creek cuts a deep canyon in the volcanic strata, but Battle Creek cuts a greater and deeper one. About ten miles east, on the north side of Battle Creek, rose a regular volcanic cone, which we resolved to visit.

Monday, October 6, Rémond and I started, our pistols carefully loaded, for hostile Indians sometimes lurked about—we saw none, however. The road leads up a gentle ascent for ten miles, rising about 150 feet per mile—a table of lava all the way, in places thickly strewn with bowlders, in others more weathered into a soil supporting straggling bushes and trees. On nearing the cone, there is more soil, and as a consequence, more trees. The cone rises, a round steep hill, steep on all sides, the top apparently cut off, thus:

It is 2,500 feet high, and in the top there is a perfectly formed old crater, a funnel-shaped basin, as is shown by the dotted line in the outline sketch above. It is perfectly round, about 900 yards around it, and 250 feet deep, its sides steep, and covered with bushes. No water ever stands in it, owing to the porous nature of the rock, and large pine trees grow down in there, their tops not reaching nearly to the rim. It appeared to be a favorite resort of bears—the "signs" were everywhere. The rim was perfect on all sides. A tolerably good figure of this is given in one of the plates of the Pacific Railroad Survey, in Doctor Newberry's report.[1]

From the cone we descended into the canyon of Battle Creek, thinking that possibly it was cut in deep enough to cut through the lava. It is in places eight hundred or one thousand feet deep, but entirely cut in these sheets of lava. It was a terrible place to get mules down and back. We found it impassable, and had to get out again, which took us several hours.

The view from this cone was peculiarly fine. The great Sacramento Valley, the Coast Range beyond in the west—in the northwest, the rough mountains beyond Shasta, for a hundred miles—in the north, Mount Shasta, looming up grandly, all white and spotless in a fresh coat of snow—in the east, the rugged, broken volcanic chain of Lassen's Peak, Black Butte, etc.—while just about us was the desolate volcanic region just described.

The country to the south of Battle Creek was too rough for us to follow down on the east side of the river, so we crossed over, passed down to Red Bluff, recrossed, and camped the next night at Antelope Creek.

Here we were in the Sacramento Valley again, a plain with majestic oaks and fertile land. We camped at Mr. Dye's ranch, an old pioneer. He was a fine old man—had come into the state in 1832 and had settled on that ranch in 1842. What changes he has seen on this coast! It seemed like a romance to hear him tell his truthful story. But he has fallen into the hands of sharpers, and the sheriff was just attaching some of his property. I felt sorry for him. Only one other man north of San Francisco has so long occupied one ranch. Mr. Dye had a fine nursery of fruit, and we luxuriated in peaches, pears, and grapes.

We spent a day in examining the canyon of Antelope Creek, which is precisely like that of Battle Creek, and the Tuscan Springs, which I described before.

October 9 we came on to Deer Creek, Lassen's old ranch, originally owned by the man[2] whose name is given to Lassen's Peak. He was murdered about two years ago by the Indians. We got into camp about noon, and in the afternoon went back to the canyon of Deer Creek.

All the way from Red Bluff the plain on the east side of the river is from five to eight miles wide. Then the volcanic hills begin—all the streams emerge from these by deep canyons. Deer Creek runs in such a canyon for over a hundred miles, often for long distances impassable. We found no fossils, for the lava was not cut through. At the mouth of the canyon we found a cabin in which lived a white man and a squaw, but we saw no other Indians. We were back to camp by dark.

October 10 we were off in good season and came down the valley twenty miles and camped at Chico, on Chico Creek. As we come south the valley becomes more fertile, and more highly cultivated. Here it is ten miles from the river to the hills, of which about eight is most excellent land and produces immense crops.

Chico is a thriving little place. We camped in the private yard of Major Bidwell, the principal citizen of the place, and while there ate at his house. We had a pleasant time.

By the way, I have forgotten whether I have given you the height of Mount Shasta—it is 14,440 feet, *the highest land yet measured in the United States.*[3] I feel proud that I took *first* accurate barometrical observations to measure the highest point over which the Stars and Stripes hold jurisdiction.

San Francisco.
November 2.

My last letter brought me up to Chico, and here I will begin again. Here, as elsewhere, one man is often the town in heart and soul. Major Bidwell left the "States" in 1840, and arrived here in 1841. He *is* Chico. He is very wealthy now, very public spirited—owns a ranch of five leagues (over twenty-two thousand acres) of fine land in the Sacramento Valley, a large mill, store, etc., and is the spirit of the growing town of Chico. Unfortunately, he was not at home; but, knowing that we were coming, he had left orders for our entertainment.[4]

Around his yard cottonwood trees had been planted ten years ago—these trees have now an average diameter of two feet—showing how trees grow here with care. In this shady yard we pitched our tent, the most pleasant place we have seen for a long time. Back of us was a fruit garden of some thirty acres, teeming with peaches, figs, grapes, etc., of which we were invited to partake *ad libitum—ad nauseam* if we chose.

Chico Creek seems on the map a little short stream. It is not so—it heads back in the hills many miles, in the volcanic table-land so often spoken of before. Somewhere on this creek certain Cretaceous fossils had been found, which were expected to throw much light on the geology of the region—it was absolutely necessary to find and examine them. I only knew that some had been seen ten miles up the creek from Chico. On Saturday, October 11, Rémond, Schmidt, and I started on mules to visit this locality, with a young Indian from the ranch as guide. Four miles across the plain brought us to the hills. Here the creek emerges from a canyon cut into the volcanic rock. Up this we made our way for about three miles, sometimes by a cattle trail, oftener without, over rocks, through thickets of chaparral—all volcanic rock, no sign of any place where fossils could possibly be obtained. Moreover, the canyon became more abrupt, and our Indian pretended to know nothing more of the road and wanted to stop. I urged him on.

At last he stopped and told us that he did not wish to go any farther, that he was afraid of Indians, that four persons had been murdered in that immediate vicinity within a few months, that his own brother had been shot in the arm, that Indians might be lurking anywhere, and that he was afraid to go any farther. We found, indeed, that what he said was true. A teamster, on a wood road near had been shot in his wagon, and his horses killed. Two girls had gone up blackberrying, on horses, with a little brother; the girls were murdered, each one pierced by over thirty arrows; the boy was carried off, and his remains were found two weeks later, sixty miles distant, where they had tortured him. We were entirely without arms, for, supposing ourselves out of danger, we had not even our revolvers. Trusting, however, that the severe punishments the Indians had received after their last murders had driven them off—a band of "volunteers" had followed them for a hundred miles, and, after finding the remains of the poor tortured boy, had killed indiscriminately all the wild Indians they could find, male or female

—I resolved to push on, and after various mishaps, at last found the coveted fossils in the bottom of the canyon.

The volcanic deposits here were about eight hundred feet thick, lava and ashes interstratified. The stream has cut entirely through into the sandstones beneath, which teem with shells. They are fossils, but are apparently as fresh as if left on the beach but a few years ago—only imbedded in sandstone. Large masses seemed half made of shells. What convulsions of nature that locality must have seen since those animals lived in that ancient sea!

We climbed out of the canyon and took our route back over the hills, sometimes through dense chaparral, at others over tables of lava, which supported a scanty vegetation of cragged bushes or more cragged trees of the nut pine. Around the latter our guide kept a sharp lookout for signs of Indians, who gather the seeds or nuts of this species for food. Whenever he found where the Indians had been, he scanned every clump of bushes very anxiously. The tame Indians that live on the ranches, or among the whites, are much afraid of the wild ones, who treat them with terrible cruelties if they catch them; moreover, like most Indians, they are very cowardly.

When near the plain again, a fine gray squirrel ran up a pine. Our Indian got off his horse, and took a "sling" out of his pocket—much like those we used to play with when boys, a piece of leather suspended between two strings about two and a half feet long. He selected a pebble nearly as large as a hen's egg, placed it in the sling, and poised it over his head, holding the stone in the leather with his left hand, his right holding the string, so that the string was over his head. Suddenly letting go with his left, he twirled it twice around with his right hand. The stone flew like a bullet and knocked the squirrel out. The Indian stood on the lower side of the tree, so the animal must have been at least seventy or eighty feet above him. It fell among the bushes, however, and got away.

We got back all safely—tired, however. The next day, Sunday, we spent very quietly in camp, and luxuriated on fruit from the garden. It was a sad day for the grapes and figs.

Monday, October 13, we went on to Pence's Ranch, about twelve miles north of Oroville. First, we went down the plain, crossed the Butte Creek, turbid with miners' dirt, and, after about ten miles, struck east.

Pence's Ranch lies just back into the first tier of foothills, and we found it on some accounts the most interesting place to the geologist in the state, owing to the proximity of several different geological formations. Here we spent two days, days most important in results.

There is a peculiar feature about the landscape that I must attempt to describe, although I fear that I cannot well make you understand it. Several of these volcanic ridges terminate here, and before they sink to the plain curious table-lands are left. The whole surface of the country has been washed away in places, except where protected by patches of lava. A hill is thus formed, a perfect "table." This would be the profile of many hills seen, thus:

a-a and *b-b*, top of hill, level lava, very hard; *c-c*, plain; *d*, valley; *f-f*, soft strata of ashes lying under the lava.

These hills are sometimes long ridges, at others, near camp, mere round hills, the lava washed away on all sides, leaving those isolated tables. The tops are bare rock, or nearly so. They are nearly level, with a slight descent toward the valley, the same slope down which the lava once flowed. You cannot imagine what a peculiar feature these hills, or "table mountains," impart to the scenery of the landscape.

And while I am at it, I may as well enlarge on this and tell you more about that great feature, the table mountains of this state. The kind I have just been describing are caused by great sheets of lava having flowed over great districts, thousands of square miles together, and then being partially worn away in later times by the denuding actions of our terrible winter rains.

There is another kind, however, still more remarkable, more noted, and of immense pecuniary importance. The Sierra Nevada, as I told you, is a broad chain, from 60 to 150 miles wide, and from 6,000 to 13,000 feet high. Its *width* is the great feature. It appears to have been upheaved, and then furrowed by water into great canyons, valleys, and "gulches." The gold is disseminated thinly through the slate rock. This slate (or quartz veins in it, with gold) has been worn down, powdered by rains and streams, the lighter materials washed into the great valleys below, the gold left in the gulches because of its greater weight; and here it is found, concentrated as it were, the gulch being a natural sluice that has been worked by nature for ages—man has only to "clean up" the gold. These are the "placer mines" of the state.

Now, the lava flowed over this country *after* much of this denudation had taken place. Immense districts, which would otherwise be gold bearing, are barren because covered up by these lava deposits.

All through the Sierra there have been immense craters, or "vents," from which enormous quantities of lava have flowed, the streams streaking the slopes, and forming table mountains. This may be taken as a sample:

a-a-a, table of lava stretching from the mountains back toward the valleys to the west; *b-b-b*, the gold-bearing slates, always standing more or less on edge and dipping east; *c*, valley of the interior of the state.

But these are *narrow* hills, and if cut across, their shape is thus:

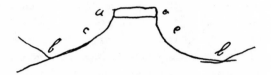

a-a, the flat, level top of the hill, naked rock, the edges abrupt precipices, from one hundred to six hundred feet thick; *b-b*, valleys of the present rivers and streams; *c-c*, steep sides, of gold-bearing slates, covered more or less with chaparral and bushes.

Now these are old lava streams, long and narrow. There is a table mountain in Tuolumne County that is *ninety miles long*, and never over a mile wide, generally less than half a mile! How could it have been formed? you ask, and the answer is the extraordinary part of the story. This lava, now the top of a mountain, ran in a stream in the bottom of a valley! Those hills of softer slate have been worn down until they are the valleys, while the harder lava has withstood the elements and forms now the top of the mountain. Most extraordinary fact! Once the outline was the line *a-a-a-a* in this sketch; now it is the line *b-b-b-b:*

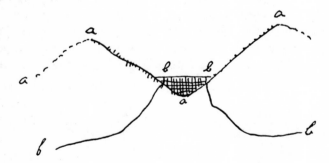

Now, upon this fact is founded an important element in one kind of mining—"tunneling"—which proves that these streams of lava flowed down the valleys of old rivers. The mountain is tunneled, often at an enormous expense, to find the old river bed, and from that get the gold. The following sketch will show my meaning:

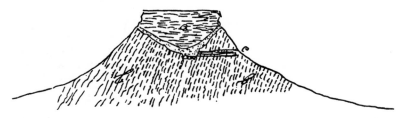

a, represents the lava top or "table"; *b-b,* the underlying gold-bearing slates, called the "bed rock," and also the "rim rock."

Between the table and the rim rock is a bed of gravel, once the soil of the surface before the molten streams of lava flowed on. This gravel is thickest in the old river bed, where they find the gold mixed with sand, gravel, and cobblestones, the whole generally cemented together into a hard, firm mass, like rock.

To get it, they pierce the rim rock by a tunnel, as at *c,* and herein is the great labor and risk. If the tunnel is begun too high, they strike above the old river bed, and hence must work at an immense disadvantage. If too low, they may pass through the mountain and not find the bed at all. This bed sometimes is beneath the middle of the lava table, at others near one side, so that should the tunnel be begun on the wrong side it must be carried to great length.

Under these mountains, you find all the evidences of the former river—not only sand, pebbles, and gold, but sometimes pools which have been filled with clay, most beautifully stratified, and containing leaves, bits of wood, etc. Some of these leaves are as plain and beautifully distinct as if their impressions were taken but yesterday in wax or putty. The bones of animals—mastodons, elephants, etc.—are also found; these, however, are rare. California, like eastern America, was once the home of these huge animals, and their remains are often found in the placer and hydraulic diggings. We have the perfect skull of a horse (extinct species) with teeth all in, found thirty-five feet underground—the teeth of elephants, mastodons, tapirs, camels, etc., and a few days ago the underjaw of a rhinoceros nearly perfect!

We remained at Pence's Ranch two and a half days, then went to Oroville. This is a smart mining town, purely mining, surrounded by claims on every side. We climbed a table mountain about three miles north, 1,300 feet high, the table of lava on top bare of all bushes, perfectly flat, but not perfectly horizontal, bounded by precipices on all sides, up which one could climb in only a few places.

Saturday, October 18, we came on to Marysville, quite a smart city, where we stayed over Sunday and attended church. I had decided that my best course now was to return immediately to the city, as we had settled certain geological points which otherwise would have required a longer stay. Monday I sent the party on, and on Tuesday I followed, arriving here the same night, the rest of the party getting in Friday, October 24. Here we disorganized the party, although there is some field work to do yet.

NOTES

1. *Pacific Railroad Reports*, Vol. VI, Pt. II, 26.
2. Peter Lassen, born in Denmark in 1800, came to America in 1829. After a residence of ten years in Missouri he emigrated to Oregon and the following year came to California by boat. In 1843, while working for Sutter, Lassen became familiar with the Sacramento Valley. He applied to the governor for a grant of land on Deer Creek and became one of the first settlers north of Sacramento. Returning from a visit to Missouri in 1848 he brought the first Masonic charter to California. Lassen later sold his Deer Creek ranch and moved to Plumas County, where, in April, 1859, he was killed by Indians. Lassen Peak and Lassen County are named for him. The name was commonly pronounced "Lawson," judging from contemporary spelling.
3. The figure 14,440 was current for many years. It was occasionally written as 14,444 for popular impression. It was superseded by the figure 14,380, computed from observations of the United States Geological Survey in 1883, which, in turn, has given way to the United States Coast and Geodetic Survey figure, 14,162 (see note on page 323).
4. John Bidwell, born in Chautauqua County, New York, in 1819, moved

westward by successive stages until, in 1840–41, he became one of the organizers of the first party of Americans to cross the continent for the purpose of settling in California. Soon after his arrival in California Bidwell went to work for Sutter, first at Bodega and Fort Ross, later at the Hock farm in the Sacramento Valley. In 1849 he purchased Rancho Chico, on the Sacramento River, where he lived until his death in 1900. General Bidwell (brigadier-general, California Militia, 1863) was a Member of Congress (1865–67) and was the nominee of the Prohibition Party for the presidency in 1892. (Articles by John Bidwell in *Century Magazine*, November and December, 1890; Bidwell: *Echoes of the Past*, The Lakeside Press, Chicago, 1928.) See, also, *The Journal of John Bidwell* (note 4, page 189).

CHAPTER VII

CLOSING THE YEAR—A MISCELLANY

Tomales Bay—An Irish Spree—Tomales Point—A Long Walk to San Rafael—The Alameda Hills—Penitentia Canyon—Summary of the Year—Continued Financial Troubles—Wreck of the Paul Pry *—Earthquake—Christmas Amenities—Contrasts.*

San Francisco.
November 30, 1862.

NOVEMBER has been a month of most lovely weather—no rain here, save one or two slight showers, air generally clear, although often foggy in the morning, so warm that I have not had any fire in my room over three or four times, free from the intense winds of summer—in short, lovely weather.

I have been in the office a part of the time, and the rest occupied in excursions in the regions about the Bay of San Francisco. My health has been most excellent. My sickness in the summer, although more severe than that which afflicted any of the other members of the party, was not so lasting in its effects. I have long been entirely well. But Hoffmann came back to the city in a sad plight; he looked bad when we returned—pale, weak, haggard—but looks much better now. Schmidt was taken down again, and went to the hospital, but he has got out again. He has been out a week, but does no work yet.

I have made several foot trips in the region about the bay since the breaking up of the party. October 27 Professor Whitney and I left here and went to Benicia for a short trip in Napa and Sonoma counties. We footed it over the hills for two days, and returned to this city October 30. There was little to interest you on that trip.

November 3, with Rémond, I started for another in Marin County, northwest of here. We went up to Petaluma, and the next day went across, west, to a place known as Tomales, near the mouth of Tomales Bay. This is a region of low, rounded hills, near the sea, therefore cool. During the summer it is subjected to intense winds, so it is nearly treeless. It is the greatest place in the state for potatoes, both as regards quality and quantity. The number raised here is enormous and, as a consequence, Irishmen abound. General Halleck has a ranch near this place, which we were on.

At Tomales there are several houses, but the only one where we could get "accommodations" was a very low Irish groggery, kept by a "lady." The place was filled with the Irish potato diggers, all as lively as the poorest whiskey could make them. One Irishman had just made some two hundred dollars by a contract for digging, and was celebrating the event, freely treating—in fact, he was just at the culmination of a three days' spree. The "rooms" of the house were far from private, the beds not highly inviting, and the customers twice as many as the accommodations. Drunkenness, singing, fighting, and the usual noise of Irish sprees were kept up through the night. Much to my disgust I had neither "bowie" nor "Colt" along, so could not command the exemption from meddling which those companions would have insured. Now, I don't mind the discomforts of the *field*, of sleeping on the ground, of diet, dust, lizards, snakes, ants, tarantulas, etc., but from drunken Irishmen, from Irish groggeries, from "ladies" of that description, "Good Lord, deliver us!"

The next morning we were off early, went to Tomales Bay, and crossed to Tomales Point. The bay is a long narrow arm of the sea that runs up into the hills, surrounded by picturesque characteristic Californian scenery. The bay is pretty, and the number of waterfowl surpassed belief—gulls, ducks, pelicans, etc., in myriads.

We were set across in a small boat, after getting some bread and meat for a lunch, which we carried along. The point is a long ridge which rises to the south, where it forms mountains, cut by deep canyons, and covered with an almost impenetrable chaparral. We footed it up south all the afternoon. The whole of this point, as well as Punta de los Reyes, is one ranch. There are a few houses on it—dairy farms, which are rented—for the chaparral peaks are only a small part of the whole, the rest is mostly fine pasturage.

Here let me give you a morsel of California ranch-land history. A Spanish grant had covered two leagues (less than nine thousand acres) which was confirmed by the United States. Although the grant was for but two leagues, mention was made of certain privileges on the *sobrante* or "outside lands." Well, the lawyers (those curses of Californian agriculture), after getting the grant of 8,800 acres, then went to work and got all the rest, the *sobrante*, of the entire point, amounting to *over forty-eight thousand acres additional*, and now hold it! Before this last was confirmed to them, squatters had settled on various parts of it, especially along the bay, near its upper end. They were ejected, and the houses are now unoccupied.

This we did not know at the time. We attempted to get to a certain ranch, which at 4 P.M. was three miles distant. We got on a wrong trail in the woods and chaparral, which at night brought us down to the bay, about six miles from the head. All this part is skirted by high hills, which come down to the water, covered with an almost impassable chaparral and furrowed by steep ravines.

We came to a cabin—it was deserted and locked up. We started up along the shore—were in decidedly a bad fix. It got dark; we could not get back, for we had come over bad places we could not pass in the night. We knew not how far we could get on, but resolved to go as far as we could, for we could not camp. The tide was down, and when it rose it would cover the whole available space to the bluff. Sometimes in the soft, black, treacherous mud and water, sometimes on the sand, we pushed on. At last we struck the mouth of a canyon, saw a light in it, and found three men, woodchoppers, who had just come there, and had got into a deserted cabin. They gave us some cold potatoes and cold meat, all they had to eat, and we slept on the floor. A miserable night, but better than to have been out, for a heavy fog rolled in from the sea, and then it rained violently.

We were up long before daylight—to take advantage of the tide which was then low, for we could not get out if it rose—followed up the bay, and at nine o'clock struck a house and asked if we could get breakfast. You may well imagine that we were very hungry. The answer of the woman was, "Yes, but it will be a hard scratch." She, however, did the best she could (who can do more?), and our appetites made up for any defect in the dishes.

We concluded to strike across for San Rafael, twenty-two miles distant. Soon after leaving the head of the bay we struck up a deep, wild canyon, exceedingly picturesque, the bottom filled with heavy timber—magnificent redwoods, often ten or twelve feet in diameter, almost shutting out the light of the sky with their dense foliage; the nutmeg tree (*torreya*)[1]; and the laurel or California bay tree, its foliage so fragrant that the whole air was often impregnated with it. But notwithstanding the beautiful way, we grew both tired and hungry, at last got some bread and pie of a Chinaman, and pursued our way. I had a heavy bag of specimens, and I was more tired when I got to San Rafael, long after dark, than I have been any other day this summer. Two nights with poor rest, three days without a decent meal, and that day we had walked from before daylight in the morning until after dark at night—no wonder that I was tired, or that I ate so much supper that I had most horrible dreams afterward.

The next day we returned to San Francisco. I then made three trips in Contra Costa County and Alameda County on the east side of the bay—one with Hoffmann, two alone. I will only tell of the last one, taken last week.

Last Tuesday I left here early; a dense fog hung over everything, but it cleared up before noon. I crossed to Oakland, and then rode to Haywards, near the "San Leandro" of your map, then struck across the hills to the southeast toward the Alameda Valley.

I stopped that night at a solitary cabin in the hills. A portion of a ranch, about eight hundred acres here, of hill land, cut by deep and steep canyons, is fenced in one field. On this about ninety head of cattle are kept, mostly cows.

A man is placed here to take care of them, and in the summer make butter and cheese. He has a horse to drive the cows with, a large corral, where he milks them in the season, a milk house, and a cabin to live in. This last is about twelve feet long by ten feet wide, of boards, shingle roof, one window and one door. There is a bunk on one side, of boards—a sort of crib, as it were—for a bed. Everything looks cheerless, dirty, comfortless. Here he lives, alone—he might sicken and die and no one be the wiser for weeks. Here, let me say, the tales of adventures, of hardships, and more than that, and of discomforts, that I have heard related by such men would make a big book. This man was a Norwegian, could not read English, could get no books in his own language except a Bible, so it is no wonder that he said that "te nights pe tampt long, and tis life tampt lonesome." He got me some supper, bread of his own baking (without butter or sauce), and eggs. We talked a spell, then "retired." As he had no extra blankets, he went out and got an old ragged piece of canvas that had been used to cover up hay with until too rotten and ragged. It was very dusty and dirty, but kept most of the cold out.

We were out before light, and I went on my way. I got into a deep canyon, then climbed a hill about two thousand feet high, commanding a most grand view, then sank into a deeper canyon, and about noon emerged into a lovely little valley— the Suñol Valley—a little plain of perhaps 1,500 or 2,000 acres, studded with scattered oaks, large and of exceeding beauty.

Here I struck another cabin, containing a woman. This cabin was, of course, far neater. She was old, but clever, and got me some dinner. The family had lived near the mouth of the canyon near, but last winter's floods had carried off the house and all they had in it. Alameda Creek drains a large extent of country, and rose to a great river. The house was carried away in the night. The family, two men, two women, and a little girl, got into a tree, in the night, the fearful torrent roaring beneath, the rain falling in torrents. Here they remained until the next day in the afternoon when they were discovered and a rope was got to them. They were rescued by being dragged sixty yards through the water. The old lady described it as a fearful time.

Alameda Creek breaks through the hills by a canyon about six or seven miles long, and I think about 1,500 or 1,800 feet deep. The sides are very steep, rising to mountains on each side. I followed down this canyon and emerged on the plain, then footed it to Mission San Jose, and the next day returned to this city. It was Thanksgiving, and all the principal shops were closed—a great holiday. I took my dinner at my boarding house.

The fine weather still continues, but it is not probable that the rains will hold off much longer. I shall probably make one or two excursions more, but all the intervals are filled up in office work. I am now in a boarding house, but my old last winter's quarters will soon be vacant, I hear, and I may return there, but have not decided yet.

Mr. and Mrs. Ashburner have come out, and are now here; it is pleasant to see them. He is in the employ of a banking house here, to look after their interests in various mining speculations.[2]

We have our old bother about getting our pay, and in the meantime we are put to much expense in consequence.

San Francisco.
December 19.

IT is a dark unpleasant night, and I am blue enough over the late news from Fredericksburg. We have meager news of our terrible defeat there. We are, indeed, in a terrible struggle. Like the rest of our defeats, perhaps it arose from *politicians*. Whatever was the cause, it is sufficient to make one blue and sad.

The first week of this month I took a trip in the mountains east of San Jose, walked nearly a hundred miles, spent one night in a cabin where a family lived on a mountain ranch, but one room, where we all slept—man, wife, five children, and myself. I slept on the floor, the rest had beds. Surely, "half the world does not know how the other half lives." The food was yet poorer than the other accommodations. The next day, on my way to San Jose, I attempted to explore the Penitentia Canyon, got in, could not get out except by climbing a bank one thousand feet high and fearfully steep. I had to walk twelve miles after it, with a heavy bag of specimens, and the exertion of the climb and the walk after lamed me for a week.

Well, the field work for the year has ceased. I have been adding up my "peregrinations" in this state since I arrived twenty-five months ago, and the following are the figures:

Mule back	3,981 miles
On foot	2,067 miles
Public conveyance	3,216 miles
Total	9,264 miles

Surely a long trip in one state! This has been over an area 625 miles long in extreme length, and has been nearly all in the coast ranges. Probably no man living has so extensive a knowledge of the coast ranges of this state from personal observation as I have, but I have seen very little of the grand features of the Sierra. When our report comes out I anticipate that it will attract much attention in the scientific world, probably more than it will here.

I do not think the Survey will be continued another year, although we have the work only fairly begun. Various influences are at work that will give us much trouble to keep it alive. The two principal ones are that the work is in advance of the intelligence of the state, and is, therefore, not appreciated; and, a more potent one, that several prominent politicians have hoped to use the Survey for personal, private speculations in mining matters and have failed—they will oppose us. You have no idea of the political corruption here. If the Survey "goes in" this winter, I shall endeavor to get into a position here, at least for a time.

San Francisco.
December 27.

WHEN I wrote my last letter and sent it by the last steamer, I thought that I closed my journal for the year, but since that time new and stirring events have occurred, worthy of being recorded in my chronicles.

The first, and most important, turned up soon. I have told you of our troubles in money matters. Well, $5,000 had been promised surely in December, then the twentieth fixed. All looked forward to this day—all were in debt, all had much salary due. My salary at the end of this month if not paid, would be about $1,760 behind—a snug little sum. In the meantime, as I had got no money for a long time, I had run about $500 in debt, on part of which I had to pay 2½ per cent per month interest. The twentieth came—but instead of the money, a letter came from the scoundrel who is comptroller, telling us that we would not be paid until May! To say that we were *indignant* would be expressing it mildly. Indignation is not a strong enough word, but the matter can't be helped.

This so deranges our affairs that I have changed my plans, and may be home in March or April, if I can raise the money.

I set out immediately to see about a professorship in a college here. I find that I can, in all probability, get it, so will probably go home in the spring, work up my report, and return here in June or July—this is my present plan—and then keep in the service of the state if the Survey is continued and the state pursues a different policy. Otherwise, I shall accept a professorship. The latter will not pay so well, nor be so conformable to my tastes, but it is well to be on the lookout for emergencies.

The next thing is something else!

Monday, December 22, was a cold, raw, blustery day, with some wind and rain. I am well acquainted with a Mr. Putnam, an officer in the California Steam Navigation Company, a monopoly that controls all the central and coast steamer business. I often visit in his family—he has four lovely little girls, the eldest thirteen or fourteen years old.[3] The company was to launch a new and fine steamer, the *Yosemite*, which lay upon the stocks at the Potrero, about five miles distant. A little steamer, the *Paul Pry*, was to go down to see the launching and carry invited guests. I was invited, and managed to have invitations extended to Averill, Gabb, and Hoffmann. We left at half-past ten, with about 150 or 200 guests.

Owing to the bad day, only half a dozen or so were ladies; the rest were gentlemen, numbering some of the most noted men of the city, public officers, etc.[4] It was said that Montgomery Street was represented by the owners of over five millions of capital in that little party. Katie Putnam was placed in my charge, but her father came on board just as we sailed. The launching was a beautiful one, the first sight of the kind I had ever seen—it was successful, and all were delighted. The *Paul Pry* was headed for home again, and we sat down to a most sumptuous lunch, where cold turkey and champagne suffered tremendously.

In the harbor is a little island, called Alcatraz, a mere rock of perhaps three or four acres, rising in cliffs from the sea about 150 feet, crowned with a fort, the defense of the interior of the harbor and a most picturesque object. To increase the pleasure of the trip and allow longer time for the lunch, we steamed around this island, keeping a short distance off. Lunch was finished. The wind blew a stiff breeze and whitecaps rolled. I was on the after part of the boat. Suddenly there came a crash that startled everyone. "What is that?" "What is that?" Everyone stared, but the suspense was short. We had run on a sunken rock, and stove a tremendous hole in the steamer, through which the water rushed into the hold, and she began to sink astern. Now came such a scene of excitement as I never saw before. The steamer rapidly settled—if she slid off from the rock, she would sink in less than five minutes; if she stuck on, she might break in two in less time.

One of the lifeboats was immediately lowered and I went below and got Katie in the boat, into which several other ladies were placed. I then stood back. The next lifeboat was run alongside and the two remaining ladies were placed in that. The excitement began to be intense. The boat had its quota of passengers instanter, when one excited individual, a prominent official, in his excitement and anxiety for his safety, jumped into the boat, fell overboard, and in getting in again upset the boat.[5]

Now came the most intense excitement, the boatload of passengers struggled in the water, several others jumped in, excited men were rushing frantically about, and, to increase the confusion, the boat took fire in the hold, and to extinguish it the steam from the boiler was blown there, which filled the whole vessel. Most luckily the lifeboat was righted and all were rescued from the water, the boat was bailed out, the wet passengers were put in, and she pushed off.

We were scarce three hundred yards from the end of Alcatraz Island. Soldiers were seen running around and soon boats put out. A whaleship also, seeing our signals of distress flying, sent off a whaleboat for our relief. The steamer, meanwhile, settled astern, and her bow raised high out of the water, but she stuck together. Boats soon arrived, and in about an hour all were safely landed at the fort.

I really felt ashamed for my sex, for manhood, when I saw what arrant cowards some of the men were. About two-thirds were as cool as if nothing had happened, but some of the remainder showed a cowardice most disgraceful.

After I saw Katie safe in the first boat,[6] and saw the second boat rescued, I went back on the upper deck, away from the excited men who clustered around the gunwale to rush into the boats that should first arrive. If the boat sank, I did not want to be among such men. All of our party showed that they had seen Californian camp life; not one showed the least trepidation or concern—coolness would hardly express it, and yet that was it. While coats, hats, umbrellas, furniture, etc., were seen floating away from the wreck, none of us could tell of such losses. We laughed at Averill because he carried away a glass tumbler in his pocket as a memento of the occasion, and the rest in turn laughed at me because I came ashore with not only my old cotton umbrella, but also Katie's that she had left.

As we left in the last boats (in fact, I was in the last but one), I had a good chance to see the whole affair. I was all prepared to swim if the boat went down. Many of the scenes were ludicrous in the extreme. Men who were the loudest to talk, here were pale as death and rushed for the earliest boats, even before the women were in. I sat on the upper deck and watched two excited men. One took out papers from his inside coat pocket, pulled off his coat and threw it overboard—off with his boots, etc. I afterward saw him in the boat without coat or hat. Another stripped to his drawers; others kept near someone who they thought could swim!

Hoffmann, Gabb, Putnam, and Averill had gone before I left. All excitement had died down—the cowards had left in the earlier boats. It was raining furiously and the water had partly filled the cabin and was sweeping on the deck with every wave. As I went below to get into the boat, I met the reporter of the *Daily Bulletin*, with whom I was well acquainted.[7] He stopped me to get "the exact time" and other "items." He was scribbling away as if there was no danger of the boat either going to pieces or sinking. He came in the next, and last, boat. A steam tug, however, soon came and got off a part of the furniture.

We were some two hours on the island; it stopped raining and we had a good chance to see the fortress. The officers treated us with every attention, and we were finally brought back to the city on a steam tug. Thus ended our mild "shipwreck."

The steamer did not go to pieces as was expected—we last saw her with her bow high in air, her after cabins entirely under water, which came above her upper deck. She was got off the next day, and hopes are entertained that she may be repaired for $20,000 or more.

But excitements were not to cease. The next morning, December 23, at half-past five, the city was visited by the severest earthquake that has been felt for some seven years. To say that houses rocked like ships in a tempest would be putting it on too thick, but they did pitch about in an uncomfortable way. The sensation is so peculiar that I cannot describe it. Three shocks followed in quick succession, the whole lasting not over six or eight seconds. The first awakened me, and with it came a slight nausea. The bed seemed lifted, then it shook, and the house rocked, not as if jarred by the wind but as if heaved on some mighty wave, which caused everything to tremble as it was being heaved. A few feminine screams were heard in other parts of the house. I got up and noted the time, but it was an hour before I could sleep again. No serious damage was done—we hear of walls cracked, plastering shaken down, clocks stopped in many houses, furniture thrown down, and such things—nothing more. These mysterious quakings and throes of Mother Earth affected me as no other phenomenon of nature ever did.

Christmas Eve I went to midnight Mass in one of the Catholic churches. Christmas was a most lovely day, the city seemed alive, all seemed happy. I took dinner with Mr. Putnam,[8] and in the evening attended a large concert with Mrs. Ashburner. Yesterday I dined with Professor Whitney, along with Baron Richthofen,[9] a distinguished young Austrian nobleman and scientific man, now in this city. He has lately been in the East Indies, China, Japan, etc., and came from there here. We all went to T. Starr King's in the evening and had a most pleasant time.

Sunday, December 28.

IT is one of the most lovely days of the season. The sky is bright, and the air of matchless purity, the mountains fifty or sixty miles distant seem as clear as if but half a dozen miles away. All the city is in excitement over the capture of the California steamer, *Ariel*, by the British pirate, *Alabama*, or "*290.*"

I was at church this morning, an Episcopal church, all decorated with evergreens, and this afternoon it seems as if all the city was in the street.

The customs of Europe and of the East are transplanted here—churches are decked with evergreens, Christmas trees are the fashion—yet to me, as a *botanist*, it looks exotic. With us at home, and in Europe, the term "evergreen" seems almost synonymous with "cone-bearing" trees, and so the term is used here. Churches are decked with *redwood*, which has foliage very like our hemlock—it is called evergreen, but it is hard for the people to remember that nearly *all* Californian trees are *evergreen*. While at Christmas time at home the oaks and other trees stretch leafless branches to the wintry winds, *here* the oaks of the hills are as green as they were in August—the laurel, the madroño, the manzanita, the toyon, are rich in their dense green foliage; roses bloom abundantly in the gardens, the yards are gaudy with geraniums, callas, asters, violets, and other flowers; and there is no snow visible, even on the distant mountains. Christmas here, to me represents a *date*, a *festival*, but not a season. It is not the Christmas of my childhood, not the Christmas of Santa Claus with "his tiny reindeer," the Christmas around which clings some of the richest poetry and prose of the English language. I cannot divest my mind and memory of the association of this season with snowy landscapes, and tinkling sleigh bells, and leafless forests, and more than all, the bright and cheerful winter fireside, the warmth within contrasting with the cold without. So do not wonder if at such times I find a feeling of sadness akin to homesickness creeping over me, that my fireside seems more desolate than ever, and my path in life a lonelier one.

NOTES

1. California nutmeg, designated by Jepson as *Torreya californica* Torrey (*Silva of California* [1910], p. 167), but by Sudworth as *Tumion californicum* (Torrey) Greene (*Forest Trees of the Pacific Slope* [1908], p. 191), is not related in any way to the nutmeg of commerce. California laurel (*Umbellularia californica* Nuttall) is found at its greatest and most profuse development near Tomales Bay.

2. Ashburner continued for a number of years to be a consultant for the Bank of California.

3. Miss Marion O. Hooker, of Santa Barbara, writes, September, 1929: "The family of Samuel Osgood Putnam has lived in or about San Francisco ever since 1853 when Mrs. Putnam and her three-year-old daughter joined Mr. Putnam there. My mother's first experience of a wreck was at this age on the *Tennessee*, in process of arriving at her new home. Of the six children, three are living: my mother [Mrs. Katharine Hooker]; her sister (Mary), Mrs. Morgan Shepard, in New York; and Edward W. Putnam, in San Francisco. (Miss) Caroline Rankin Putnam, (Miss) Elizabeth Whitney Putnam, and Osgood Putnam have died."

4. "There were the Collector of the Port, Naval Agent, Surveyor of the Port, Postmaster, Captains Baby, Seeley, Barclay, Pease, one of the newly elected Pilot Commissioners, and other gentlemen familiar with salt water, besides Captain Winder and other officers from Alcatraz, leading merchants, lawyers, etc., and the ubiquitous representatives of the press—all anticipating a delightful excursion, and each evincing a disposition to cultivate those social amenities which are so indispensably necessary to real enjoyment on festive occasions of this character" (*Daily Alta California* [San Francisco, Tuesday, December 23, 1862]).

5. "Just as the boat was pushing off, the newly elected Pilot Commissioner aforesaid, at once perceived that there was no person aboard suffi-

ciently skilled to assume the management of the craft. He sprang to the rescue, and in doing so capsized the boat, pitching all hands, and heads, and bodies into the foaming briny deep" (*ibid.*).

6. Mrs. Katharine Hooker writes, September, 1929: "My father insisted on dropping me down into the first boat, among the agitated ladies, much against my will as I wanted to remain with him. Then and there a singular fact came to light—I was tucked in by the side of Mrs. Chenery with whom I had been wrecked once before, at the age of three, when the steamer *Tennessee,* headed for San Francisco from the Isthmus of Panama (1853), ran on the rocks just north of the opening of San Francisco Bay, having missed its bearings in the fog."

7. The San Francisco *Bulletin* (December 23, 1862) contains a spirited account of the wreck, not quite so facetious as that in the *Alta.*

8. Mrs. Hooker writes: "Professor Brewer . . . was a great friend of our family and a familiar visitor at our house. We were all very fond of him. Besides the long talks with my father, he was devoted to the younger children and I often picture him with a little girl on each knee, while they sang, and he trotted them with a vigor that never failed."

9. Baron Friedrich von Richthofen worked from time to time with the California State Geological Survey and was a lifelong friend and correspondent of Whitney's (Brewster, *Life and Letters of Josiah Dwight Whitney* [1909], p. 240).

BOOK IV
1863

CHAPTER I

IN AND ABOUT SAN FRANCISCO

Lecturing at San Jose—A Chinese Theater—The College at Oakland—A "Grand Explosion" at the Beach—Chinese New Year's—A Speculator—The Quicksilver Mines Again.

San Francisco.
January 11, 1863.

WELL, the holidays have passed. The day before New Year's was a lovely day, but New Year's itself was rainy enough. I made some calls, but the rain interfered seriously. We had little parties at our boarding house, both New Year's Eve and New Year's night—informal, pleasant little affairs.

The Emancipation Proclamation was hailed with gladness by a large majority of our loyal population. It is looked upon not only as a military necessity, but also as a grand step in civilization, as the great movement to unite the nation again by laying the foundation for the early removal of the grand bone of contention.

I was invited to deliver a lecture in San Jose on Saturday evening, January 3, and went up that day, arriving at 3 P.M. I expected to stop with Mr. Hamilton, but found his household in some slight confusion owing to the addition of a young daughter to the family that morning. All parties were doing well at the last accounts. I had a very good audience and I believe it was considered that the lecture was a very good success. My subject was "The Mountain Scenery of California." I was invited to the house of a wealthy citizen, where I stopped until Monday morning, and had a very pleasant time. Mr. Belden, at whose house I stopped, is an old pioneer—came here in 1841, while the territory was under Mexican rule, and his wife followed five years later. The changes they have seen in this state in the twenty years they have been here must seem like romance. Like other pioneers he made money, and unlike most of them he has saved a snug little sum of it and now lives in very comfortable circumstances, indeed.

I returned on the fifth in order to attend a meeting of the California Academy of Sciences, of which I am the recording secretary. On Wednesday evening, January 7, I was at Rev. T. Starr King's, and had a most pleasant time indeed.

I have told you much about the Chinese here—that there are thousands in the city, that whole streets look like a Chinese town so far as people and wares are concerned. Well, last Thursday evening, I attended the Chinese theater. Now, if I could correctly describe this it would make a most interesting letter, but unfortunately I cannot. Words would entirely fail. Whether it was opera, tragedy, or comedy, or a mixture of the three, I have no idea—I think it was perhaps a mixture—but it was all comical enough, and yet intensely interesting because of its extreme singularity, so very unlike anything I have ever seen before.

The place is extremely low, frequented by the lower classes of Chinese, and is in a rather poor building. A stage extends entirely across one side, raised about three feet above the floor, covered with mats, and without drop curtain in front or side scenes. Stage furniture stands upon the two sides, a table in the middle, and two doors closed by fancy Chinese curtains lead to two rooms behind the stage—the "behind the scenes" of the place.

The orchestra, five or six persons, sits at the back part of the stage. One holds a small metallic instrument, a sort of cross between a small gong and a flat bell; another has several blocks, of different sizes and shapes, upon which he beats with two sticks, making a *noise* but surely not *music;* another beats a drum, looking and sounding like a half-barrel tub covered with leather. A large gong hangs beside him, which he pounds in the "terrific" portions of the play. Another plays most of the time on a stringed instrument, in principle something like a fiddle with two strings, but entirely unlike a fiddle, or anything else describable, in both shape and sounds. In the more noisy parts of the play he beats a pair of huge cymbals—about as musical as would be the clashing together of two pieces of sheet iron. Another plays on a guitar-like instrument, or, by way of variety, lays this down and blows a sort of shrill clarinet. There is *system* in their music, but neither melody nor harmony. It could not be expressed at all by the characters we use for writing music. Such is the orchestra. It keeps up a sort of accompaniment to the whole play, but some of the interludes are "awful." Imagine a room in which one man is mending pots, another filing a saw, another hammering boards, another beating a gong, and two boys trying to tune fiddles, and you will have some idea of some of their grand efforts in the music line.

A strong odor of burning opium pervades the room, for a few are smoking that narcotic. Against the wall at one side of the stage is the idol, with some pictures, Chinese letters, etc., and a lamp burning in front of it. The acting is the most comical part of the whole, but I could only tell which part was considered funny and which pathetic by watching the effect on the audience; it could never be perceived from the play itself.

San Francisco.
February 7.

SINCE I last wrote I have received your letters of December. You ask about the pronunciation of San Joaquin. It is pronounced *San Waugh-keen'*—accent on the last syllable. The Spanish names in this state are generally pronounced in nearly Spanish style, thus: Marin (County) is known as *Ma-reen'*, San Jose as *San Ho-say'*, San Juan is called *San Huan'*, or *San Wan'*, etc.

The month of January is past, and lovelier weather could not be—but three or four rainy days—its contrast with the last year's January is a good commentary on the uncertainties of the Californian climate. In January, 1862, there fell in this city 24.36 inches of rain; in January, 1863, only 3.63 inches, or only *one-seventh* as much. There had fallen last winter, up to the first of February, over 38 inches; this winter only 6.5 inches.

Last Saturday I went across the bay to Oakland, where I visited the college and saw the president.[1] The college is a small-fry affair, but it is the best in the state, and the largest except a Jesuit college.[2] They are more anxious to have me unite with it than I had expected. I shall probably be offered a professorship, in which case I shall accept it conditionally. I had an offer a few days ago to go into Mexico and examine a silver mine for some capitalists. It went hard to refuse it, but I could not well go, and I did not wish to assume the responsibility of advising in the matter, where I have had so little experience, and where so much money might be lost if my conclusions should not be correct. Therefore, I declined.

Sunday, March 1.

NEARLY a month has elapsed since I wrote the beginning, and I have continually neglected finishing. I find it even harder to write letters in the city than in camp, for after writing all day in the office I feel but little like writing in the evening. But I will go on with my story.

A company is constructing a carriage road directly west of the city to the ocean, and at the ocean beach it passes around a high bold cliff, which it was necessary to blast. Directly under this, on a level with the beach, a cavern was found extending back some fifty feet. The papers announced that on Friday, February 6, a large quantity of powder would be put in this, the mouth closed, and "the whole cliff torn off." The "grand explosion" was announced to come off at 3 P.M., and hundreds went out to see it—in carriages, on horseback, on foot. Our whole party walked out. It was a tedious walk, some five or six miles of it being over sand hills, in which one sank to the ankles at every step. The point was reached, and all waited for hours. The great hole had to be filled nearly full of sand, and at last nearly all the crowd left. I, however, remained. Fifty kegs of powder were placed in the cave, the mouth filled with sand and rocks for some thirty feet, a long fuse laid, and at last, long after dark, the train was fired. The explosion was a failure so far as effects were concerned. It threw out an immense quantity of sand and rocks from the mouth, but nothing more—an immense volume of flame, a very heavy sound, and a disgusted crowd who got lost in the sand hills on their way back to the city. I caught a bad cold which I have not got over yet.

The next excitement was the Chinese New Year's, which came off the third week in February. The festival lasts two weeks, but the police grant them the privilege of firing firecrackers only three days. I do not know the reason for their burning so many firecrackers, but I believe it has some religious significance. I thought I had seen firecrackers before, but became convinced that I had not. All day Tuesday, February 17, there was a continuous roar of firecrackers. About sunset I strayed through the main Chinese street, where the wealthier merchants live and have their places of business. From the roofs of the houses the "crackers" were in progress. At home we see Chinese crackers only in small packs about four inches square and one inch thick, the crackers all of a size and red. Not so here; they have not only these small packs, but immense ones, containing vastly more. I have seen them over a foot long, with partly small and partly large crackers—the latter yellow and large and thick as a stout man's thumb, exploding with a noise like a musket. Most of the crackers are in bunches about three times as large as those in vogue among boys at home about July 4.

But to get back to my story—one scene will describe many. On the top of a store is a crowd of twenty or thirty men (Chinese)—packs of crackers are lighted, hurled in the air, and allowed to fall in the street. A part of the time twelve men are lighting and throwing out the packs—a hundred crackers in explosion at each instant, making a continuous roar that can be heard over the whole city. As twilight comes on, the night becomes more picturesque. The roar, not only of this place, but of a hundred other places in the city, the dense volume of smoke that rises from the burning powder, the crowds of Chinese in the streets below—all conspire to produce a grand effect. Some of the wealthier houses spend as high as $600 for firecrackers alone.

Wednesday, February 18, was still worse. The exploded husks accumulated so thickly in front of some of the houses that they took fire and the engines came out to extinguish them, the fire bell of the city giving the alarm! This I know seems a big story, but it is *true*. The police only allow the firing of crackers for three days, but I will venture to say that it would take many thousands of dollars to foot the bills. On one street a hundred bushels of exploded husks could have been shoveled up—another big story, but also true.

We have a lot of "Secesh" at my boarding house—among them a sister of Van Doren. The last night of carnival the boarders at the house (Virginia Block) proposed to have a little private masquerade. All the preparations were made, when some wag went to the Chief of Police and told him that "a grand Secession masquerade ball" was coming off. He stopped the whole affair—decidedly "rich," but it shows how loyal this city is.

Affairs of the Survey are very uncertain. I shall be home sometime during the year, but when I can't tell. I have been nominated to the Professorship of Chemistry at Oakland and will probably be elected in June. If I am I shall accept.[3]

San Francisco.
March, 1863.

WE had been desiring to examine a region of mountains a little beyond New Almaden, the great quicksilver region. A gentleman came into the office March 4 with some fine specimens of cinnabar (quicksilver ore) and was greatly excited thereat. A new mine has just been found, ore *very* rich, and very abundant—the old story—the farm upon which it occurs can be got very cheap, not over one-hundredth of what even a tolerably good mine is worth—the whole thing could be got for $6,000. He was intensely excited, wanted me to go right down and examine it, would pay me $150 for three days' work. Now, we are not allowed to take fees from private individuals, but I told him I would go if he would pay my expenses and those of an assistant for a week's trip in that region. He was delighted.

Thursday, March 5, early, found Hoffmann and me on the steamer for Alviso, our speculator along. He thrust five twenty-dollar pieces into my hand, begged me to accept it as a present, which I most virtuously refused, except thirty dollars to pay our expenses on the proposed trip. He had all the papers necessary to make the purchase. The company was formed—twenty-six shareholders, the stock to be —— dollars. He had the specimens of ore in his pocket, a lawyer engaged in San Jose to search titles, even pen and ink to make the Frenchman, on whose land the rich mine was, sign immediately, and blank paper on which I was to write a favorable report to the company. All the way up he could talk of nothing but "cinnabar," "quicksilver," and "ledges."

Alviso was reached and we took the stage to San Jose, where we hired a private conveyance to the mines. We found the farm and the Frenchman, but *not* the mine—no trace of one. The ore had come from the Guadalupe Mine. It was evidently a plot by which he expected to sell his land (for which he had only a squatter title) "unsight and unseen" to persons in this city, whose mining zeal should outrun their discretion. Yet he pretended that he had a mine—he would show it the next day. The end was that the speculator, after talking hard French, harder Spanish, but the hardest English, left most disgusted, his bright visions of sudden wealth gone, and we took up our way to the mountains near, promising to return in a week and examine a mine he would *then* show us.

This was near the Guadalupe Mine, and the next day, Friday, March 6, we visited that. As I have told you before, three quicksilver mines lie within a distance of six miles, the Guadalupe, the Enriquita, and the New Almaden. Of these, only the last has proved very valuable. We saw Doctor Mayhew, the superintendent and engineer of the Guadalupe; he said that the company had spent upward of $400,000 in prospecting. This mine is a good illustration of the uncertainty of mining quicksilver. The ore is found in three different conditions: as fine threads of the brilliant red cinnabar in the harder rock, called *jilo;* or as a red looking earth, known as *terres;* or in great chambers of solid ore, called *labors.* Now, of course, the occurrence of these last is the most desirable, but they are very capricious, following no regular law in their distribution. A year and a half ago a large *labor* was discovered in the mine, which is not yet exhausted, although they have taken out over 100,000 pounds of metal from it; yet, until the discovery, its presence had not been suspected, although one drift had passed within eleven feet of it and another had been worked to within *four inches of it* and then stopped. Years later this *labor* was discovered by accident in cutting an air passage from one part of the mine to another.

We visited the Enriquita Mine. It, too, is doing but little. The Dutch superintendent and his Irish wife received us kindly and treated us to lager beer. We pushed on our way and stopped at New Almaden, a mine of real value. Here we remained from Friday until the next Tuesday, exploring the region. We had intended to work south of New Almaden, but the very broken country and dense chaparral prevented us. A large region is thrown into high ridges and very deep canyons, the ridges from 1,500 to 3,700 feet high, but mostly about 1,800 or 2,000 feet, covered with a dense growth of almost impenetrable chaparral. We reached a few elevated points, from which we could map out the topography of the country.

NOTES

1. The College of California at Oakland was the beginning of the University of California, now at Berkeley. Samuel H. Willey, D.D., was at this time acting president.
2. Santa Clara College, founded 1851.
3. The following letter is quoted in Willey's *A History of the Colleg of California* (San Francisco, 1887):

"San Francisco,
Apr. 1, 1863.

Rev. S. H. Willey
Dear Sir:
YOUR favor of yesterday is received, informing me that the Board of Trustees of the College of California at a recent meeting had honored me with the election to the chair of the Professorship of Natural Science in the College. In reply, I am happy to accept the appointment, subject to the condition that during my connection with the Geological Survey, my first duty will be to serve that, and that the time I may devote to the instruction in the College shall be regulated by the wants of the said survey. I will at all times endeavor to advance the interests of the institution according to my abilities and opportunities. I am, sir, your obedient servant,

Wm. H. Brewer."

CHAPTER II

TEJON—TEHACHAPI—WALKER'S PASS

Excitement at Clayton—Hill's and Firebaugh's—Desperadoes—Fresno City—Kings River—Visalia—Desolation and Misery—Crossing Kern River—Storm on the Plains—Cañada de las Uvas—Fort Tejon—San Emidio Ranch—Indian Reservation—Liebre Ranch—Yucca Plants—Tehachapi Pass—Walker's Basin—Keysville—Walker's Pass—Indian Wells — Desert Discomforts — Greenhorn Mountain—Tule River—Millerton—Hornitos.

On the San Joaquin Plain.
Easter Sunday, April 5, 1863.

I sent my last letter in such a hurry that I did not have time to finish the account of our trip in the Santa Cruz Mountains, and now so long a time has passed that I will let it go. Suffice it to say, that we had a trip of ten days with some exceedingly pleasant features, had lovely weather, examined a region we had long been anxious to visit, measured the height of a number of mountains, some of them over three thousand feet high, had some most lovely landscape views in the clear spring atmosphere, stopped among the lumbermen and "bachelor rancheros," had dirty beds with fleas and bugs, and bad feed and fatigue, and finally got back to San Francisco safe and sunburned. There we had some work to do, and then a rapid preparation for the long trip on which we are now started.

It is desirable to get our Survey in such a form that at the close of the year it can be wound up and the results be in good shape, if such action be considered advisable at that time. The great metallic wealth of this state lies in the Sierra Nevada, and as yet we have not satisfactorily determined the most important problems relating to the geology of that mountain chain. To that we propose to devote this summer's labors.

This chain and the coast ranges unite, as you will see by the map, in the vicinity of Fort Tejon, some three or four hundred miles southeast of San Francisco. It is desirable to visit this region, which is supposed to throw some light on the structure of both chains. Owing to insufficient rains, the pasturage is *very* poor and we considered it impracticable, or at least very expensive, to get in there with our wagon and a larger party, so Gabb and I have undertaken to do the work on horseback, trusting to such accommodations as we can find or carry with us; and on this trip we have started. We have a few most indispensable articles, such as blankbooks, paper, extra shirt and socks, packed in our saddlebags; a few instruments, a blanket and oilcloth rolled up and carried behind our saddles—such is our outfit for a trip to extend over a thousand miles in the aggregate and to take us at least two months.

Wednesday, April 1, we left the city and went to Martinez. It was a beautiful evening—the hills about the bay were green and lovely. Thursday we went by stage to Clayton near Mount Diablo, where our mules were. The plain on the north was covered with young grass of the softest green; the majestic oaks scattered over the plain are just coming into leaf.

At Clayton we found all in excitement. Supposed copper mines and quicksilver mines had just been discovered—worthless—and great is the excitement. The town was filled, the one "hotel" crowded, speculators from all the adjacent country were on hand—one heard nothing but "stocks," and "feet," and "ledges," and "claims." We remained there that day, getting our mules shod, etc., and had the full benefit of the excitement. Among the crowd was one man on his way to the San Joaquin to buy a drove of hogs. He had been for ten years a hog drover for the butchers of San Francisco, and what mines were to other people hogs were to him. He talked "hog" all the time, and it was a relief to find one man not crazy about copper. While we were sitting in front of the hotel a Mexican came in with two small hogs *packed on a horse*—our drover was interested and admired the horse that would carry live hogs on its back. We left him the next morning, still talking about hogs, totally regardless of copper mines or quicksilver leads.

Friday, April 3, we were off in good season. I rode the horse and Gabb rode my mule, Kate. The horse was inclined to show his Californian nativity when first mounted, but he soon cooled down. It was cloudy. We passed over the low pass at Mount Diablo and struck down the canyon east. It began to rain. There were no houses to stop at, so we kept on and soon got beyond the rain. We struck out on the great San Joaquin plain, stopping at "Dutch Fred's," a shanty in a grove of oaks, dignified by the title of "Public House." Saturday, April 4, we pushed on. It was clear, and at times the snowy Sierra showed dimly in the distance on the east, while Mount Diablo and its range towered grandly on the west.

But this plain is the great feature. The mountains are so dim in the east that they are not conspicuous, so that the plain seems as interminable as the ocean, except on the west. The ground is already dry, the grass very scanty and seldom over two or three inches high. As it grew hotter a mirage flitted around us—a great lake of clear water continually before us for hours. Once the appearance was so deceptive that we took the wrong road, supposing that it led around the water we thought was ahead of us. It took us out of the way, so that we did not strike the right road for some miles. You can have no idea of this extraordinary optical and atmospheric phenomenon. The eye is so deceived that the understanding refuses to dissent from the apparent truth.

Both of us were somewhat stiff after our two days' riding, so a halt here over Sunday is very acceptable. It is Easter, but it does not seem so here—a quiet house on the river bank, far from the hills, on this monotonous plain.

Visalia.
Sunday, April 12.

MONDAY, April 6, we started on and rode thirty-one miles without any incident. We stopped that night at Hill's Ferry, a dirty place, where a drunken "Secesh" made an uproar the whole night. A few *vaqueros* were managing a lot of wild Spanish cattle, to get them on board the ferryboat, and gave us some specimens of horsemanship and lassoing that were decidedly fine.

Tuesday we came on to Firebaugh's Ferry—a long fifty miles, a weary ride, and we had nothing to eat the entire day. Nor was this all. In crossing a slough early in the morning, a mixture of very muddy water and very sticky mud, Kate and Gabb got mired. They got out safely, but both cut a sorry figure, wet and covered with sticky mud, the former from head to tail, the latter from head to foot. We pushed on, the sun came out and dried them, and the mud was brushed off.

Firebaugh's was even a harder place than Hill's. I ought to have mentioned that near our Sunday's stopping place a murderer had just been arrested, and that at Hill's four horses had just been stolen. When we got to Firebaugh's we found more excitement. A band of desperadoes were just below—we had passed them in the morning, but luckily did not see them. Only a few days before they had attempted to rob some men, and in the scrimmage one man was shot dead and one of the desperadoes was so badly cut that he died on Monday. Another had just been caught. Some men took him into the bushes, some pistol shots were heard, they came back and said he had escaped. A newly made grave on the bank suggested another disposal of him, but all were content not to inquire further. This semi-desert plain and the inhospitable mountains on its west side are the grand retreats of the desperadoes of the state.

It is needless to say that knife (newly ground) and six-shooter are carried "so as to be handy," but I trust that I will ever be spared any actual use for them.

Wednesday we came on to Fresno City—only eighteen miles, but there was no other stopping place for forty miles, so we had to stop. The country had been growing more and more desolate. We had left the trees behind at Hill's, except occasional willows along the sloughs, and this day, for sixteen miles we rode over a plain of absolute desolation. The vegetation that had grown up last year, the wet year, was dead, and this year none has started. Sometimes no living thing cheered the eye, nothing in sight alive for miles. We crossed the slough ten miles above Firebaugh's, where there is a ferry, and a solitary shanty stands, the only house for the eighteen miles. We then struck across the plain in the direction of Fresno City, without track or guide, and at last that place loomed up.

Fresno "City" consists of one large house, very dilapidated, one small ditto, one barn, one small dilapidated and empty warehouse, and a corral.[1] It is surrounded by swamps, now covered with rushes, the green of which was cheering to the eye after the desolation through which we had passed. These swamps extend southeast to Tulare Lake. We got into the place after much difficulty, but our animals had to content themselves with eating the coarse rushes that grew on the edges of the swamp. The cattle and horses that live on this look well.

Thursday we came on to Kings River,[2] forty miles. Two men came with us, and in taking us across the swamp by a "better way," the horses came through safely, but Kate got mired again and Gabb had to get off and wade it and lead her. A mule mires much easier than a horse, having so much smaller feet. This day's ride was even more tedious and monotonous in its features. The land was wetter, so was green, but we were nearer the center of the plain, which is here not less than sixty miles wide! We rode upward of thirty miles without any tree or bush—except once a single small willow was visible for two hours, but we passed nearly two miles from it; it was a mere speck.

Hour after hour we plodded on. The road was good but a dead level—no hill or ridge ten feet high relieved the even surface—no house, no tree—one hour was but like the next—it was like the ocean, but it depressed the spirits more. The coast ranges were dim in the haze on the west; the snows of the Sierra Nevada eighty to a hundred miles distant on the east were hardly more distinct; while north and south the plain stretched to the horizon.

About six miles from Kings River we struck a belt of scattered oaks—fine trees—and what a relief! For, except a few cragged willows, shrubs rather than trees, in places along the sloughs, we had seen no trees for the last 130 miles of the trip! We crossed Kings River, a swift deep stream, by ferry, and stopped at a house on the bank, the most like a *home* of anything we had seen for two hundred miles. The owner was a Massachusetts Yankee, and his wife a very intelligent woman—I noticed an atlas of the heavens hanging up in the sitting room—but she was a Kentuckian and smoked *cigarritos* as industriously as we did pipes.

Friday, April 10, we came on here, twenty-five miles, crossing an open plain of nearly twenty miles. The morning was clear, and the view of the snowy Sierra most magnificent. Tomorrow we push on, and anticipate a rough time for the next four or six weeks.

Visalia is a little, growing place, most beautifully situated on the plain in an extensive grove of majestic oaks. These trees are the charm of the place. Ample streams from the mountains, led in ditches wherever wanted, furnish water for irrigating. We have stopped here two days to allow our animals to rest and get inspiration for our trip ahead. The Indian troubles east seem quelled, so we apprehend no trouble on that score. The ringleader was killed near here yesterday morning, and another detachment of soldiers has just left to further punish those in the mountains.

One sees many Indians here in various stages of civilization, but generally rather low. On Kings River we visited a camp of the Kings River tribe. They were hard looking customers. They lived in a long building, made of rushes carefully plaited together so as to shed the rain, making a long shed perhaps sixty or eighty feet long. The children were entirely naked; the adults all with more or less civilized clothing. All have very dark complexions, nearly as black as negroes.

Fort Tejon.
May 5.

MONDAY, April 13, we left Visalia and came on to Tule River, about thirty miles, still along the plain, here very barren, but the Sierra rose quite near us. We stopped at a little tavern and store, with miserable accommodations, but poor as it was, it was to be our last for near a hundred miles. I spent the evening in reading *Les Miserables*—a returned miner was reading *Shakespeare*. I question if the next generation here will care for Shakespeare, or any other author, growing up in ignorance, far from school, church, or other institution of civilization.[3] The Tule River is a small river, easily forded, with wide stretches of barren sand on either side.

Tuesday we came on thirty miles and stopped at Coyote Springs, about six or seven miles from White River. The road this day was through a desolate waste—I should call it a *desert*—a house at Deer Creek and another at White River were the only habitations. The soil was barren and, this dry year, almost destitute of vegetation. A part of the way was through low barren hills, all rising to about the same height—in fact, a tableland washed down into hills. We stopped at a miserable hut, where there is a spring and a man keeps a few cattle. He was not at home, but his wife was, and she gave us something to eat, and we slept out upon the ground near our mules. The woman was a fat, ignorant thing, with four little girls who looked like half-savages; in fact, they were scarcely better than Indians.

Wednesday we came on thirty-five miles to Kern River, the most barren and desolate day's ride since leaving Fresno, and for thirty miles we saw no house. We continued among the low barren hills until we came near Kern River—here we had to leave the road and go down the river nine or ten miles to find a ford. We followed a few wagon tracks, left the hills, and struck down the plain. The soil became worse—a sandy plain, without grass, in places very alkaline—a few desert or saline shrubs growing in spots, elsewhere the soil bare—no water, no feed. We saw some coyotes (wolves) and antelope. Night came on, and still we found neither grass nor river ford. Long after dark, when we began to get discouraged and to fear we would have to stop without water or feed for ourselves or animals, we heard some dogs bark. Soon we saw a light and soon afterward struck a cabin. Here we found some grass, went into the house, made some tea, and then slept on the river bank. Here in a cabin lived a man, wife, and several children, all ragged, dirty, ignorant—not one could read or write—and Secessionists, of course.

Kern River is a wide swift stream, here about twenty or twenty-five rods wide, with a treacherous sandy bottom. I dared not risk our animals in it without seeing what it was, so I hired the man to cross it first with his horse. We crossed safely, but it was up to the horses' sides most of the way. That day, Thursday, we came on thirty miles to the mouth of this canyon. From Kern River we saw the mountains on all sides of us—the Sierra on the east, the Coast Range on the west, the two joining on the south—the high peaks, some of them capped with snow, rising like a vast amphitheater. The morning was cloudy and clouds hung over the highest peaks. In about six miles we crossed the last slough of Kern River, then struck south for the mountains, across a complete desert. In places there were patches of alkali grass or saline desert shrubs, in others it was entirely bare, the ground crusted with salt and alkali, like snow—the only living thing larger than insects for many miles was a rattlesnake that we stopped and killed.

The storm came up black behind us; it stretched like a wall across the valley from mountain to mountain; the clouds grew lower on the peaks. We hurried up our tired and jaded animals, who seemed almost worn out. The storm struck us; but instead of rain it was wind—fierce—and the air was filled with dust and alkali. It was fearful, like a sandstorm on the desert. Sometimes we could not see a hundred yards in any direction— all was shut out in clouds of dust. In the midst of this storm my tired horse fell flat with me and got away; luckily he was too tired to run, and I easily caught him again.

At last the rain set in and wet us thoroughly. It was night when we struck the mouth of the Cañada de las Uvas,[4] and found a house. They would not keep us at first, no accommodations, but at last we prevailed and stopped. We got some supper, shivered in the cold, then went to bed. Our host, a young man, with Indian servants, generously gave us his room. There were but two rooms in the house. He, a Spaniard, and two squaws occupied the other room—with a pack of cards and a bottle of whiskey to heighten the pleasure of a social evening.

The last few miles the plain had risen rapidly, until the mouth of the canyon where we were had an elevation of over a thousand feet. The country had improved on this slope, with some grass, while on striking the mountains all was green. All changed, and what a relief! We had ridden on this plain *about three hundred miles!*

Friday, April 17, we got up, put on our wet clothes, got some breakfast, saddled our wet animals, and started. It still rained some. We came up the canyon four miles to this place, Fort Tejon. The canyon is wild and picturesque and rises rapidly. At about four miles from the entrance it widens into an open, wide valley, and here, surrounded by mountains, is the old Fort Tejon. A pretty grassy valley, at an altitude of about three thousand feet above the sea, covered with scattered oaks of venerable age and size and of great beauty, with mountains rising around to the height of perhaps seven thousand feet, the valley and hills green—picturesque, beautiful, quiet—such is the site of old Fort Tejon.

The buildings, some fifteen or twenty, are neatly arranged, all built of *adobe*, but many are already falling to ruin. The old stables are crumbling, the corrals are down. After Uncle Sam had located here and had built buildings costing at least $60,000 to $80,000, spending here in all over half a million, of course it turned out that it is on a Spanish grant, and is private property. Under the virtuous reign of Buchanan the proof was got, and the grant was confirmed to a political friend—now a rank Secessionist—a Mr. Bishop, who got the buildings all built to his hand. As there was no further use for the fort, it was vacated two or three years ago.

Here we remained three days. One day we climbed the mountains near and got a grand view of the surrounding region—the coast ranges, the Sierra, the great valley and plain on the north, the distant desert on the east. Sunday I spent quietly here, and remained over Monday.

Tuesday, April 21, we started on a trip in the mountains west. It rained a little. A party was prospecting for silver some twenty miles west; one of the party was here and piloted us to their camp by a wild trail through the mountains. The miners' camp was at an elevation of perhaps six thousand feet, by a spring, under a grand old live-oak tree. This tree was a beauty, over twenty-eight feet in circumference in the smallest place, with wide spreading branches. High, rugged, and rocky peaks rose on all sides, with more or less snow. We had a hearty dinner of venison, and then sat by the bright and cheerful camp fire, reminding us of old times. The night was cold, and ice froze an inch thick, but we slept warmly.

April 22 was still cold. We rode down the canyon about five miles to see a "copper lead"—found it no copper at all worth noting. A fierce snowstorm set in, so we returned to camp and there spent the rest of the cheerless, comfortless day. The squalls came on every little while. The stormy day was followed by a cold night. April 23, the morning was very clear, the peaks about us intensely white in their new garb of snow. We then rode down the canyon ten miles to San Emidio Ranch, at Mr. Alexander's.

April 24 we went up the canyon again to further examine the region. The higher mountains are of granite; outside are sandstones, lying in thick strata highly elevated, standing at a high angle. Some of these strata rise in steep, bare ridges, thrust up as it were 1,500 to 2,000 feet above the canyon, or 4,000 to 5,000 feet above the sea, forming grand but peculiar features in the landscape. The next day, April 25, our mule got away and we spent most of the forenoon in catching her, so we could not leave. We spent the rest of the day at the ranch house.

San Emidio Ranch is a large and valuable one, near the mouth of the canyon, commanding a lovely view of the plain, of Buena Vista Lake, and of the coast ranges. Two large springs water the gardens—fruits and grapes grow to perfection—yet nothing would tempt me to live in such isolation. We found three ladies there. Mr. Alexander was absent, but we had a letter from him to the ladies, and we had a hearty welcome, for visitors are scarce there. The ladies of the place were a sister and niece of Mr. Alexander and a visitor from a neighboring ranch (forty miles distant). The ladies had been there three years, had come out from New York for the romantic life on a Californian ranch, but were heartily tired of it now.

Mr. Alexander is a rich man—a single business transaction of his, three years ago, amounted to $90,000. He was now absent, but we met him on his return. He had just refused

$15,000 for a thousand cattle—his cattle were numbered by thousands and his horses by hundreds. How did he live? In what style? His house was surrounded by the houses and huts of his Indian and Mexican servants and *vaqueros*. The residence was of *adobe*, the floor of rough stones, the furniture rude. The ladies had been eight months without seeing any other white woman. They were twenty-eight miles from their nearest neighbors and forty from their next nearest. They were a hundred and forty miles from their post office (Los Angeles), and the same distance from a doctor—in fact, from any of the refinements of civilization. No wonder that even with all the wealth of their relative they thought California decidedly a humbug.

Sunday, April 26, we bade goodbye to our friends and came on here, twenty-eight miles, a part of the way by the plain. The only incident of note was our killing a very large rattlesnake—his rattles were broken and part gone, but he had eight left. The next day was rainy, and I spent it in writing up my notes and a letter to Professor Whitney.

Tuesday, April 28, we started and rode to the Tejon Indian Reservation, about twenty miles, and stopped at the house of Mr. Boschulty, the Indian agent. The reservation is in a corner of the great valley, the extreme southeast corner. The land rises in a gradual slope to the hills, a regular inclined plane, to the height of some 1,800 or 1,900 feet. It is the prettiest spot I have seen in this part of the state, with plenty of water and ample means of irrigation. The grass is green, the water good. The next day we rode up into the Tejon Pass about fifteen miles, a wild pass, but little traveled now. We killed another big rattlesnake, with nine rattles left, the rest broken off. We saw a large wildcat. I ought to have told of the game we saw in the San Emidio. We saw many deer, ten in one group at one time; another time one came so near that we got a shot at it with our pistols.

To return to the reservation. An Indian woman died, and

that night the Indians burned her hut, her clothing, and all her things in the fire—for her use in the next world. We stayed at the reservation until Friday and were most hospitably entertained. In the meantime the superintendent of Indian affairs came down with his secretary. His name is Wentworth, and he seems entirely unfitted for his responsible office.

We returned here on Friday; Saturday we visited a region ten miles northeast. We have now finished here, at least all that our time will allow, and tomorrow we shall leave. We are a hundred miles from Los Angeles, the post office where I will send this letter, if I can, to be mailed.

One more item about the Indian reservation. Lieutenant Beale had it placed there; last year he bought up a Spanish claim and now has it confirmed as a Spanish grant—a tract of twenty-two leagues or 87,750 acres, over 137 square miles! —that is the way to get land.

Murphy's, Calaveras County.
June 7.

MAY 6 we left Fort Tejon and crossed through the mountains south, to the Liebre Ranch. The pass is a very picturesque one, 4,256 feet high, with peaks on each side rising several thousand feet higher. The valleys are green, the region beautiful, but all changes on crossing the chain. We passed down a valley, dry and alkaline. Two little salt lakes were dry—the salt and alkali produced by the evaporation covering the ground like a crust of ice. For several miles we followed a line of earthquake cracks which were formed in 1856. The ground had opened for several feet wide, no one knows how deep, and partially closed again. We hear that these cracks extend nearly one hundred miles. In the valley we passed down a woman was killed by her house falling in the earthquake.

The Liebre Ranch belongs to Lieutenant Beale.[5] It is eleven leagues (about eighty square miles), and controls nearly all the water and, consequently, feed, for three times that amount, in fact for a hundred miles east. We had a letter from the owner to his Spanish *major-domo* (head man in charge), and we were hospitably entertained according to the Mexican fashion. We examined the region, crossed the valley into the mountains north, which command a wide view of the rough mountains south and the barren desolate deserts stretching east to the Colorado. This occupied us until May 9, when we bade adieu to Don Chico and followed down the edge of the mountains until near Lake Elizabeth, then struck across the desert north.

Now, I wish I could describe this desert so that you might really appreciate what it is—a great plain, rising gradually to the mountains on each side, sandy, but with clay enough in the sand to keep most of it firm, and covered with a scanty and scattered shrubby vegetation. It does not look so naked as much of the San Joaquin and Tulare plains that are not desert. The shrubs are of cragged growth, and belong to species which can stand the severest drought, for in some years there is scarcely any rain at all during the rainy season—a year may elapse and not an inch of rain fall.

The most noticeable shrub of the region is a species of yucca (*yucca gloriosa*); the Mexicans call it "palm," the Americans call it "cactus." It is neither.[6] It comes up, a single stalk, four to six inches in diameter, bristling with stiff leaves of the size and shape of a sharp bayonet. These die below as the plant increases in height, only a tuft remaining at the end. Sometimes it thus stands a single column fifteen to twenty feet high, with a tuft of leaves on the end. More often, however, it branches, and one sometimes sees a tree with numerous such branches, each with its tuft of leaves at the end, the trunk bare and without bark, apparently as dead as a fence rail. The plant is truly magnificent when in flower, spikes of flowers terminating the branches, with an immense number of blossoms, each nearly as large as a lily.

Another species of yucca, much smaller, grows on the drier

mountains and is now in flower. It has a tuft of bayonet-like leaves at the bottom, which live and grow many years, like the century plant. At last, a flower stalk eight or ten feet high springs up from the center in an incredibly short time—this bears a pyramid of several hundred delicate greenish-white flowers, each of the size and somewhat the shape of a tulip.[7] Of far less interest on this desert is the creosote bush,[8] every part of which stinks, making the whole air offensive. Sage bushes (*artemisia*) make up most of the vegetation, however.

This plain stretches off to the east, and many volcanic knobs rise from it, each perfectly bare of vegetation. At the base of one of these, fifteen miles from any other water, there is a nice spring of pure water. We found no feed, however, so pushed on. It became dark and we entered the mountains at Frémont's Pass, east of the Tejon Pass. The wind came down fearfully through a gap. At a late hour we struck water and a little grass at Oak Creek, and there we camped and spent a most cheerless night. The wind was cold and it howled, and our poor animals fared nearly as badly as we, for there was scarcely any feed.

Sunday, May 10, very early, we got some breakfast at a house not far off in this miserable spot, and then pushed on about eighteen miles until we struck fine grass and good water and camped about noon. We crossed the ridge to Tehachapi (on your map *Tah-ee-chay-pah*) Pass. From the ridge we had a wide view out on the desert. The Tehachapi Valley is a pretty basin five or six miles long, entirely surrounded by high mountains, and lies over four thousand feet above the sea. It is covered with good pasturage, and several settlers have located there.

We camped under some magnificent oaks; the one over our heads was over six feet in diameter, with a most magnificent head. We had tea, sugar, two tin cups to make our tea in, crackers, and jerked beef. We made our dinner and supper on this. Strange as it seemed to us, there had been religious service in the valley that day. A Methodist South preacher was visiting a settler and held "meeting" that day in a house, and all of the half-dozen families of the valley turned out. The settlers are "Pike" (Missouri men), and it was delicious to see the women dip snuff.

Monday, May 11, we kept on in a general northerly direction over a terribly rough trail, over mountains and across canyons. From some of the heights we had magnificent views. We worked west nearly to the great valley, then struck east again and camped in a wild picturesque spot on Pass Creek. With my revolver I shot a hare, which made us a supper and breakfast.

May 12 we went up the canyon a few miles, then over a high ridge to Walker's Basin. This is "The Park" on your map, and lies in the center of the chain about fifteen miles south of Kern River. The trail for the first twelve miles was horrible, a hill 1,000 or 1,500 feet high was so steep that we could scarcely lead our mules up it. The ridge is not sharp, but opens out into pretty little valleys with scattered trees and grass. The view of Walker's Basin from this ridge is lovely in the extreme—a pretty basin of three or four square miles, perhaps more, green as emerald, surrounded on all sides by very high, rugged, granite mountains, whose gray, bleak, and barren sides are in striking contrast with the green basin below. The creek on which the basin is, flows out through an impassable canyon. We met in the basin a band of Indians, mostly women and children, the widows and orphans of the "braves" killed in the recent battles on Kern River. They were a hard looking set. On passing out of Walker's Basin we fell in with a man on foot, and we camped together on a stream. Company is desirable in a hostile Indian country, where, sleeping out, the lonely traveler is liable to be "picked off" in the night.

May 13 our provisions were gone, and our supper the night before and breakfast that morning were even more meager than usual. We came on to Keysville that day. We had some grand scenery—on the east, high granite mountains, with enormous precipices of naked rock, while on the opposite sides were mountains nearly as high. On crossing the last high ridge, before sinking into the valley of Kern River, from the summit, a wide view burst upon us, all of rugged, barren, granite mountains, with but few scattered trees—a scene of wild desolation. At the forks of the Kern River is another little basin, like Tehachapi and Walker's basins—all lying in the middle of the chain, all surrounded by high, granite mountains.

Near these forks is Keysville—you can scarcely see the name on the map. It is the largest place within ninety miles—much more on the west, south, and east—yet it contains but eight houses all told. But it was the largest place we had seen in a month's travels. We got hay and barley for our hungry and jaded animals, and "square meals" for ourselves, so we stopped over a day. A store, with no floor but the ground, a saloon and "hotel" with ground floor and not a chair about the establishment, are the accommodations which we consider *princely* after our much harder fare. We have a bunk of boards to sleep on (using our own blankets). Our bedroom has no floor but that made by nature. Such were our accommodations—the best we had seen for some time, and the best we were destined to see for some time to come.

The spot is picturesque—the granite mountains are steep and high, and the Kern River runs through a wild, picturesque canyon. The place derives its importance from a few gold mines near, and from being on the road to the Slate Range, Coso, Owens River, and other mining districts. It was a peculiarly lively time there. Indian troubles in the spring had expelled the miners from some regions. The Indians were now nearly quelled and the miners were returning; but revolvers, rifles, and carbines formed a conspicuous part of every equipage, and I saw no other party so small as ours trusting itself alone. I resolved to go through Walker's Pass and return, as the great work of this year is to examine the geological characteristics of the Sierra Nevada, and it was thought best to cross the chain at as many points as is possible. We thought the Indians were quiet, and it was considered safe.

May 15 we started and rode up the pass twenty-five miles, following up the Kern River, which for this distance flows in a valley often a mile wide. The bottom is barren granite sand, however, and produces but little feed. A few houses along the way were deserted since the Indian troubles, but at twenty-five miles a man by the name of Roberts holds his place in spite of the Indians, and here we stopped. He has a field fenced with some feed, and we got supper and breakfast, sleeping out in the field near the house.

May 16 we came on to the Indian Wells, through the pass, twenty-eight miles. The valley narrows above Roberts', but the slope is gradual, although we rise so high. It grows more and more barren, the granite mountains grow less and less timber as we work east. The summit is reached, 5,302 feet above the sea, where a strong breeze draws through, and, although so high, the air is as dry as in a severe drought at home—no wonder that a desert stretches so far on the east. This desert basin is very high, from three thousand to four thousand feet. As we look out on the plain, mountains rise from it in every direction, but they are but detached chains—one can travel all the way to the Colorado River on the plain, which is everywhere a terrible desert. One of the mountain chains, fifty miles distant, is the Slate Range, now creating much excitement because of silver mines discovered there.

This desert is but a continuation of the one described in my last letter. It has less yucca, but like that, is covered with scattered bushes, such as can stand a dryness you cannot appreciate East. Doctor Horn, of Camp Independence, in Owens Valley, a perfectly reliable man, was stationed in that valley last August. He has kept a rain gauge, and from that time to this, *the rainy season,* there has fallen *in the aggregate less than a quarter of an inch of rain!* None can fall now until next winter, and possibly not then, and yet these shrubs can live in such a climate, if they get a good wetting every two or three years. A view, comprising a field as large, or nearly as large, as the state of Connecticut, *has not a single tree in sight.* Such are the Californian deserts.

We descended gradually and curved around the hills for about fourteen miles from the summit to the Indian Wells, where we stopped. A more god-forsaken, cheerless place I have seldom seen—a spring of water—nothing else. Here was a cabin, where we got our suppers. There was not a particle of feed. A little grass (bunch grass) that can grow in a dry climate, occurs four miles distant—it is cut with a sickle, tied up in little bundles, and sold at ten cents per pound—and cracked corn at fifteen cents per pound. This we got for our horses. You can well imagine that the inhabitants of such a place are not the most refined, but they hold a claim four miles distant— "ranch," they call it—where there is water and *some* feed, and keep horses that get used up on the desert.

On arriving at this miserable hole the first greeting was inquiries about Indians. Two men had reported seeing a band in the pass where we had come through. Several horses and mules had been discovered shot near there that day, and one man was missing. We had seen no Indians, although we had been well on the lookout.

We spent a most uncomfortable night there. The wind blew and filled the air with sand—sleeping on the ground, it seemed as if we would be stifled at times. I arose at half-past three, my teeth gritty with sand, face black, eyes full. The sand lay half an inch thick in places on my coat, which I had used on my saddle for a pillow. We saddled and came on twelve miles to a poor spring, keeping a brisk lookout for Indians. At the spring we got some breakfast, then pushed on—returned by the same route and reached Keysville on Monday night, May 18.

May 19 we left Keysville and came on over Greenhorn Mountain by the hardest wagon road I have ever seen that was much traveled. In places the road is so very steep that I cannot see how loads are got over it at all. We saw some government teams, where they had to double their six-mule teams to get an *empty* wagon up the hills. Most of the freight is packed over on mules.

The morning was cloudy, and it began to rain. The rain was mingled with snow, which almost froze us. It thawed as fast as it fell, wetting us thoroughly. We stopped at a place called Greenhorn—a place of two or three miners' cabins and a store—having made the fourteen miles in four hours. We stopped there over the rest of that day.

Wednesday, May 20, the rain stopped and we rode from Greenhorn to Tailholt (a good sample of Californian names)[9] over a very rough road, up and down, sometimes by wagon road and sometimes by trail.

On the west slope of the Sierra there is more timber—scattered oaks and pines—and some of the slopes are covered with chaparral. There is *some* more grass over the hills, and all tells of a moister climate. From some of the ridges we had wide and magnificent views of the foothills and the great valley on the west, but the coast ranges were shut out by the hazy air. Tailholt is on White River, among the hills, but we struck a very comfortable place to stop.

All these streams flow out into the valley through wild, impassable canyons, but in the mountains the valleys often spread out into little basins, some of which are of exceeding beauty; but being made of granite sand, none are very fertile. We passed two such, the Posé Flat and Little Posé Flat, both little basins, green on the bottom, with a clear stream of water, and fine old oak trees.

May 21 we came on to Tule River, out on the plain, and uninteresting—stopped there one day—then came on to Visalia and stopped over Sunday. At Tule River we were overtaken by some men with a band of horses and mules, which they were bringing from Indian Wells, where we had been. Several had been killed by the Indians, others wounded, and for safety they were bringing them to the Tulare Valley.

Our funds were getting low—no remittance—and we started for San Francisco, intending to inquire for letters at Millerton and Hornitos. We found that we must live economically to get back.

Monday, May 25, we came on thirty miles—all the way on the plain, so nothing of interest. The next day, thirty miles more to Millerton, which is on the San Joaquin near where the old Fort Miller was.[10] We struck into the hills just before reaching the place. There is some mining done there. The hotel where we stopped showed a truly Californian mixture of races —the landlord a Scotchman, Chinese cooks, Negro waiter, and a Digger Indian as stable boy. The San Joaquin is here a swift, clear stream, scarcely contaminated by the miners' washings. The country about Millerton is picturesque. Back of the town a high table mountain of lava is conspicuous for a great distance.

Two days more, all the way through the low foothills, brought us to Hornitos, near Mariposa. Here we found a letter from Professor Whitney instructing Gabb to go to San Francisco and me to go on to Columbia and meet him. Gabb left that night. So we can be said to have terminated that trip. Our poor animals were jaded, for we had ridden them *over a thousand miles* in less than two months—in fact, I rode 1,100 in the two months. I stopped a day at Hornitos to examine some copper mines there, all of which proved to be very poor; but the people are excited there on the matter of copper, and much money is lost and won in copper speculations.

NOTES

1. Fresno City was at a considerable distance from the site of the present city of Fresno, which was founded ten years after the date of this journey.

2. The members of the Whitney Survey, and many others at that time and since, fell into the error of writing the name of this river in the possessive singular—"King's." If the possessive indication is used, it should be in the plural, but the practice recommended by the United States Geographic Board is to omit it altogether. The river derives its name from the Spanish *Rio de los Santos Reyes,* "River of the Holy Kings," a designation given in 1805 or 1806 (Francis P. Farquhar, "Spanish Discovery of the Sierra Nevada," in *Sierra Club Bulletin,* XIII, No. 1 [February, 1928], 58).

3. Yet this region today has some of the finest schools in the state.

4. "Canyon of the Grapes," the present Tejon Pass.

5. Edward Fitzgerald Beale (1822–93), co-hero with Kit Carson at the Battle of San Pasqual, bearer of first California gold to reach the East, superintendent of Indian affairs for California, importer of camels to the West, surveyor-general of California and Nevada, minister to Austria, etc.

6. It is the tree yucca or Joshua tree. Jepson cites it as *Yucca brevifolia* Engelmann; Sudworth, as *Yucca arborescens* (Torr.) Trelease.

7. A form of the Spanish bayonet (*Yucca whipplei* Torrey).

8. *Larrea tridentata.*

9. "There are a number of stories about the naming of the second village on White River. I believe that the following account is correct. When the stage road was built across the mountains it left Dogtown to the south and passed near a mining claim. The miner who owned the mining claim had built a small cabin. On the door of his cabin he had nailed a cow's tail for use as a handle to pull the door open and closed. Because of this he called his mine Tailholt. Then the stage station, and later the village which grew up around it was called Tailholt.—The name Tailholt was used as a nickname for White River many years after the post office was established." (*San Joaquin Primeval—Uncle Jeff's Story,* arranged by F. F. Latta, printed at the press of the *Tulare Times,* Tulare, California [1929], pp. 55–56.)

10. Fort Miller was established during the local Indian troubles in 1851, but was soon abandoned. Millerton was for a time the county seat of Fresno County, but in 1874 the seat was moved to the new settlement at Fresno and Millerton became deserted.

CHAPTER III

THE BIG TREES—YOSEMITE— TUOLUMNE MEADOWS

Columbia—The Calaveras Grove—Big Tree Statistics —Sonora—Big Oak Flat—Bower Cave—Crane Flat—Descending into the Yosemite—Features of the Valley—Measuring Yosemite Falls—To the High Sierra—Mount Hoffmann—Lake Tenaya— Tuolumne Soda Springs—Mount Dana—Vegetation of the Sierra—Attempt on Mount Lyell— Fourth of July at Tuolumne Meadows.

Yosemite Valley.
June 21, 1863.

SATURDAY, May 30, I rode to Knight's Ferry, thirty-eight miles, crossing the Merced and Tuolumne rivers, both streams muddy from mining operations. Knight's Ferry is quite a pretty mining town, on the Stanislaus, and there is a fine bridge over the river. Sunday, May 31, I rode to Columbia, most of the way near Table Mountain, which forms the great feature of the landscape, its crest level, or with a low slope toward the plain, its top of bare rock, its sides steep, often in cliffs near the top.

Monday, June 1, I stayed all day at the quiet little place of Columbia, a very pretty mining town, entirely unlike anything in the East. It grew up at a rich placer region. Ditches sixty miles long bring water to wash the gold and irrigate the gardens. The gold has been mostly washed out, many miners have left, so many houses are empty of inhabitants. Many of the houses are embowered with climbing roses, now in full bloom, and the place is lovely. The underlying rock is limestone, which is worn very rough—knobs, poles, pinnacles, thirty feet high—once all filled in with soil, making a level flat. Here were rich placers, and much of the soil has been removed, leaving these ragged rocks bare. The effect is very peculiar.

Here our plans were formed for a trip across the mountains to Aurora, *via* Yosemite. Hoffmann is sent to Clayton for more animals, and in the meantime Professor Whitney and I will visit certain silver mines three days' ride distant. Accordingly, on June 2, we left Columbia and went to Murphy's by a very picturesque road, crossing the Stanislaus where it flows in a valley a thousand feet deep. At Murphy's we met Mrs. Whitney, and Professor Whitney heard of the severe sickness of a sister in San Francisco, so he dared not go on—so I rode to the Big Trees, the celebrated Calaveras Grove, fifteen miles from Murphy's.

There is a fine stage road over the hills, abounding in rather picturesque views. The way is mostly through an open forest and there is nothing to indicate the near presence of any such vegetable wonders—one is even inclined to doubt the truth of the guideboard which proclaims "$\frac{1}{4}$ mile to the Big Trees." One sees nothing to indicate any such thing, until crossing a little hill, one enters the valley and the grove. The first two trees, "The Sentinels," stand one on each side of the road, like two faithful sentinels, truly, and huge ones they are.

There are about ninety trees of this species in this grove, which is in a valley, sheltered from the winds. The prevailing trees are sugar pine, pitch pine, false cedar (called here *arbor vitae*),[1] Douglas spruce, and silver fir—all of which grow to a large size, often over two hundred feet high and ten feet in diameter, so the "big trees" always disappoint the visitor. They do not seem as large as they really are, but on acquaintance they grow on the mind, so that in a day or two they can be appreciated in all of their gigantic proportions. I measured over a dozen and found that the measurements popularly given are about correct. I will give the circumferences at three feet from the ground of some that I measured: Pride of the Forest, 60 feet; another, 65; Pioneer's Cabin, 74; General Scott, 51; Mother and Son, 82; Old Bachelor, 62; Empire State, 67; and so on. Many of them are over three hundred feet high.

The largest trees have fallen. The "Father of the Forest" is prostrate—it is said to be 116 feet around it—it was probably 400 feet high, although it is generally estimated that it was 450 feet high. It is burned in two, affording a fine opportunity to measure it in places. At 195 feet from the base the wood is 9 feet 9 inches in diameter inside of the bark! These measurements I made myself.

One gets the best idea from a prostrate tree. It lies like a wall fifteen to twenty feet high—a carriage might be driven on the trunk. One prostrate tree was hollow; it had burned out and the cavity was large enough for a man to ride through eighty feet of the trunk on horseback! This was called "The Horseback Ride." Only three days before I arrived it split in pieces and caved in. No one will ride through it again. A man rode through three days before and his horse tracks were fresh on the inside when I arrived.

A tree was felled a few years ago. It took four men twenty-seven days to get it down. It was cut off by boring into it with long augurs. This tree lies there still. The stump is six feet high and about twenty-four feet in diameter inside of the bark. A house is built over the stump to protect it. Stairs of twenty-seven steps carry one up on the prostrate trunk. At thirty feet from the base the diameter is considerably less, or only some thirteen and a half feet. The wood is perfectly sound to the very center. Professor Whitney carefully counted the annual rings four times over at this place (thirty feet from the base) and found the tree there 1,255 years old. It is remarkable that the wood should be sound that was already over eight hundred years old when Columbus set out on his voyage of discovery. The wood is much like red cedar in color and texture, only coarser, and is very brittle.

There is a fine hotel there, well-kept, and it is a most charming place to visit.² I stayed two days, with barometer, and found the height to be 4,800 feet above the sea. The bark of some of the larger trees is two feet thick, it has been measured *over* two feet. There has been much error written about these trees—that there were no young trees, etc. There are trees of all ages and sizes. Six or seven groves are now known, all large, all in valleys at altitudes of 4,500 to 6,000 feet. Large quantities of seeds have been sent to Europe, and one nursery in England has over 200,000 young trees. Probably more groves will be found in this state.

I returned to Murphy's, rejoined Professor Whitney, and examined the immediate vicinity. Murphy's is a pretty and thriving little mining village, with sluices and ditches on every side—everything telling of the only interest there.

Saturday, June 6, we rode to Cave City, over a very rough trail, visited a cave there and returned again to Murphy's. The cave is in limestone and is quite extensive, running at least half a mile underground. It was like all limestone caverns, with chambers and galleries, and stalactites, etc., but all was very dirty. Cave City is a dilapidated little village, the placers worked out, and decay is on every side.

I spent a quiet Sunday at Murphy's, and Monday, June 8, we rode to Sonora, returning through Columbia, killing a big rattlesnake on the way. At Sonora I found a tremendous pile of letters, the accumulation since April 1.

Sonora, like Columbia, is a mining town, but is a large village, having perhaps three or four thousand inhabitants. It is but three miles from Columbia, and there are several other mining villages very near—all owing to the rich placers of the region. Water for washing this dirt is brought from the Stanislaus River, over and through a very rough country, in a ditch over sixty miles long! And here is a Californian history. A ditch supplied water, but the miners thought the water rates entirely too high, so they would build an opposition ditch. It was estimated to cost $300,000; they built it, but it cost over *one and a half millions,* or over five times the estimated cost. It was scarcely finished before it was sold at mortgage sale for $150,000, *and bought in by the old company,* the one that this was to run opposition to, so both fell into the same hands. It was destroyed the winter of 1861–62 by the high water.

Hoffmann was to meet us at Sonora with extra animals, with which to cross the mountains. Leaving a letter for him with instructions for him to join us, we went on to Big Oak Flat, twenty miles, between the Merced and Tuolumne rivers—a rough but picturesque road, the last part running up a hill over three miles long.

Big Oak Flat is a little mining village, on a little flat. The "big oak" which gave name to the place has been undermined and killed, although attempts were made to protect it by town ordinance. It was nearly ten feet in diameter, and in the days of its glory must have been a grand tree.

Thursday, June 11, we spent in that neighborhood. From a hill north of the town I had a fine view—the surrounding mountains, steep, rugged, and covered with scattered trees, the distant higher Sierra capped with snow, Table Mountain for thirty or forty miles of its course, its top a gentle slope and apparently as regular as if graded by a race of Titans. The great plain, and even the coast mountains, were in view. We took tea with Mr. St. John, a lawyer of the place, and to our disgust received a letter from Hoffmann that his letter and instructions had been sent off by mistake in the express. To hasten matters I returned for him to Sonora, which took two days, getting back Saturday night.

Our plans were to cross the Sierra *via* Yosemite Valley, by the Coulterville Trail, and return by another, the Sonora Trail. This was expected to take us five weeks. Our party consists of four: Professor Whitney, Hoffmann, and myself, and John, a hired man. Besides our riding-horses, we have two pack-mules, which carry provisions and such blankets as we cannot carry behind our saddles. Of course our personal baggage is cut down to the very lowest figure—only what little we can carry in our saddlebags. So, too, our cooking arrangements must be primitive—four knives and forks, four tin cups, a coffeepot, a tin pail to cook beans in, a pan to wash dishes in, and a frying pan in which we fry meat and bake bread—no tent, no shelter, fewer blankets than we used to have in our wagon, compass, two barometers, and other instruments.

We left Big Oak Flat on June 14 and struck east by the trail into the mountains. We passed the flume of a ditch which brings water for the miners from the river forty miles above. This crosses a valley in a flume that is really stupendous, being 2,200 feet long and 280 feet high. Twenty-two wooden towers, the highest 288 feet, support the wire cables which suspend the flume. These towers, built of timbers, are wonders in themselves.

We camped that night on the North Fork of the Merced River, where there is a remarkable cave.[3] I have seen a number of caves, but this is the prettiest—a cave in the limestone, where there has been a large chamber, and the top has caved in, leaving it open to the sky. Trees grow in there, and there is a pool of very clear blue water. It is 109 feet down to this water. The effect is charming—the clear sky overhead, the trees in there, the ferns on the walls, the clear waters below, the dark chambers running from the main portion far back into the limestone.

Although the day had been hot, ice froze on our blankets that night. The lady owner of the cave and the house near, a French peasant woman, young, whose clear blue eyes did not look "spunky," was the heroine of a tragedy some months ago. She had a brutal husband, much older than herself, who, when drunk, often threatened to kill her, and one day attempted it, when she took his pistol from him and shot him. He lived two or three days, while she, womanlike, carefully nursed him.

June 15 we came on to Crane Flat, on the Yosemite Trail, a very rough country, through open forests of enormous trees, with some grand views of the mountains. In crossing a mountain over seven thousand feet high, we had a grand view of Castle Peak, and the higher, snowy Sierra. After crossing the summit we struck a pretty, little, grassy flat, called Crane Flat, where we camped—a pretty place, a grassy meadow surrounded by forests, and lying at an altitude of over six thousand feet.

June 16 Hoffmann and I visited another grove[4] of big trees, which is near—grand old trees, but not equal to the Calaveras trees—we saw no tree over twenty feet in diameter, but a stump was twenty-three feet inside the bark.

The trail grew rougher, across canyons, over hills. At last we crossed a hill, and a view of a portion of the valley burst upon us in all its beauty—the valley far below us, its peaks rising still higher than we were. We descended into the valley, passed the Bridal Veil Fall, and camped beneath the enormous precipice called Tu-tuc-a-nu-la.[5] The next day we crossed the river and went up the valley about three miles, and camped directly in front of the great Yosemite Fall. We see it continually when we are in camp; it is the first thing we see in the gray dawn of morning, the last thing at night.

I will give a general description of the valley, without following the order in which the points were visited. Yosemite (pronounced *Yo-sem'-i-tee*) Valley is not only the greatest natural curiosity in this state, but one of the most remarkable in the world. It is on the Merced River, about sixty miles from the plain. (It is on a branch and marked *Yohamite* on your maps.)[6] The stream is so large that it cannot be forded here. I tried it two days ago, but although in the best place, my horse was carried off his feet. We had to swim, and got a good wetting, so you can see that there is plenty of water. Below, the river runs in a narrow canyon; here, it widens out and is the Yosemite Valley proper—six or seven miles long, and about a mile wide, and over four thousand feet above the sea. The trails to it lead over the mountains.

We descend into the valley by a terribly steep hill—we have to descend nearly three thousand feet within four or five miles. We strike the level bottom, green and grassy, and pass up. The Bridal Veil Fall is in front (called also *Po'-hono*). It is a stream larger than Fall Creek at Ithaca, and falls *clear* at one leap over eight hundred feet! Some measurements make it 950 feet. The stream entirely dissolves in spray, which sways back and forth with the breeze, like a huge veil. It is vastly finer than any waterfall in Switzerland, in fact finer than any in Europe.

Now the valley begins to assume its characteristic and grand features. By the Pohono Fall rises the Cathedral Rock, a huge mass of granite nearly 4,000 feet high; while opposite is Tutucanula, a bluff of granite, rising from this plain perpendicularly 3,600 feet (we made it over 3,500 by our measurements). It rises from the green valley to this enormous height without any talus at the foot. You cannot conceive the true sublimity of such a cliff. It is equivalent to piling up *nine* such cliffs as the entire height of the walls of the ravine at Taughannock Falls.

We pass up the valley to our camp. South of us rises the Sentinel Rock, like a spire, 3,100 feet above us. There are the Three Brothers, the highest 3,443 feet above the valley, and in front the great Yosemite Fall, of which more anon.

As we pass up, the grandeur actually increases. The valley divides into three branches. We pass up the north, with the North Dome on one side, a huge dome 3,700 feet above the valley, and on the other the Half Dome, entirely inaccessible, like a huge dome split in two and only one-half remaining, its broken side rising in precipices 2,000 feet or more high, the whole rock being some 4,000 feet above the valley, or 8,000 feet above the sea. Here, in this valley, between these, is a lovely little lake, perhaps a quarter of a mile across.

The Merced River, however, comes down the central canyon. Here, between precipices of at least two thousand feet, nearly perpendicular, are two falls—the Vernal Fall, or Piwyac, and the Nevada, or Yowiye—the former four hundred feet high, the latter over six hundred feet. The quantity of water is large, so you may well imagine that they are sublime. The lower fall, Piwyac, is easily reached by a trail, and by ladders which carry us up to the top. Not so the Nevada Fall—that is more difficult—but we went up to measure it, to see if the heights previously given were exaggerated. We had a rough climb, but were amply repaid by the fine views. But few visitors ever attempt it.

But the crowning glory of the valley is the Yosemite Fall, before which I write this. The stream, fed by melting snows back, is a large sized millstream, say fifteen feet wide and two or three deep at the top of the fall. It comes over the wall on the north side of the valley, and drops 1,542 feet the first leap, then falls 1,100 more in two or three more cascades, the entire height being over 2,600 feet! We measured it yesterday.[7] I question if the world furnishes a parallel—certainly there is none known.

It was desirable to measure this, so yesterday morning Hoffmann and I started very early to climb the cliffs by a ravine a short distance up the valley. He wanted to get bearings for a map of the region. For six hours we had a terribly hard climb, and exciting from its danger. We climbed up cliffs that seemed almost perpendicular, but we measured the falls, and a cliff by its side, which is over three thousand feet high, the upper half of which is perpendicular. A large stone hurled over brought no echoes, it struck too far down for us to hear it. The view from the top was magnificent. The valley, over half a mile—in fact over three thousand feet—below us, its green plain spotted with trees which seemed flat bushes, the river winding through it, the granite domes around, and last of all, the snowy peaks of the higher Sierra just beyond, rising several thousand feet higher—all conspired to form a scene of grandeur seldom met with. I have seen some of the finest scenery of Switzerland, the Tyrol, and the Bavarian Alps, but I never saw any grander than this.

Well, we got back after fourteen and a half hours of severe fatigue, which makes my hand tremble today so that I can write only slowly and crabbedly.

There are two or three parties of visitors here, perhaps a dozen persons in all. There is a house where one gets pretty good fare at three dollars per day, but it is by no means a fashionable hotel with all of the modern improvements. The difficulty of access and the expense deter most of those who would wish to visit this place, yet a photographer[8] packed in his apparatus on mules and took a series of the finest photographs I have ever seen. A gentleman from New York, with three ladies and as many gentlemen, is here. We dined with him once. His expenses must amount to seventy-five dollars per day for animals, guides, packers, etc. But he enjoys the trip.

There is another fall in the south branch, said to be 1,100 feet, but we have not visited it. By Tutucanula is a little fall called 3,000 feet, but it does not drop clear much over 2,000 feet; then it runs over broken rocks. The stream is very small —it looks like a thread dangling from the high rocks.[9]

Soda Springs, on the upper Tuolumne River.
Sunday, July 5.

IT is a beautiful Sunday, and we are surrounded by the grandest scenery. Two of us are here alone now. I have attended to my botanical specimens, baked two loaves of bread, washed my clothes, and am at last ready for writing. My last was written in Yosemite Valley, two weeks ago.

June 23 we were on the trail very early, bade adieu to the valley, came up the terrible hill that leads out, where we rose about three thousand feet in less than six hours, and struck east. In crossing some of the ridges the views of the Sierra were sublime. We camped at Porcupine Flat, a pretty, grassy flat, at an elevation of 8,550 feet, surrounded by scrubby pines, and tormented by myriads of mosquitoes. The grand Sierra was in sight, a cluster of rugged peaks that Professor Whitney thought over fourteen thousand feet, and he became excited—could hardly sleep that night—a group of mountains so high, of which absolutely nothing is known! Subsequent measurements, however, prove that we overestimated their height.

June 24 we climbed a peak over eleven thousand feet high,[10] about five miles from camp, which we named Mount Hoffmann, after our topographer. It commanded a sublime view. Perhaps over fifty peaks are in sight which are over twelve thousand feet, the highest rising over thirteen thousand feet. Many of these are mere pinnacles of granite, streaked with snow, abounding in enormous precipices. The scene has none of the picturesque beauty of the Swiss Alps, but it is sublimely grand —its desolation is its great feature. Several little lakes are in sight. The scene is one to be remembered for a lifetime.

We returned to camp, took a hearty dinner, then fought mosquitoes with industry and built smokes beside our blankets before we could sleep.

June 25 we came on to Lake Tenaya, a most picturesque alpine lake, about a mile long and half a mile wide, of clear, cold, ice water, lying 8,250 feet above the sea. Its clear waters are very blue and very deep, no fish live in it, it is too high for the water to have air enough in it to support them.[11] Scattered pines are around the lake, or grow in the crevices of the granite. Above rise domes of granite, many of them naked, while patches of snow lie around on every side. Of course it freezes every night.

June 26 we came on to this camp, the Soda Springs, on the upper Tuolumne River, at the altitude of 8,700 feet. The river valley here forms a flat nearly a mile wide, green and grassy, while around is the grandest alpine scenery. It is a most lovely spot. Several mineral springs are here—cold water, charged with carbonic acid gas, giving it the name of "soda" springs. The waters are charged with iron and various other metallic salts, are highly tonic, pleasant to the taste, and would be worth a fortune anywhere in the old states.

A small party exploring for a road camped near us and brought us papers, with the bad news of the invasion of Pennsylvania. The night, as all our nights are here, was perfectly cloudless and clear, the moonlight on the snowy peaks around us forming a beautiful scene. The frost, as usual, settled heavily on our blankets that night.

June 27 we moved on and camped about three miles from the summit of Mono Pass, at an altitude of 9,800 feet—like the other camps, picturesque enough. I went on the summit of the pass to measure it—it is over 10,700 feet.[12] The night was cold and everything froze up hard, but a bright camp fire cheered the evening.

A high mountain rose on the east, from which we hoped to get important bearings, so on June 28 we were up early and Hoffmann and I started for it—over rocks, ice, and snow.[13] It is over thirteen thousand feet high and afforded a yet grander view than any we had had. The air was very clear, and we remained on the summit over four hours, taking bearings and barometrical observations. Hundreds of peaks were in sight, probably over fifty that are over twelve thousand feet. All north, west, and south was a scene of the wildest mountain desolation. On the east, at our feet, lay Mono Lake, an inland sea surrounded by deserts. A chain of extinct volcanic cones lay to the south of it, while alternate barren mountains and more barren desert plains stretched east to the distant horizon. It is not often that a man has the opportunity of attaining that height, or of beholding such a scene.

Professor Whitney was not well, so he did not go up, but on our return we gave him such a glowing description, and as we had found the climb so easy, he resolved to make the ascent.

So on June 29 I returned with him, thus making the ascent on two successive days. He thought the view the grandest he had ever beheld, although he has seen nearly the whole of Europe. Hoffmann measured another pass, through which it is desirable to get a trail.[14]

Here let me say that thus far we are on a trail that crosses the mountains, over which supplies are packed on mules during the summer to Esmeralda, freight being eight or nine cents per pound from Mariposa, Coulterville, or Big Oak Flat, all of which places pack over this trail. Trains of animals thus pass nearly every week.

We have found so much of interest here, among the rest finding the traces of enormous *glaciers* here in earlier times, the first found on the Pacific slope, that we have been detained much longer than we expected. Professor Whitney found that he would not have time to get around the whole trip with us, and as our provisions were nearly exhausted, he resolved to return and leave us to finish it alone.

Accordingly, June 30, we returned again to Soda Springs and made preparations for the "change of base." Our huge camp fire this night burned down three large trees, starting us from our beds. But the night was lovely indeed, the bright moon lending its charm to the scene.

Here let me make a digression and speak of the vegetation. In ascending the chain from the west the vegetation continually changes. You know that the Sierra Nevada is a very broad chain, being from fifty to one hundred miles wide, here perhaps about eighty miles. As we leave the plain, where there are but few trees, the grass is already dry and withered. In the foothills, to the height of four thousand feet, there are scattered oaks and pines. Above this come the fine forests of gigantic trees, all evergreen; the oaks have disappeared, and in their places are pitch pine, sugar pine, false cedar, and some Douglas spruce. Above this, at six thousand to seven thousand feet, are the noble fir and silver fir; we get above this, and at eight thousand to nine thousand feet a scrubby pine (*Pinus contorta*)[15] is almost the only tree—to the height of about 9,700 feet. Then, at about nine thousand feet, a low scrubby pine comes in, which extends up to eleven thousand feet or more but is a mere shrub. Its branches are very tough, and it will grow where fifty feet of snow falls on it every winter and lies on it for seven months in the year—in fact, never leaves the vicinity. Such is the *Pinus flexilis* of botanists.[16]

Above this, peculiar alpine plants come in, all very small, which extend to the summits of the highest peaks here, a little over thirteen thousand feet. Snow still lies in patches as low as 8,500 feet and is abundant above 10,000 feet. I have collected over a hundred species of mountain plants since I left Big Oak Flat.

July 1, early, Professor Whitney started for Big Oak Flat, taking John and one pack-mule for supplies, while Hoffmann and I packed the other mule and started up the Tuolumne River to see more of that mass of high mountains observed there and explore the country generally. No trail led up the valley, but we made our way up about ten and a half miles, and camped at the head of the valley,[17] at the altitude of about 9,200 feet. Here there is a grassy flat half a mile wide, which terminates just above in a grand rocky amphitheater. Sharp granite peaks rise behind to about thirteen thousand feet, with great slopes of snow, and pinnacles of granite coming up through, projected sharply against the deep blue sky. It was most picturesque, wild, and grand. And what an experience! Two of us alone, at least sixty miles from civilization on either side, among the grandest chain of mountains in the United States, whose peaks tower above us—we sleeping in the open air, although a thousand feet higher than the celebrated Hospice of the Great St. Bernard, the frost falling white and thick on our blankets every night. The high granite walls of the valley, the alpine aspect of the vegetation, all conspired to make an impressive scene. Just opposite camp a large stream of snow water came over the rocks—a series of cascades for 1,000 or 1,500 feet in height—a line of spray and foam. By our side a little rill supplied us with the purest of cold water. Such was our camp—picturesque, romantic; but prosy truth bids me to say that mosquitoes swarmed in myriads, with not one-tenth the fear but with twice the ferocity of a southern Secessionist. We "turned in" early; the bright moon lit up the snowy peaks grandly above the great rocky amphitheater, while the music of waterfalls lulled us to sleep.

July 2 we are up early. First, a hasty and substantial breakfast, then we prepare to climb the highest peak back. The frost lies heavy on the grass, and we are some distance before the sun peeps over the hill. Over rocks and snow, the last trees are passed, we get on bravely, and think to be up by eleven o'clock. We cross great slopes all polished like glass by former glaciers. Striking the last great slope of snow, we have only one thousand feet more to climb. In places the snow is soft and we sink two or three feet in it. We toil on for hours; it seems at times as if our breath refuses to strengthen us, we puff and blow so in the thin air.

After over seven hours of hard climbing we struck the last pinnacle of rock that rises through the snow and forms the summit—only to find it inaccessible, at least from that side. We had to stop at 125 or 150 feet below the top, being something over 13,000 feet above the sea, the barometer standing 18.7 inches. As we had named the other mountain Mount Dana, after the most eminent of *American* geologists, we named this Mount Lyell, after the most eminent of *English* geologists.[18]

The view from our point was the most desolate we had yet seen. All my adjectives are exhausted in my former descriptions, yet this surpassed them all for sublimity. A high precipice, perhaps one thousand feet nearly vertical, lies on the south side of the dome, forming part of a great amphitheater a mile across, of which two other similar granite needles form part of the sides.

We got back nearly used up, and were not long out of our blankets. July 3 we were in no hurry to rise after our yesterday's labors, so we lay "in bed" until the bright sun shone into the valley and melted the heavy frost from our blankets. It was a lovely morning—the sky of the deepest blue—at this great height—not a cloud in sight. We botanized, etc., during the morning, and in the afternoon returned here to Soda Springs. On our way we saw a large wolf, the only wild animal of any considerable size that we have seen here.

July 4 we celebrated by riding down the river a few miles and climbing a smooth granite dome for bearings, for we hope to work up a map of this region, of which no map has ever been made. The one you have is entirely incorrect, being made by guess. The view from this granite dome would be a grand one for a painter, although not so grand as those farther south that I have described. We are eight hundred feet above the river—for a foreground we have a series of smooth, low, granite domes, with the grassy flats by them, with a cascade of the Tuolumne in front, a dark pine forest beyond stretching up against the slopes, while beyond and above are sharp granite pinnacles destitute of trees and streaked with snow. Two peaks were especially fine—Unicorn Peak, a sharp needle over eleven thousand feet high, and Cathedral Peak, about the same height—the latter something the shape of a huge cathedral.

A great glacier once formed far back in the mountains and passed down the valley, polishing and grooving the rocks for more than a thousand feet up on each side, rounding the granite hills into domes. It must have been as grand in its day as any that are now in Switzerland. But the climate has changed, and it has entirely passed away. There is now no glacier in this state—the climatic conditions do not exist under which any could be formed.[19]

We rode back early, ate a tremendous dinner of preserved chicken, in regard to the day we celebrate. In the evening we built a tremendous bonfire of dead trees, but there were only the two of us to enjoy it. Today John returned with an ample supply of provisions.

NOTES

1. Incense cedar (*Libocedrus decurrens*).
2. This hotel was operated for many years until destroyed by fire in 1943.
3. Bower Cave.
4. The Tuolumne Grove.
5. El Capitan.
6. For origin of this and other names in the Yosemite region see Francis P. Farquhar, *Place Names of the High Sierra* (Sierra Club, 1926).
7. The figures of the United States Geological Survey are: upper fall, 1,430 feet; total falls, 2,565 feet.
8. This was C. E. Watkins, whose views of Yosemite still stand among the finest ever taken.
9. These heights are exaggerated. Illilouette Fall is less than 400 feet, while Ribbon Fall, near El Capitan, is 1,612 feet in sheer drop.
10. The United States Geological Survey map shows 10,921, minus 85 feet correction.
11. This has been disproved.
12. The Whitney Survey measurements in this region are from one hundred to several hundred feet too high.
13. Mount Dana, 13,050 feet (U.S.G.S.).
14. Probably the present Tioga Pass.
15. Locally called tamarack pine, but similar to, if not identical with, the lodgepole pine of wide range in the West—*Pinus contorta; Pinus murrayana;* or *Pinus contorta* var. *Murrayana.*
16. This is the white bark pine (*Pinus albicaulis*). *Pinus flexilis* is seldom found west of the Sierra crest.
17. Lyell Fork.
18. It is strange that they found it so difficult, for hundreds of people have since made the ascent without trouble. Perhaps they tried to scale the rock too far to the east. The altitude is 13,090 feet (U.S.G.S.).
19. Members of the Whitney Survey, led by Professor Whitney himself, were altogether too dogmatic about this. John Muir discovered glaciers in the Sierra in 1870, and later conducted experiments on the Mount Lyell glacial system. King discovered the Shasta glaciers the same year. There have since been found a number of glaciers of unmistakable character, such as those on Mount Ritter, Mount Darwin, and the Palisades.

CHAPTER IV

MONO LAKE—AURORA—SONORA PASS

Mono Pass and Bloody Canyon—Peculiarities of Mono Lake—Visit to the Islands—Aurora—Life in a Mining Camp—Walker's River—Sonora Pass—The Stanislaus—Sonora—Murphy's—Political Observations.

Lake Mono.
July 11, 1863.

JULY 6 Hoffmann and I visited a peak about four miles north of camp, to complete our bearing for this region. It is a naked granite ridge, about 10,500 feet high, and like all the rest commands a sublime view.[1] We found mosquitoes on the very summit, and on a table-land at about nine thousand feet, in an open forest, they swarmed in myriads. On this table we found a pretty lake, of very clear water, about a mile long—a most picturesque sheet of water. These little lakes abound here in the higher Sierra.

July 7 we were astir early, packed up, and crossed the Mono Pass, and descended to the plain near Mono Lake. It has now been just two weeks since we have been lower than 8,600 feet, having had a decidedly "high time." For two weeks we have slept in the open air, entirely without shelter, at altitudes sometimes a thousand feet higher than the Great St. Bernard in Switzerland, the nights clear, and freezing every night; now we descend to a valley, a lower region, but still higher than the highest of the White Mountains in New Hampshire.

After crossing the pass, the way leads down Bloody Canyon —a terrible trail. You would all pronounce it utterly inaccessible to horses, yet pack trains come down, but the bones of several horses or mules and the stench of another told that all had not passed safely. The trail comes down three thousand feet in less than four miles, over rocks and loose stones, in narrow canyons and along by precipices. It was a bold man who first took a horse up there. The horses were so cut by sharp rocks that they named it "Bloody Canyon," and it has held the name—and it is appropriate—part of the way the rocks in the trail are literally sprinkled with blood from the animals. We descended safely, and camped in the high grass and weeds by a stream a short distance south of Lake Mono. This camp had none of the picturesque beauty of our mountain camps, and a pack of coyotes barked and howled around us all night.

July 8 Hoffmann and I visited a chain of extinct volcanoes which stretches south of Lake Mono. They are remarkable hills, a series of truncated cones, which rise about 9,700 feet above the sea. Rock peeps out in places, but most of the surface is of dry, loose, volcanic ashes, lying as steep as the material will allow. The rocks of these volcanoes are a gray lava, pumice stone so light that it will float on water, obsidian or volcanic glass, and similar volcanic products. It was a laborious climb to get to the summit. We sank to the ankles or deeper at every step, and slid back most of each step. But it was easy enough getting down—one slope that took three hours to ascend we came down leisurely in forty-five minutes. The scene from the top is desolate enough—barren volcanic mountains standing in a desert cannot form a cheering picture. Lake Mono, that American "Dead Sea," lies at the foot. Between these hills and our camp lie about six miles of desert, which is very tedious to ride over—dry sand, with pebbles of pumice, supporting a growth of crabbed, dry sagebrushes, whose yellow-gray foliage does not enliven the scene.

July 9 we came on about ten miles north over the plain and camped at the northwest corner of Lake Mono. This is the most remarkable lake I have ever seen. It lies in a basin at the height of 6,800 feet above the sea. Like the Dead Sea, it is without an outlet. The streams running into it all evaporate from the surface, so of course it is very salt—not common salt. There are hot springs in it, which feed it with peculiar mineral salts. It is said that it contains borax, also boracic acid, in addition to the materials generally found in saline lakes. I have bottled water for analysis and hope to know some time. The waters are clear and very heavy—they have a nauseous taste. When still, it looks like oil, it is so thick, and it is not easily disturbed. Although nearly twenty miles long it is often so smooth that the opposite mountains are mirrored in it as in a glass. The water feels slippery to the touch and will wash grease from the hands, even when cold, more readily than common hot water and soap. I washed some woolens in it, and it was easier and quicker than in any "suds" I ever saw. It washed our silk handkerchiefs, giving them a luster as if new. It spots cloths of some colors most effectually.

No fish or reptile lives in it, yet it swarms with millions of worms, which develop into flies. These rest on the surface and cover everything on the immediate shore. The number and quantity of these worms and flies is absolutely incredible. They drift up in heaps along the shore—*hundreds of bushels* could be collected. They only grow at certain seasons of the year. The Indians come far and near to gather them. The worms are dried in the sun, the shell rubbed off, when a yellowish kernel remains, like a small yellow grain of rice. This is oily, very nutritious, and not unpleasant to the taste, and under the name of *koo-chah-bee* forms a very important article of food. The Indians gave me some; it does not taste bad, and if one were ignorant of its origin, it would make fine soup. Gulls, ducks, snipe, frogs, and Indians fatten on it.

Camp 127, on the Stanislaus River.
Monday, July 20.

On July 10, luckily, a man was going to the islands in the lake and invited us to go with him. Myriads of ducks and sea gulls live here in the summer and breed on the islands. Hens' eggs are worth from $1.00 to $1.50 per dozen in Aurora, thirty miles distant, so an old mountaineer conceived the idea of gathering these gulls' eggs for market. He dug out a neat canoe from a tree, and in the spring hired two Indians to help him collect the eggs. He drives a good business for two months in the spring.

This man wanted to go to the island to see about some things he had left there last spring. The day was nearly still—thunderclouds hung over the mountains, and occasionally there was a light breeze. At 3 p.m. we started; we had but little wind, and that in fitful puffs, so we did not arrive until nearly dark. There is fresh water in only one spot, on the larger island, in a little swamp and patch of tule a few rods in extent; here we camped and slept on the soft grass. There were two slight showers in the night, and some thunder, quite a novelty to us, for one very rarely hears thunder west of the Sierra.

July 11 we were up at dawn, a clear, calm morning. Clouds of gulls screamed around us. An early breakfast, then a tour of examination. These islands are entirely volcanic, and in one place the action can hardly be said to have ceased, for there are hundreds of hot springs over a surface of many acres. Steam and hot gases issue from fissures in the rocks, and one can hear the boiling and gurgling far beneath. Some of the springs are very copious, discharging large quantities of hot water with a very peculiar odor. Some boil up in the lake, near the shore, so large that the lake is warmed for many rods—no wonder that the waters hold such strange mineral ingredients. The rock is all lava, pumice, and cinders. At the northeast corner of the island are two old craters with water in them. The smaller, or north island, has no fresh water—it looks scathed and withered by fire. One volcanic cone, three hundred or four hundred feet high, looked more recent than any other I have seen in the state.

We sailed back to camp, stopping on the north shore, where some Indians (Pah-Utes) were gathering *koochahbee*. Along this shore many curious rocks stand up from the water, of lime tufa, made by springs in former times. They are of very fantastic shapes, often worn by the water into the form of huge mushrooms, ten to twenty feet high. I took a bath in the lake; one swims very easily in the heavy water, but it feels slippery on the skin and smarts in the eyes.[2]

The afternoon was showery and there was much thunder, and we made our preparations for an uncomfortable night. Soon after sunset the clouds came over the whole heavens, night set in intensely black and dark, the wind rose from the southeast, and the thunder roared incessantly. Soon the big drops began to fall; we took the hint and went to bed. The rain set in in earnest. The lake roared with the wind, which, with the thunder and rain and our lonely place and no shelter, made the situation peculiar. But luckily the rain was not so heavy as we anticipated—it did not wet our blankets through, yet I cannot say that it was a pleasant or comfortable night. At every movement of the body the water would run in from one's face, under the clothes, and feel *so* cold—one's hair and beard saturated—yet it was stifling to sleep with head covered. The tall grass and weeds around my bed nodded in the wind and slatted big drops of water over me. It rained at intervals all night, but in the morning cleared up after a fashion. It still looked bad, and we resolved to push on to Aurora, thirty miles, where we would find shelter.

We were soon on our way, most of the journey over a sandy desert on the north side of the lake. Mono Lake was at no very early period much higher and larger than at present. It has been gradually lessening and shrinking, leaving its old shores like terraces stretching around the present water, and great sandy desert plains covered with sagebushes, once its bed. We measured terraces 680 feet above the present lake.[3] About ten miles before reaching Aurora we struck into the hills north, the town being about fifteen miles due north of the east end of the lake, near what is called "Walker's Diggings" on your maps.

You doubtless have heard of the mining district of Esmeralda. Well, Aurora is the head of this—is a city, in fact, and the second in importance on the east side of the mountains. It has grown up entirely within two years, numbers now probably five thousand inhabitants, and is like California in '49. The town lies in a valley—canyon rather—and the first view is picturesque. The hills around are barren as a desert, with scattered scrubby pines and more scrubby cedars here and there—no grass, no anything to attract man, except the precious metals. With so large a population there are not accommodations for a fourth of the people. Thousands of "prospectors" come there poor as rats and expect to grow immensely rich in a few months—but, alas! most of them will either die here or leave still poorer. They live in such quarters as such a region will enable them to get up—hundreds of huts made of stones or dug in the earth with a canvas roof—such are the houses of the outskirts. The center of the town, however, is better—whole streets of wooden buildings, erected in the most cheap and expeditious manner, a few of brick. There are men at work, but far more are idle, for it is Sunday. We put our horses in stables at $1.25 per day each for hay alone. There is no hotel, but there are many lodging houses—we take "beds" at one and eat at a restaurant. I would much rather sleep out in my blankets if a clean spot could be found, although it is showery weather.

We got letters but, much to our disgust and disappointment, no *money*, and we have not enough for actual wants. We also got late and brilliant news from the armies. In the mountains we heard of the invasion of Pennsylvania. Here we heard that Lee's army is whipped, that Vicksburg is ours, and that gold is falling.

Aurora of a Sunday night—how shall I describe it? It is so

unlike anything East that I can compare it with nothing you have ever seen. One sees a hundred men to one woman or child. Saloons—saloons—saloons—liquor—everywhere. And here the men are—where else *can* they be? At home in their cheerless, lonesome hovels or huts? No, in the saloons, where lights are bright, amid the hum of many voices and the excitement of gambling. Here men come to make money—make it *quick*—not by slow, honest industry, but by quick strokes—no matter *how*, so long as the law doesn't call it *robbery*. Here, where twenty quartz mills are stamping the rock and kneading its powder into bullion—here, where one never sees a bank bill, nor "rag money," but where hard silver and shining gold are the currency—where men are congregated and living uncomfortably, where there are no home ties or social checks, no churches, no religions—here one sees gambling and vice in all its horrible realities.

Here are tables, with gold and silver piled upon them by hundreds (or even thousands), with men (or women) behind, who deal *faro*, or *monte*, or *vingt-et-un*, or *rouge-et-noir*, or who turn *roulette*—in short, any way in which they may win and you may lose. Here, too, are women—for nowhere else does one see prostitues as he sees them in a new mining town. All combine to excite and *ruin*. No wonder that one sees sad faces and haggard countenances and wretched looks, that we are so often told that "many are dying off"—surely, no wonder!

July 13 we visited one of the mines and were kindly shown through it and the mill. We looked a little around the region and got supplies to the limit of our means. But our want of money was *the* great drawback—we could not get all we wanted for the return trip. My pants were used up. I got a pair of cheap overalls, which I had to sew over entirely two days after, as they ripped the whole length of the legs.

Immense sums of money have been spent here in this region, an immense number of claims have been taken up, nearly twenty quartz mills have been erected, costing perhaps over a million; but whether the mines will ever pay is to me a question. I hear that *five thousand claims* have been recorded, and I question if ten have as yet paid the money expended on them; in fact, only a very small part of this immense number have ever been worked at all, only taken to speculate on. One or two mines *may* pay, the majority never will.

Where Aurora is, is as yet not known. *We* think it in California, but there is a dispute whether it be not over the line and in Nevada Territory. Most of the inhabitants wish it there, so that Uncle Sam will pay their bills of government, but like true American citizens, who will not be deprived of their *rights*, they vote in *both places*, in California and in Nevada, and their votes have thus far been accepted in *both*. Politicians may well exclaim, "Bully for Aurora!"[4]

Tuesday, July 14, we left Aurora, "dead broke." We had much calculation over our funds, to see if we had enough to get our breakfast before starting, found that we had, and two dollars over, with which we start for a trip of 130 miles through the mountains. We had not traveled two miles before we met a tollgate, which took $1.25 of our store.

We came on twenty-eight miles, mostly over a very barren country with no feed, until we struck the meadows on one of the forks of Walker's River—a pretty little basin, lying some 6,500 feet above the sea. Here we camped, and a relief it seemed to sleep in the open air. At Aurora, at our lodgings, one room was small, but it had twelve sleepers, in bunks, like berths on shipboard—here no such crowding.

July 15 we came on fifteen to twenty miles, over a rough, desolate country, of volcanic hills, with some pretty, grassy valleys in places, on the forks of Walker's River. We camped by a little lake, at the altitude of over seven thousand feet. During the day we passed a very large and copious hot spring; the water is boiling, and a stream large enough for a small mill runs away.[5] We spent one day here, going on a peak south of camp.

July 17 we came on up the pass and camped at a little grassy flat, near the summit of Sonora Pass, at the altitude of 9,450 feet—surely a high camp. The night was very clear, and it froze very hard. I slept very cold, which led to a severe rheumatic attack the next day.

July 18 Hoffmann and I climbed a peak, perhaps over twelve thousand feet. The views are not so grand here as on the other trail, yet are fine. We had hundreds of peaks in sight with snow on them. The view was very wide, extending even across the plain—the coast ranges could be dimly traced, although 150 to 180 miles distant. A severe rheumatism came on and I was glad to get back to camp. Another cold night followed, but I slept warmer. Water boils at about 192° or 193°. John complains that the beans must be very dry—it takes from eight to ten hours to cook them—he can't understand the low boiling point, nor understand the barometer sinking so low.

We had intended to observe Sunday, as we did not the last one, but both my companions wanted to come down to a warmer level and my rheumatism advised the same thing, so we came over the summit and on about twenty miles. The summit is about 10,300 feet; there is snow in banks for some distance, then we sink into a canyon on one of the upper branches of the Stanislaus River and follow down that. We leave the volcanic region and get into one of granite. The canyon rises in steep hills on both sides, two or three thousand feet above the river. We camped at an altitude of about six thousand feet, where it is warm and pleasant. Today, although Monday, we are having our "Sunday," and are spending the day in washing our clothes, writing, etc.

Murphy's.
Sunday Morning, July 26.

It is very early, not yet five o'clock in the morning, but I will write, for it is cool; soon it will be hot enough. We are back again among the haunts of civilization, and, with the open windows, I find it impossible to sleep with the many noises, the crowing of numerous roosters through the village, the barking of dogs, the lowing of cows—in fact, the many noises of dawn in this little village are in such contrast with the still solitudes we have had for the last six weeks, where one ever sleeps on the alert for the slightest sound, that all combine to prevent sleep. The effect is heightened by the fact that it is very hot—we are in a house instead of the free open air, and in a sultry, effeminate bed instead of the cool, vigorous earth. It is but just sunrise, and I have been awake for an hour and a half.

July 21 we climbed the ridge south of the canyon of the Stanislaus to get a look over the country—up a steep hill, perhaps three thousand feet above camp. The tops of all the hills here are capped with lava, with granite under, as if a rough granite country with rounded hills and long valleys had had beds of lava poured over it. Such, indeed, was the case. It covered the summits and streamed down the valleys, sometimes nearly a hundred milés, forming those remarkable features called "table mountains" that I have so often written about before.

Well, in the higher Sierra, along our line of travel, all our highest points were capped with lava, often worn into strange and fantastic forms—rounded hills of granite, capped by rugged masses of lava, sometimes looking like old castles with their towers and buttresses and walls, sometimes like old churches with their pinnacles, all on a gigantic scale, and then again shooting up in curious forms that defy description. We climbed on such a lava point, and had a wide view in all directions. We had a rather hard day's work.

July 22 we raised camp and came on near thirty miles. We left the valley, climbed a high hill, and then struck across a barren country—fine trees, but nothing else—no grass. We had some fine views and passed some camps of men at work on the road. A road is now in course of construction near the line of this trail—a great work—men are now at work on the heaviest part. Our measurements were looked for with great interest.

We descended the long western slope by following down ridges, a part of the time on table mountains or ancient streams of lava, a part of the time in valleys—at last we camped where there was water, but scarcely anything for the animals to eat. We did not have a pleasant camp, and getting into a hotter climate, I did not feel well. John complained, and Hoffmann looked and felt as if on the eve of a severe sickness.

July 23 we came on to Sonora, which we reached a little after noon. At about a dozen miles back from that place we began to reach the signs of an approaching civilization, and at last we struck a mining district. Immediately, all became life— it seemed like jumping from the wilderness solitudes to a thickly settled region. It grew hotter, the roads grew dustier— my head ached as if it would burst. It was over 100° F. when we struck Sonora. Our room, called the coolest in the house, showed a temperature of 103° F. in its coolest place. How it roasted us! Four days ago we were sleeping in the open air, with scanty blankets, where the soft mud at night would be frozen in the morning hard enough to bear our horses. Now, after so few days, in the dry dust, with this intense heat—it is most too much for flesh and blood, and no wonder that sweat streams from every pore.

At Sonora I got some money, and July 24 we came on here. I reëmployed John for another trip. We passed through Columbia and the mining region I have before described. The heat was most intense, but here, at Murphy's, we have a most excellent hotel.

Saturday, July 25, I kept quiet all day, wrote up notes, observed barometer, got horses shod, etc. Professor Whitney is at the Big Trees and came down to see me; we discussed work and formed plans. We will get ready tomorrow (Monday) and the next day start on a new trip across the Sierra. I have lost over thirty pounds since I left the city, but feel in excellent health; the rheumatism induced by the cold and exposure is entirely gone here in this hot place.

Professor Whitney on his return had a sad greeting—three dear friends dead—one, a very dear sister, in San Francisco, Mrs. Putnam, recently died, leaving six children, the eldest not fifteen, the youngest an infant of a few weeks—it is a most sad dispensation.

I have long refrained from writing any politics, and will not say much now, but a few words on affairs here may be of interest. This state is as loyal as any eastern one. She must be so. Secession would be a yet greater folly than with the southern states. With an immense territory, with a population of less than a million—one-half of which is in a district embracing only one-tenth of the state, the remainder scattered over a territory of over 160,000 square miles, with over 600 miles of seacoast—she would be as an infant; a tenth-rate power could annoy her and crush her resources. Yet, there are many Secessionists—enough to fill the minds of loyal citizens with just cause for anxiety. These may be divided into three classes: the first, small yet formidable, of desperadoes, who have nothing but their worthless lives to lose, and might gain something by robbery in case of an outbreak; second, a class of southern descent, whose sympathies are with the South, who do not wish to see civil war, yet who would glory in the fall of the Republic.

The third, and last, is the largest, and comprises a considerable party, mostly the Breckenridge part of the Democratic party, who at present control and really represent the Democratic party in this state. These *call* themselves Union men, but deny that the government has any power to put down rebellion constitutionally, that in fact the United States was always a "confederacy," but never a nation. Some of these are active Secessionists, but most are only talking men, who wield some power. Judge Terry, who killed Broderick, you remember, and is now at Richmond, is an example of this class, and many other men who once held office. Were they in power now it is not probable that they would commence active hostility against the Union, but they would throw every means in their power against the general government. Some of their papers openly rejoice over southern victories or northern defeats, and all of them put the worst possible light on all northern matters, such as praising the bravery of the southern generals and men, and implying the cowardice of the northern ones.

But the Union element is vastly in the majority, unconditionally loyal. This state has had so many southern scoundrels in office that the people are afraid of them.

NOTES

1. Ragged Peak (10,858 feet).
2. Mark Twain, in *Roughing It,* describes the effect of Mono Lake water on a flea-bitten dog.
3. See Israel C. Russell, "Quaternary History of Mono Valley, California," *8th Annual Report of the United States Geological Survey, for 1886–87.*
4. "September 16, '63, Aurora was found to be upwards of three miles inside the State (or Territory) of Nevada, instead of in California, as previously supposed" (Joseph Wasson, *Bodie and Esmeralda* [San Francisco, 1878], p. 49). For a lively description of this region see J. Ross Browne, *Adventures in the Apache Country, with Notes on the Silver Regions of Nevada* (New York, 1869); also articles by J. Ross Browne in *Harper's New Monthly Magazine* (August and September, 1865). There is a bibliography of the Mono Lake and Aurora region in the *California Historical Society Quarterly,* Vol. VII, No. 2 (June, 1928).
5. Fales Hot Springs.

CHAPTER V

TO CARSON PASS AND LAKE TAHOE

News from Yale College—Off on Another Journey—Silver Valley—Silver Mountain—A New Mining Town—Mining Fever—Volcanic Lava Caps—Hermit Valley—Winter in the Sierra—Volcano—Tragedy Springs—Silver Lake—Carson Pass—Hope Valley—The Placerville Road—Slippery Ford—Pyramid Peak—Lake Valley—Lake Tahoe—Trout—Mushroom Towns—Dealing in "Feet"—American River Trails.

Volcano, California.
August 13, 1863.

I was to start on a new trip Tuesday, July 27, but Hoffmann was again taken sick—too sick to travel—which gave me much anxiety. I got supplies, then rode up to the Big Trees, fifteen miles, where Professor Whitney was stopping, to consult him. As I have before said, there is an excellent, quiet hotel at Big Trees, and Mrs. Whitney and child are boarding there this summer. The house was full, with a pleasant company. The next day Professor Whitney and I repaired our barometers in preparation for a new trip. He decided upon plans, and I returned to Murphy's and sent Hoffmann to San Francisco to recruit.

That night I received letters from Professors Brush and Johnson, of Yale College, that took me all aback—an unofficial notice that in all probability next year I shall be invited to a professorship in Yale College—a position I would of all others prize. The college has recently been designated as the recipient of the Morrill Land Fund, donated by Congress to schools for the advancement of agriculture and the useful arts. If the country does not go to pieces Yale will receive some five or six thousand dollars per year from this and new professors will be required to fill the quota. If the place is offered me I shall surely accept it, gladly, and next year will turn my back to the Pacific and my face eastward and homeward once more. It is not yet a certain thing, however, but within the range of possibility.

July 29 I was up at half-past three and in the saddle by four, rode out of town in the early dawn, a lovely quiet morning, and arrived at Big Trees in time for breakfast.

Nora, Professor Whitney's child, was unwell, so he resolved to stop another day. I, however, went on and camped some fifteen or twenty miles east, killing a tremendous rattlesnake on the way. They grow huge here in the foothills of the Sierra. I took John and our pack mules along. A Doctor Hillebrand goes with us this trip as visitor. He is a resident of Honolulu, Sandwich Islands, where he has lived fourteen years. He is a man of much ability and importance there—a doctor, has charge of botanical and horticultural gardens there, a good botanist.[1] He and his family are spending a few months in this state and will return in a few weeks to the Islands. He carried a package of papers with him and collected plants on the trip. We remained there over the next day, and in the meantime Professor Whitney joined us, Nora being better.

July 31 we came on a few miles to Silver Valley, where we stopped—a pretty, grassy, mountain valley. We are already up to a considerable altitude. We examined and climbed a steep volcanic hill lying on the granite near, where volcanic explosions have been reported. We found no traces of such, but had a fine view, commanding a wide extent of country—the snowy Sierra lying along east of us, a wild, rocky, and desolate landscape. The family who live in Silver Valley leave in the fall.

Here let me anticipate. Recent reputed discoveries of silver ore at Silver Mountain, just east of the crest, on the headwaters of the Carson River, near Ebbetts Pass on your maps, has caused much excitement. An old emigrant road over the mountains, *via* the Big Trees, runs within ten or twelve miles of it, and now, suddenly, travel is pouring over this route. A stage runs part of the way, until the road becomes very rough; then a "saddle train," with a few pack animals, takes the passengers and their luggage to the promised land. So horses in these mountain valleys all at once become important, and at Silver Valley the stages stop and saddle trains start.

August 1 we came on to near the summit—first over a high ridge, then down a long, rough, and rocky hill to Hermit Valley, on the Mokelumne River, then up again to near the summit, camping at an altitude of over eight thousand feet, where we found grass, water, and wood. We had some fine views on the way and our camp was surrounded by rocky, volcanic pinnacles, picturesque and grand. We had a late supper and good sleep again in the cool air, for the frost settled heavily on our blankets.

August 2 we climbed one of the volcanic pinnacles near camp, over nine thousand feet high, from which we had a grand and very extensive view. We spent the afternoon in camp, and had a bright camp fire in the evening. The night was clear and there was no frost, although the temperature sank to 25° F. August 3 we came on a few miles, crossed the summit, and camped in a canyon about three miles from Silver Mountain. John rode into town, and again we luxuriated in fresh meat. August 4 we went into the town. Professor Whitney examined the mines. I took observations for altitude.

Silver Mountain (town) is a good illustration of a *new* mining town. We arrive by trail, for the wagon road is left many miles back. As we descend the canyon from the summit, suddenly a bright new town bursts into view. There are perhaps forty houses, all new (but a few weeks old) and as bright as new, fresh lumber, which but a month or two ago was in the trees, can make them. This log shanty has a sign up, "Variety Store"; the next, a board shanty the size of a hogpen, is "Wholesale & Retail Grocery"; that shanty without a window, with a canvas door, has a large sign of "Law Office"; and so on to the end. The best hotel has not yet got up its sign, and the "Restaurant and Lodgings" are without a roof as yet, but shingles are fast being made.

On the south of the town rises the bold, rugged Silver Mountain, over eleven thousand feet altitude; on the north a rugged mountain over ten thousand feet. Over three hundred claims are being "prospected." "Tunnels" and "drifts" are being run, shafts being sunk, and every few minutes the booming sound of a blast comes on the ear like a distant leisurely bombardment.

Perhaps half a dozen women and children complete that article of population, but there are hundreds of men, all active, busy, scampering like a nest of disturbed ants, as if they must get rich today for tomorrow they might die. One hears nothing but "feet," "lode," "indications," "rich rock," and similar mining terms. Nearly everyone is, in his belief, in the incipient stages of immense wealth. One excited man says to me, "If we strike it as in Washoe, what a town you soon will see here!" "Yes—*if*," I reply. He looks at me in disgust. "Don't you think it will be?" he asks, as if it were already a sure thing. He is already the proprietor of many "town lots," now worth nothing except to speculate on, but when the town shall rival San Francisco or Virginia City, will *then* be unusually valuable. There are town lots and streets, although as yet no wagons. I may say here that it is probably all a "bubble"—but little silver ore has been found at all—in nine-tenths of the "mines" not a particle has been seen—people are digging, *hoping* to "strike it." One or two mines *may* pay, the remaining three hundred will only *lose*.

It was a relief to meet Mr. Bridges, an old rambler and

botanical collector, well known to all botanists.[2] For twenty years he rambled in South America and explored the Andes for plants for the gardens and herbariums of Europe. He first sent seeds of the great Amazon water lily to Europe. He spent three months on the island of Juan Fernandez, came to California, and has been supplying the gardens of England and Scotland with seeds, and the herbariums with plants, from this coast, for the last few years. It was a relief to meet him and talk botany; yet, even he is affected—he has dropped botany and is here speculating in mines.

"Mining fever" is a terrible epidemic; when it is really in a community, lucky is the man who is not affected by it. Yet a *few* become immensely rich.

While visiting the region we kept our camp up the canyon some three miles from town. August 5 Professor Whitney and I climbed the Silver Mountain, a long walk of twelve or fifteen miles, and a very rugged climb. It is over eleven thousand feet high, the highest in the region, and commands a very extensive view. In the north we look into the Carson Valley and even see the mountains of Washoe; in the west is the summit of the Sierra Nevada, the volcanic crests here worn into fantastic outlines; on the southeast are the mountains near Mono Lake and about Aurora; and in the east chain after chain, extending far beyond the state line into the territories. I collected some fine alpine plants and, curious enough, among the stones on the top are myriads of red bugs—beetles—red and brilliant—a pint could easily be collected. We have seen them on many of the peaks, even on Mount Dana, at over thirteen thousand feet.

We got back to camp tired enough, but a hearty supper and sound sleep brought us out all straight again. August 6 we returned again to town and further examined the mines. From Scandinavian Canyon we had some views of the peak of Silver Mountain that are truly grand.

We had spent all the time we had to spare, so August 7 we

came up and camped at a little lake at the summit, to examine the pass. It is a most picturesque spot—a small lake of clear water, with green grass and trees around it, and snow banks lying here and there, while on the north of us are a series of volcanic ridges, rough and jagged in outline. Near camp was a most picturesque volcanic cone, its sides made up of strata of basalt, curved and inclined—the whole most beautifully columnar. There are several little lakes in depressions near the summit, all very picturesque, but *the* feature of the region is the volcanic cap to the mountains—those pinnacles in lava. Were I to see them truly represented on canvas or paper in views of any other country, I should have pronounced the views unnatural and grossly exaggerated. We examined two or three gaps along the summit—a matter of much importance here, for they are building a wagon road through—then finished our work and prepared to return.

August 8 we were off early and started on our return. In about six miles we struck the old emigrant wagon road, where I left the party and went up on the summit to measure the height of that pass, the rest going on. I got back to Hermit Valley that night, and stopped at a house there. The house is a mere cabin, although now a "hotel." Twelve men slept in the little garret, where there were ten bunks, called "beds." Two men have wintered here, and in the winter have killed several rare animals—two gluttons, stone martens, silver-gray foxes (so rare that their skins are worth fifty dollars each), large gray wolf, etc.

In winter the climate is truly arctic. A smooth pole in front of the house is marked with black rings and numbers to indicate the depth of snow. The deepest last winter was eighteen feet, but it falls much deeper at a greater elevation in the vicinity. And herein consists the marvel of this alpine region— the immense depth of snow in the winter (on the Placerville Pass the winter of 1861-62 the aggregate depth of snow was fifty feet at an altitude of seven thousand feet) and the little rain in summer, so the snow melts under the clear sky even where fifty or sixty feet fall nearly every winter. Stranger still, trees flourish, of several species, all of large size—there are at least five species of cone-bearing trees that grow where the snow lies abundantly for seven or eight months each year and it freezes nearly every night of the summer; and four of these species grow to trees four feet or more in diameter even under these conditions.

In winter the only way of getting about is on snowshoes, not the great broad Canadian ones that we see sometimes at home, but the Norwegian ones—a strip of light elastic wood, three or four inches wide and seven to ten feet long, slightly turned up at the front end, with an arrangement near the center to fasten it to the foot. With these they go everywhere, no matter how deep the snow is, and downhill they go with frightful velocity. At a race on snowshoes at an upper town last winter the papers announced that the time made by the winner was half a mile in thirty-seven seconds! And many men tell of going a mile in less than two minutes.[3]

The next day I rode on to Big Trees, a long ride, and overtook the rest of the party there. We remained there one day, then Professor Whitney went to San Francisco on business and sent John and me to go over the Amador road to Lake Bigler. We are on our way there now. My watch needed repairs so I have been detained here at Volcano over today. We had a hot, dusty ride through several mining towns. Why the place is called Volcano I do not know. Bayard Taylor, in his travels, speaks of several craters near, but no one else has ever seen them—they probably exist only in the poet's imagination.[4]

There was a terrible accident here yesterday afternoon. The premature explosion of a blast horribly burned two workmen; one died this morning, the other will perhaps follow. The poor fellows were brought into town last night and are horrible objects to look at.

We leave in the morning, cross to Hope Valley and to Lake

Bigler; then I don't know where—perhaps to Plumas and Sierra counties. We shall skim the cream off from the geology of the state this summer, and, as a consequence, of the sight seeing—so I will have less reluctance about leaving it should I come home next summer, as I have intimated a possibility of doing.

Lake Tahoe (or Lake Bigler).
August 23.

AUGUST 14 we left Volcano and came on up a new road just built, or rather building, across the mountains near the route of the old emigrant road. Soon after leaving Volcano we struck a table mountain of lava and followed it for two days, up a very gentle grade. Deep valleys lay on both sides, and from many points we had wide views of the foothills and the great valley, all bathed in a blue haze of smoke and dust, like our Indian summer intensified, in which all objects faded in the distance and those on the horizon were shut out entirely.

The next day, still following up the table, we got partially above the sea of dust and haze and at times we could even see the coast ranges. At Tragedy Springs we were up over seven thousand feet. This place took its name from a fearful tragedy. Four men were killed here by the Indians and their bodies burned. Their names are carved on a large tree by the spring, their only monument.[5] Here we descended into the valley of Silver Lake, a lovely little sheet of water, very deep and blue, resting in a basin of granite, high, picturesque peaks of volcanic rocks rising on the east. We camped at the foot of the lake, where a large house is going up. Log hotels are going up all along this road, in anticipation of custom when the road shall be finished. We are now up to where we have cold nights—the thermometer sank to 28°.

August 16 we came on nine or ten miles to where we found a little feed. My funds were running so low that I dared not stop over Sunday where we had to buy hay, as we had been obliged to thus far, the cheapest price being four cents per pound.

We rose from Silver Lake and crossed the Carson Spur, a high ridge of stratified volcanic ashes and breccia, capped by hard lava. The South Fork of the American has worn a very deep canyon through this cap down into the granite below. The road winds around the side of this canyon, and here we had the most picturesque scenery of the route. Below us, a thousand feet, dashed the river over granite rocks, the cliffs worn into very fantastic shapes—old castles, towers, pillars, pinnacles—all were there; while above, rugged rocky peaks, volcanic, of fantastic shapes, rose a thousand feet more, fearfully steep, the snow lying in patches here and there. We crossed this spur, behind which lies a pretty lake, called Summit Lake, but "Clear Lake" on your maps. Near this we camped, at an altitude of near eight thousand feet, the snow lying just above us and about us—of course a cool night, temperature down to 26°.

August 17 I climbed alone a high peak south, over 10,500 feet high—a steep, heavy climb. I had to carry barometer, bag with lunch, thermometer, levels, hammer, etc., a canteen, and my botanical box. We have had high winds for several days, and in the morning clouds enveloped the summit of the peak, but they fled before I reached it. The last eight hundred or a thousand feet rise in a steep volcanic mass, so steep as to be only accessible in one place, around cliffs and up steep slopes on the rocks, but there was no serious difficulty or danger. A cold, raw, and fierce wind swept over the summit, but the day was very clear and the view sublime. The peak is just south of Carson Pass, and some twenty to twenty-five miles south of Lake Tahoe (Bigler). The lake was in full view its whole extent—my first sight of it—its waters intensely blue, high, bold mountains rising from its shores. I am higher than the great crest of the Sierra here and hundreds of snowy peaks are in sight. Hope Valley[6] lies beneath, green and lovely—high mountains, eleven thousand feet high rising beyond it. Besides Lake Tahoe, there are *ten* smaller lakes in sight, from two miles long down to mere ponds (which were not counted)—all blue, of clear snow water. It was, indeed, a grand view.

I descended by getting on a steep slope of snow, down which I came a thousand feet in a few minutes where it had taken two hours' hard labor to get up. I had some time so I returned *via* the summit of Carson Pass, which was but two miles from camp. The pass is about 8,800 feet high. I got back to camp tired enough, and, of course, slept well after the fatigue of the day, although it froze in the night.

August 18 we crossed the pass and descended into Hope Valley, at the head of Carson Canyon—a beautiful basin, surrounded by high mountains on all sides—itself high, over seven thousand feet. We were hungry enough, and as we were getting dinner some Indians came to camp—two squaws, who had been out digging nuts. The younger, perhaps seventeen or eighteen years old, had a papoose on her back, tied to a willow frame, flat, with a willow flap above to keep the sun off. An old Indian, husband of the elder squaw, soon came up with his boy. He had his bows and arrows and a few squirrels he had shot. They were better looking and much better formed than the Diggers west of the mountains—better features, noses not so flat, mouths not so large, skin lighter—the younger squaw even had well-formed legs and small ankles, both very rare among the Indians west of the mountains. Just before we finished our dinner the old Indian sent the squaws away and gallantly ate up the remains of our dinner, while the squaws looked wistfully back.

We were nearly out of money, but I figured that we could get to Lake Tahoe, with tight calculation, the next afternoon; so I rode up the valley about four miles to visit a copper mine, and, to my great astonishment, found an old acquaintance there, who had been buying into the mine with another partner. He was anxious to get my advice in regard to it, and after I had seen all and given the advice he was only too glad to lend me twenty dollars, which relieved my anxiety immensely.

Now I resolved not to go directly to Lake Tahoe, but to cross the summit and climb Pyramid Peak, which had been in plain view for a long time—a very high and conspicuous point, which had never been measured. So we packed off, crossed a hill, sank into Lake Valley and out of it again, crossed the summit and struck the Placerville road, the grand artery of travel to Washoe. Over it pass the Overland telegraph and the Overland mail. It is stated that five thousand teams are steadily employed in the Washoe trade and other commerce east of the Sierra—not little teams of two horses, but generally of six horses or mules, often as many as eight or ten, carrying loads of three to eight tons, on huge cumbrous wagons. We descended about eight miles and camped at Slippery Ford.

This great road deserves some notice. It cost an immense sum, perhaps near half a million, possibly more. A history of this road would make a good Californian story. First an Indian trail, then an old emigrant road crossed the mountains; when, seeing its importance, the state and two counties, by acts of legislature and appropriations, at a cost of over $100,000 (I think), made a *free* road over on this general line. But the engineers, honest men, had neither the time nor means given them to do their part of the work well—as a consequence, it was not laid out in the best way. The mines of Washoe were discovered, and an immense tide of travel turned over the road. Men got franchises to "improve" portions of the road and collect tolls for their remuneration. Grades were made easier, bridges built, the road widened at the expense of private companies, who thus got control of the whole route. In other words, the *state* built a road that these private companies could transport their materials *free* over to build their *toll* road. Now, the tolls on a six-mule team and loaded wagon over the road amount to thirty-two dollars, or thirty-six dollars, I am not certain which sum, and it has paid immensely. In some places the profits during a single year would twice pay the expense of building, repairs, and collection of tolls! With such strong inducements men could afford to "lobby" in the legislature and get the franchises. A portion of the road, which is *assessed* as worth $14,000, last year collected over $75,000 in tolls! It takes a legislature elected for *political* services to grant such franchises.

The trade to Washoe, being so enormous, other roads are being built across the mountains. The Amador road, described in the first part of this letter, will be to some extent a rival road, but their pass is higher and all the passes south are higher. This pass is less than eight thousand feet.

Clouds of dust arose, filling the air, as we met long trains of ponderous wagons, loaded with merchandise, hay, grain—in fact everything that man or beast uses. We stopped at the Slippery Ford House. Twenty wagons stopped there, driving over a hundred horses or mules—heavy wagons, enormous loads, scarcely any less than three tons. The harness is heavy, often with a steel bow over the hames, in the form of an arch over each horse, and supporting four or five bells, whose chime can be heard at all hours of the day. The wagons drew up on a small level place, the animals were chained to the tongue of the wagon, neighing or braying for their grain. They are well fed, although hay costs four to five cents per pound, and barley accordingly—no *oats* are raised in this state, barley is fed instead.

We are at an altitude of over six thousand feet, the nights are cold, and the dirty, dusty teamsters sit about the fire in the barroom and tell tales—of how this man carried so many hundredweight with so many horses, a story which the rest disbelieve—tell stories of marvelous mules, and bad roads, and dull drivers, of fights at this bad place, where someone would not turn out, etc.—until nine o'clock, when they crawl under their wagons with their blankets and sleep, to be up at early dawn to attend to their teams.

Pyramid Peak lies about four or five miles north of this point. No one was inclined to accompany me on the climb, all dreading the labor. So the next morning, August 20, I started for the ascent alone. It was very early and cool, frost lying on the grass by the river, but not on the hillside. I climbed a steep hill; in fact, it was all climb, but not so hard as I had expected, for in four hours I was on the summit with barometer, bag with thermometer, hammer, lunch, and botanical box. The day was fine, not a cloud in sight, the air very clear, though of course hazy in the distance. I remained on the top over three hours.

The view is the grandest in this part of the Sierra. On the east, four thousand feet beneath, lies Lake Tahoe, intensely blue; nearer are about a dozen little alpine lakes, of very blue, clear, snow water. Far in the east are the desolate mountains of Nevada Territory, fading into indistinctness in the blue distance. South are the rugged mountains along the crest of the Sierra, far south of Sonora Pass—a hundred peaks spotted with snow. All along the west is the western slope of the Sierra, bathed in blue haze and smoke; and beyond lies the great plain, which for 200 miles of its extent looks like an ill-defined sea of smoke, above which rise the dim outlines of the coast ranges for 150 miles along the horizon, some of them over 150 miles distant. It is one of those views to make a vivid and lasting impression on the mind.

I was back at the house by sunset. All were surprised to find me no more tired, but the fact is, I have never felt in more vigorous health and my weight is reduced to good walking condition. I am now less than 140 pounds.

August 21 we came on up the road, crossed the pass, sank into Lake Valley, and are camped now at the south side of the lake. From the summit there is a fine view of the lake and of Lake Valley, which stretches south from the head. We camped here over yesterday and today, to get barometrical observations and let our animals rest in the good pasture they find here. We are camped in a pretty grove near the Lake House and a few rods from the lake. It is a quiet Sunday, the first we have observed for four weeks.

The lake is *the* feature of the place. A large log hotel is here, and many pleasure seekers are here, both from California and from Nevada. I was amused at a remark of a teamster who stopped here for a drink—the conversation was between two teamsters who looked at things in a practical light, one a stranger here, the other acquainted:

No. 1. "A good many people here!"

No. 2. "Yes."

No. 1. "What they all doing?"

No. 2. "Nothing."

No. 1. "Nothing at all?"

No. 2. "Why, yes—in the city we would call it *bumming* (Californian word for *loafing*), but here they call it *pleasure*."

Both take a drink and depart for their more practical and useful avocations.

Lake Tahoe, once called Lake Bigler and as such is on your maps, is the largest sheet of fresh water in the state.[7] Recent measurements in connection with the boundary survey make it twenty-three miles long and ten wide, but it has always before passed for much more. It has been recently sounded; its greatest depth is 1,523 feet, and most of it is over 1,000 feet deep. It lies at an altitude of over six thousand feet, while around it rise mountains four thousand feet higher. Its Indian name, Tahoe, was dropped and it was called after Governor Bigler, a Democratic politician. He was once of some notoriety here, but since he has turned "Secesh" all the Union papers have raised the cry to have his name dropped, and the old Indian name has been revived and will probably prevail.

The purity of its waters, its great depth, its altitude, and the clear sky all combine to give the lake a bright but intensely blue color; it is bluer even than the Mediterranean, and nearly as picturesque as Lake Geneva in Switzerland. Its beautiful waters and the rugged mountains rising around it, spotted with snow which has perhaps lain for centuries, form an enchanting picture. It lacks many of the elements of beauty of the Swiss lakes; it lacks the grassy, green, sloping hills, the white-walled towns, the castles with their stories and histories, the chalets of the herders—in fact, it lacks *all* the elements that give their peculiar charm to the Swiss scenery—its beauty is its own, is truly Californian.

The lake abounds in the largest trout in the world, a species of speckled trout that often weighs over twenty pounds and sometimes as much as *thirty pounds!* Smaller trout are abundant in the streams. An Indian brought some into camp. I gave him fifty cents for two, and they made us two good meals and were excellent fish. He had speared them in a stream near. We were eating when he came; when we finished he wanted the remains, which I gave him. Rising satisfied, patting both hands on his stomach, he exclaimed, "Belly goot—coot-bye." Many of these Indians, like the Chinese, cannot pronounce the letter R, substituting an L.

Camp 142, on the Truckee River.
August 27.

On Monday morning, August 24, we started on. I got funds from Professor Whitney and felt all right again. We kept east three miles, to the southeast corner of the lake, then struck north along the shore, now in Nevada Territory. We kept near the shore all day, often crossing ridges where rocky points jut out into the water. We camped at a pretty spot on the shore, the lake in front, and high granite ridges rising behind us— had a cold night.

August 25 we came on, turning around the northeast corner of the lake and striking over ridges to the northwest. Beautiful as Lake Tahoe is from the south, it is yet more so from the north—from having a finer background of high, rugged, black mountains, some of them eleven thousand feet high, or near it, their dark sides spotted and streaked with snow. This end will eventually become the most desirable spot for persons in pursuit of pleasure.

We struck over a ridge, came to the lake again at its north end, then left it entirely, crossing a high volcanic ridge and sinking into a new mining district which is just starting—a new excitement, and people are pouring in. As we went down a canyon we passed numerous prospecting holes, where more or less search has been made for silver ore. Since the immense wealth of the Washoe mines has been demonstrated, people are crazy on the subject of *silver*.

We passed through the town of Centerville, its streets all staked off among the trees, notices of claims of town lots on trees and stumps and stakes, but as yet the *town* is not built. One cabin—hut, I should say—with a brush roof, is the sole representative of the mansions that are to be. Three miles below is Elizabethtown, a town of equal pretentions and more actual houses, boasting of two or three. We stopped at the main *store*, a shanty twelve feet square, made by driving stakes into the ground, siding two sides with split boards, and then covering with brush. Bacon, salt, pepper, tobacco, flour, and more than all, poor whiskey, are kept. The miners have camps —generally some brush to keep off the sun and dew; but as often nothing. Some blankets lying beside the brook, a tin kettle, a tin cup, and a bag of provisions, tell of the *home* of some adventurous wandering man. We passed the town and camped two miles beyond, in Tim-i-lick Valley. The day had been warm, but the night was cold enough to make it up, the temperature sank to 20° F., twelve degrees below the freezing point.

August 26 our animals eloped in the early morning and it took us until ten o'clock to find them and pack up. In the meantime we were joined by a boy thirteen or fifteen years old, who was coming this way and is with us now. Only such a country as this can produce such boys. He was from Ohio—came here in '53—went back *alone* in '58—had his pocket picked twice, but was keen enough to have his money in his boot—came back *alone* last year—now, with his horse, takes a trip over a hundred miles from home among these mountains.

Well, we struck over the mountains for the Truckee River, to this place, where new mines have "broken out"—at least, a new excitement. We crossed a high volcanic ridge, very rough trail, all the way through an open forest of pines and firs, as one finds everywhere here, and camped on the river about Knoxville. Here I have been examining the "indications" today. Six weeks ago, I hear, there were but two miners here; now there are six hundred in this district. A town is laid off, the place boasts of one or two "hotels," several saloons, a butcher shop, a bakery, clothing stores, hardware and mining tools, etc.—all in about four weeks.

I would give twenty-five dollars for a good photograph of that "street." A trail runs through it, for as yet a wagon has not visited these parts. The buildings spoken of are not four-story brick or granite edifices—not one has a floor, not one has a chair or table, except such as could be made on the spot. This shanty, in the shade of a tree, with roof of brush, has a sign out, "Union Clothing Store." I dined today at the "Union Hotel"—a part of the roof was covered with canvas, but most of it with bushes—and so on to the end of the chapter. The crowd—only men (neither women nor children are here yet)—are all working or speculating in "feet."

Let me explain the term "feet" as used here. Suppose a hill has a vein of metal in it; this is called a "lead." A company "takes up" a claim, of a definite number of "feet" along this vein, and the land 150 to 200 feet each side. The length varies in different districts—the miners decide that themselves—sometimes 1,000 feet, sometimes 1,500, at others 2,000 feet are allowed. In a mine that claims 1,000 feet, a foot sold or bought does not mean any particular foot of that mine, but one one-thousandth of the whole. Thus, the Ophir Mine has 1,500 feet, now worth over $4,000 per foot. One never sees shares quoted in the market, but feet; and feet may be bought at all prices, from a few cents in some to over $5,000 in the Gould & Curry Mine in Washoe. A man speculating in mines is said to be speculating in feet.

There is great excitement here—many think it a second Washoe. Some money will be sunk here before it can be known what the value will be. I have but little faith in it myself. I surely would not invest money in any mine I have seen today, and I have visited eight or nine of the best.

Forest Hill.
Sunday, August 30.

WHERE I last wrote our camp was on the side of the Truckee River, about six miles below the lake. We had a very comfortable camp. Many small squirrels, a very small species of chipmunk, swarmed about camp—most beautiful little animals. Our boy—his name is Mehafey—rigged up a trap made of our dishpan set on a T-bait, and caught seven of them. By the time we got ready to start they had all got away but one; that one John has brought along—a beautiful little animal, tame already, it rides in his pocket or on his shoulder. I will try and get him to the city if I can. They live only in the high mountains.

August 28 we were up and off early. It rained a little the evening before, but not much. We struck up Squaw Valley, a pretty little grassy valley, then rose the steep ridge to the pass, which is between eight thousand and nine thousand feet. We stopped and lunched there, while I took observations for the altitude of the place. We had a grand view of the mountains south, spotted with snow, and the dim ones east, far in Nevada Territory, and of the western slope fading into blue haze.

We crossed the summit and sank into the canyon of the Middle Fork of the American River, then rose a high hill. We had heard that there was abundant grass on the road, but we found it only partly true. For the rest of the day we followed a volcanic lava ridge—in places only a sharp ridge, with a canyon a thousand feet deep on each side. It was grand, but a very rough trail, in fact, "awful bad," and, what was worse, no water. We traveled until camp time, then three hours later —for water to camp by. The sun set and the full moon rose long before we struck water. At last, however, we found it—it was nearly nine o'clock before we got our supper. Of course it tasted good after our fatigue and fast. We stopped hungry, thirsty, fatigued, and, as a result, in bad humor; after supper and a big beefsteak we were in fine humor again. The moon was peculiarly bright, the night warm and balmy, just right to sleep well.

August 29 we were off early, but were directed on the wrong trail and it cost us much labor. We followed on down the volcanic table, with a deep canyon on the south, the air very hazy and thick, the foothills becoming lost in the haze in a few miles. We were rapidly getting into a warmer climate. At noon we struck a mining town, Last Chance—hot, dusty in the extreme. Here we found we were on the wrong trail and had to cross three deep canyons. A trail is cut down the steep sides. We descended some 1,500 feet, then rose another volcanic table as high as the first—the top of this canyon, from table to table, is not over three-quarters of a mile to a mile, its depth about 1,500 feet. We crossed this table, passed the little place called Deadwood, and then we had the El Dorado Canyon to cross—still worse—nearly or quite two thousand feet deep, its sides still steeper. Here is a toll trail, very narrow—often a misstep on the narrow way would send the horse and rider, or mule and pack, down hundreds of feet, to swift and certain destruction. It was fearful, yet we had to pay $1.50 for the privilege of passing it. There is a cluster of mining cabins in the canyon. A nugget has just been taken out that weighs seventy-eight ounces (over eight pounds[8]) and worth some $1,500.

Well, we came out of that and stopped last night at Michigan Bluffs, a mining town. The town is supported by claims in "washed gravels" that form bluffs nearly two thousand feet above the bottom of the canyon, yet stratified by water. Our horses cost us two dollars each for keep over night. I was anxious to get on, so came to Forest Hill this morning, six miles, once more in a wagon road, but hot and dusty—temperature over 90°.

Since writing the above I have received a letter from Professor Whitney calling me to San Francisco immediately. I will start in the morning at three o'clock and mail this there.

Some of you ask about when I am coming home, and if I have the same old trouble about getting money. First question —I don't know, probably next year. Second question—the same old trouble. The state now owes Professor Whitney (including our unpaid salaries) about $25,000, and in his letter received this afternoon he says he doesn't know what he is to do unless he gets money within a month; he has borrowed until he cannot raise any more. My salary is now back to the amount of $2,800, or for one year and two months, and I have to borrow for my personal expenses. I am tired of it. We may possibly get some money in September, but most probably not until December.

It is infamous—political hacks get their money more regularly. We must wait, as our bills have less "political significance," as the comptroller calls it.

NOTES

1. William Hillebrand was born at Nieheim, Westphalia, 1821. After studying at Göttingen, Heidelberg, and Berlin, he practiced medicine for a time. On account of his health he sought other climates, visiting Australia, the Philippines, and California before settling in the Hawaiian Islands. During a residence of twenty years in Honolulu he made exhaustive studies and

collections of Hawaiian flora. He was the private physician of King Kamehameha V and a member of the Privy Council. In 1871 he left Hawaii and resided in Germany, Switzerland, Madeira, and Teneriffe. He died about 1888 (Preface to Hillebrand, *Flora of the Hawaiian Islands* [1888]).

2. "Thomas Bridges came to California in 1856, and for the next nine years collected on the coast, much of the time in this State, his collections going mostly to Europe. After his death, in 1865, his wife presented the California collections then on hand to the National Herbarium at Washington. They were distributed by Dr. Torrey" ("Geological Survey of California," *Botany*, II [1880], 558). A handsome flower, *Pentstemon bridgesii*, named for him, is found in the Yosemite region. A variety of lupine, also, bears his name.

3. In *Overland Monthly* (October, 1886), Dan De Quille tells the story of "Snowshoe Thompson," who for many years carried the mail across the Sierra on Norwegian snowshoes (skis).

4. Bayard Taylor, *Eldorado, or Adventures in the Path of Empire* (1850), chap. xxiii.

5. The story of Tragedy Springs is told in Serg. Daniel Tyler, *A Concise History of the Mormon Battalion in the Mexican War, 1846–1847* (1881), p. 337. A party of the Mormon Battalion, returning from southern California to Utah by way of the San Joaquin Valley in the summer of 1848, sent scouts ahead to find a way across the Sierra. About the middle of July the main party, about thirty-seven in number, advanced into the mountains. Because of the rarity of the book, the incident is here quoted in full:

"Some four or five miles took them to what they named Tragedy Springs. After turning out their stock and gathering around the spring to quench their thirst, some one picked up a blood-stained arrow, and after a little search other bloody arrows were also found, and near the spring the remains of a camp fire, and a place where two men had slept together and one alone. Blood on rocks was also discovered, and a leather purse with gold dust in it was picked up and recognized as having belonged to Brother Daniel Allen. The worst fears of the company: that the three missing pioneers had been murdered, were soon confirmed. A short distance from the spring was found a place about eight feet square, where the earth had lately been removed, and upon digging therein they found the dead bodies of their beloved brothers, Browett, Allen and Cox, who left them twenty days previously. These brethren had been surprised and killed by Indians. Their bodies were stripped naked, terribly mutilated and all buried in one shallow grave.

"The company buried them again, and built over their grave a large pile of rock, in a square form, as a monument to mark their last resting place, and shield them from the wolves. They also cut upon a large pine tree near by their names, ages, manner of death, etc. Hence the name of the springs."

6. Hope Valley was named by the Mormon party after crossing the Sierra from Tragedy Springs, "the spirits of the explorers who first discovered it reviving when they arrived in sight of it" (*ibid.*, p. 239). Frémont had passed this way in February, 1844.

7. Lake Tahoe was discovered in 1844 by Frémont, who called it first Mountain Lake, later Lake Bonpland. In the early fifties it was given the name Lake Bigler, for John Bigler, Governor of California (1852–58). In

1862 the name Tahoe, said to mean "big water" or "water in a high place," was proposed and soon became generally adopted. An attempt to revive "Bigler" was made in 1870, when the California State Legislature passed "An act to legalize the name of Lake Bigler" (*California Statutes 1869–70*, chap. lviii, p. 64). This was not repealed until 1945, when the Legislature made "Lake Tahoe" official.

8. This should be six pounds, six ounces (troy weight).

CHAPTER VI

THE NORTHERN MINES AND LASSEN'S PEAK

A Railroad Ride—Gardner and King—Election Day in San Francisco—Return to Forest Hill—Grass Valley—The Yuba Mines—To Feather River—Genesee Valley—Big Meadows—Lassen's Peak—View from the Summit—Volcanic Activity—Hat Creek—Fort Crook.

Genesee Valley, Plumas County.
September 20, 1863.

MONDAY morning, August 31, I was up at half-past two and took stage, and was far down the foothills before sunrise, breathing clouds of dust. You can have no idea of the dust of these roads in the dry season. I took breakfast at Auburn, then staged six miles farther, where I took the cars.

It was a delight to travel by rail again, the first time since I left the states. At Sacramento I took steamer, and meeting an old friend, had a pleasant trip. On the way down two young men came up to me, asked if my name was Brewer, and introduced themselves as two young fellows just graduated last year in the Scientific School at Yale College, who this summer have crossed the plains. Their names are Gardner[1] and King.[2] Of course I was glad to see them; King I have taken with me on this trip.

We were late in the city, from a curious detention. The alarm was given, "A man overboard!" The steamer was stopped after much delay, backed, a boat sent out, when it was found that a man who lived in a cabin on the river had jumped overboard to swim home, which he did before the boat reached him. It made much excitement and detained us half an hour.

Next day I met many old friends. All the Survey had come back to the city from different directions—our first meeting all together since last February. Professor Whitney and I talked plans; and much miscellaneous business had to be attended to. In the afternoon I met the celebrated traveler, J. Ross Browne.[3] He appears a quiet fellow, not at all the one to visit so many distant lands and write such genial accounts of what he sees.

Wednesday, September 2, was Election Day, an important day in the history of California. I cast my first vote in nine years—for the "Independent Union Ticket." In the city there were three tickets: Regular Democratic, Regular Union, Independent Union—the first with too much of a sprinkling of Secesh; the second a politicians' ticket; the last, a truly people's ticket, was successful. The state Union ticket was largely elected, for the state is soundly loyal. And here let me say that the so-called "Democratic" ticket was by no means loyal. The mass of Democrats are unquestionably loyal men, but in this state the leaders of the party are many of them southern men, avowedly Secesh. The most active stump orator of the party in this canvass, Judge Robinson, who has a son in the Rebel army, cannot practice law in this state because he will not take the oath of allegiance. Weller is open Secesh, Bigler nearly as bad. All these men have tried hard to be arrested and be made martyrs, in Vallandigham style, have publicly said everything they could against the government to bring about such an end, but have not succeeded. Downey, the candidate for governor, was governor when we arrived here three years ago. He is an Irishman who was down on us (the Survey) because Professor Whitney would not use his official influence as State Geologist to aid him in mining speculations; he is not full-blooded Secesh, but about half-and-half. The candidate for state printer is Secesh, etc. Luckily for the state and country this set of broken-down politicians, some of them full-fledged scoundrels, did not get into power. The election was quiet everywhere.

A friend tells an election anecdote, which he says is true—true or not, it is good. Two Irishmen meet at the polls, one accosts the other:

No. 1. "Mike—hev yer vowted?"

No. 2. "Yes."

No. 1. "Vowt agin fur Downey, for the dommed Yankees are staleing the counthry away from us."

The next two days were spent in various matters in the office—plans of operations were decided on—very busy, yet not much done. I got a letter from Schmidt, our kind German cook last year—he is safely in his home in Hamburg, in Germany. I resolved to take King with me; Gabb goes on a trip to Oregon, Hoffmann and Whitney to the Sierra, Cooper to Lake Bigler.

Saturday, September 5, in the afternoon we took steamer for Sacramento. All the time I was in the city the weather was cool—fog came in from the sea every night—but on getting away from the coast it changed entirely. At Sacramento I sent King to Grass Valley, where we got an extra horse for him, while I rode to Forest Hill, where I had left John and the animals. It is striking to pass inland from the sea at this season. From fog and cold air to a cloudless sky, heat, and dust.

At Forest Hill I found all right. I settled bills, and September 7 we packed up and came on. I had two barometers to carry. We passed several mining towns, and crossed the very deep canyon of the North Fork of the American River; but a horse got sick and we had to stop at Iowa Hill, a long row of neat houses perched on the apex of a very sharp ridge, with placer diggings all around. This was once a very important town, but now is much smaller, its placers being mostly worked out.

September 8 we came on to Grass Valley, a pretty place of

three thousand or four thousand inhabitants. We crossed the canyon of Bear River and passed several mining towns—the air hot and the roads dusty in the extreme. There we spent one day. I wanted to establish a barometrical station, so left the instrument with a friend, Mr. Blake, who has a quartz mill near.

This is a rich region and quartz mills abound. The gold-bearing quartz runs in veins from a few inches to several feet thick. The quartz is mined, the same as any other mineral, then crushed to a very fine powder by mills. This powder then runs in water in shallow troughs, over quicksilver, which takes up the gold. Sometimes, instead of quicksilver, it runs over blankets, the fibers of which catch the gold; the blankets are washed out every ten or fifteen minutes in a tank of water which has quicksilver at the bottom. Hundreds of men have been ruined by unprofitable quartz mining, and others have become immensely rich very easily.

Here is the famous Allison Ranch. The owners were ignorant Irishmen who could neither read nor write. They spent their money as such men will. One at one time went to San Francisco and bought sets of diamond jewelry to the amount of $12,000, which he presented to the women in a house of ill fame in a noted part of the city. He drank himself to death last year.

Here also is the Rocky Bar Ledge. Two poor men, brothers, common laborers, discovered it and in eighteen months made, clear of all expenses, $750,000. They are now spending their money very profusely in fast horses and similar luxuries.

Walter Frear lives here, the pastor of a small congregation. I called and found him very pleasantly situated. He has a rather pretty wife and two children. Strange enough, I met here a lady, a widow, now Mrs. Baker, whom I knew twelve years ago in Lancaster, New York, as a Miss Mills. She had entirely passed from my mind, but brought back old memories again.

I found King and the horse all right, so September 10 we started on our way—first to Nevada, a few miles, a fine town in a rich mining region, then to San Juan North (there are several other San Juans in the state), then to Camptonville, a miserable, dilapidated town, but very picturesquely located, with immense hydraulic diggings about. The amount of soil sluiced away in this way seems incredible. Bluffs sixty to a hundred feet thick have been washed away for *hundreds of acres together*. But they were not rich, the gold has "stopped," the town is dilapidated—but we had to pay big prices nevertheless.

September 11 we passed Galena Hill, with extensive hydraulic diggings; then a deep canyon of the North Yuba, to Brandy City, with *tremendous* hydraulic diggings; then up a long volcanic ridge, ten miles, to Eureka. At places we attained an elevation of over five thousand feet and commanded wide views of the Sierra, the great tables of lava in the south, the rugged Downieville Buttes in the east, Pilot Peak, Table Mountain, etc.

At Eureka we came upon the slates again, with gold—a mining town. Then a vile trail, nine miles, *very* rough, across canyons and ridges, with high peaks with patches of snow. We at last sank into a very deep canyon, perhaps two thousand feet deep, to Poker Flat, a miserable hole—but what we lacked in accommodations was made up in prices. Ten white men and two Chinamen slept in the little garret of the "hotel." Our horses fared but little better, and our bill was the modest little sum of fifteen dollars.

September 12 we were off early, passed several little mining towns, Whiskey Diggings, Potosi, Rowland Flat, etc. We had a very rough trail, being in the very heart of the Sierra, which here sinks to an altitude of six thousand to eight thousand feet. We crossed a high ridge by Pilot Peak, stopped, unsaddled our horses, and went up it, as we were within five hundred feet of the top. It is a little over seven thousand feet high, and commands a grand view from the top. Lassen's Butte in the northwest is a truly grand object. On the northeast is a rough region, cut by very deep canyons; on the southeast are the rugged peaks between us and Downieville.

We then descended into a tremendous canyon, over three thousand feet deep, of the Middle Feather River, at Nelson's Point. It was nearly sundown, but it was such a miserable hole, and as it was Saturday night we resolved to push on to Quincy, ten miles farther—a trail much better than we had had for some time. Nelson's Point is very picturesque—the deep, steep canyon and the rocky slates standing on edge, are peculiar. It was long after dark when we got to Quincy, where we struck a good hotel and rested over Sunday—my first for three weeks.

Quincy is a pretty place, in a most charming valley, has nice houses, a pretty courthouse, etc., but no church. I heard that there is not a church or schoolhouse in the county, which has a population represented by over fifteen hundred voters.

Our expenses for one week had been over $150 in getting through these mining towns; we now struck a cheaper region.

In Camp, at Lassen's Peak.
September 28.

MONDAY morning, September 14, I laid in supplies and went to Genesee Valley, about twenty-three miles northeast. We crossed the lovely American Valley, then a high range of hills about ten miles wide, then sank into Indian Valley, one of the loveliest valleys in the state. It is a basin, elevated about 3,500 feet above the sea, entirely surrounded by mountains, which rise several thousand feet higher, sheltered from winds—in fact, containing all the elements of beauty. It is about ten miles long and three or four wide, and covered with fine farms.

From this we passed up the canyon of Genesee Creek, which after a few miles opens out into another pretty basin, but smaller and less fertile than Indian Valley. It is also higher and colder. My object in coming here was to look for fossils said to occur here in the auriferous slates. We found them, a most important matter geologically.

This was my birthday—I am thirty-five—half my "three-score years and ten" are past. As I lay in my blankets that night, long I reflected on it as I watched the stars in the cloudless sky.

We stopped there three days, examining the region, looking at some copper mines just found there, and collecting fossils. We camped on the ranch of a Mr. Gifford. I found that he was from Aurora, New York, and knew many of my old acquaintances, and we revived old memories by spending an evening talking about them.

One meets here people apparently out of their station. I met a Mr. Wilson—evidently a man of intelligence—dressed in dirty duck pants, dirty gray shirt, with shoes and hat to match. Yet he was once in the navy, a graduate of the Naval Academy at Annapolis, was on General Somebody's staff, had spent three years in the Mediterranean, had seen much of the world, had resigned and had gone into railroad making. He became rich, but lost all in unfortunate speculation, and is now here, in this out-of-the-way place, among uncongenial companions, away from family and old associations, prospecting for copper and hoping to retrieve his fallen fortunes.

We were here four nights, and each night it froze—the temperatures were 27°, 27°, 21°, and 22°.

September 18 we moved down the valley about five miles and camped at Mormon Station, where I found more fossils. They were in a rough, rugged canyon. I had several miles to walk. Both King and John were unwell, so I had it all to do. The next day I collected more, killed a rattlesnake, and packed my specimens. It clouded over and sprinkled some. We turned in early and were hardly well in our blankets when it began to rain. You cannot imagine how cheerless and uncomfortable it is to lie out in the rain—how one looks up at the black sky, lets the rain patter on his face, saturate his hair and beard, as he thinks of home and its cheerful fireside and luxurious comforts. Happily it did not rain much, not enough to wet our blankets through. Sunday, September 20, we spent quietly in camp, and I wrote my last letter home. It was clear and cool.

September 21 we packed up and came on to the Big Meadows, twenty-four miles. We went back to Indian Valley and followed down that for ten miles, then over low hills for ten more, when we sank into the valley of Big Meadows. As we struck this we came in sight of Lassen's Peak, rising gloriously, scarcely forty miles distant, a grand object indeed. We camped by the Feather River, here a large, cold stream, abounding in trout. The night was intensely clear, and the temperature sank to 20°. There is a store here and we laid in more provisions, for it is the last we shall find for many a weary mile.

September 22 we came on toward Lassen's Peak, some twenty-two miles farther. First, for fifteen miles, we followed up the valley, the Big Meadows, a level valley with high mountains on all sides; but *the* grand feature is Lassen's Peak, rising beyond the head of the valley. In front, the level meadow; then the dark forests along the base of the mountain; while beyond, and above all, rises the bare and desolate peak of snow and rocks, against the blue sky, which is streaked here and there with thin clouds.

We struck the road that runs between Red Bluff and Honey Lake, followed up it for some miles and camped at a little grassy flat. Soon some teamsters came along and camped with us, and made the evening merry with their jokes and songs. The night was clear, and the thermometer sank to 19°.

September 23 we came on but about two miles, to the end of our trails or roads, to the tent of an old hunter named Loveless, who knows all about the region, and of whom we hoped to get information as to how we could reach the peak. Here we found camped some men from the Sacramento Valley, who had come up to escape from the heat of the valley, to hunt and have a good time. Judge Walsh, a very wealthy ranchero, and his family, were the main ones. He greeted us warmly, and gave us a quarter of venison. We camped near them, and I spent the rest of the day in writing up my notes. We dined with them and had a pleasant time. Mrs. Walsh, a Scotch lady, was lively and was enjoying the gipsy life. They were well provided with tents, a cook, servant, etc. Here King met an old friend who had crossed the plains with him, and they had a jolly reunion. I slept cold that night, for the thermometer sank to 12°.

September 24 our friends showed us some hot springs and other curiosities of the region. We rode through the woods and chaparral for some miles. First came the Boiling Lake, a lake of about four acres of hot water, boiling furiously in many places, and clouds of steam rising from it. Around it the steam issued from hundreds of jets. The rocks have been decomposed by these agencies, and the bed of the lake is of thin, fine clay, which bubbles and sputters like some titanic mush kettle.

Two miles from these are the Steamboat Springs, where steam and hot water issue from hundreds of places. There is a pool of boiling water in the canyon, about two rods across, in which there is a mass of water and steam rising, in jets, often six to eight feet high, splashing and roaring incessantly, while clouds of steam roll away up the canyon. Wherever you climb over the bowlders around this you feel hot steam puffing out around you and hear the hissing and gurgling everywhere beneath your feet. We have been in a volcanic region for some days and these springs show that it is not entirely cooled underneath yet.

Thence we went to Willow Lake, a pretty little sheet of water embosomed in the hills. It abounds in trout. The rest had come there in the morning and had caught near two hundred of them. We stopped a little while and I caught my *first trout*. We returned from our rough ride, tired and hungry, and dined again at the camp of our friends. We made great havoc in the trout and venison, then spent a lively evening around their bright camp fire of huge logs. We returned to our camp, and the thermometer sank that night to 15°.

September 25 we were up early and off. We had prevailed on our friends to pilot us to the base of the last cone and make the ascent with us. So Mr. Walsh, Keating, and Eastman, with a pack horse, came with us.[4] We rode about sixteen miles and camped here, where we are yet, at the base of the last peak, at the highest grass, at an altitude of about eight thousand feet. We had no trail, but went through woods and chaparral, across canyons and through swamps. We passed another cluster of boiling springs, even more extensive than those of the previous day. Hot water, steam, gases, and sulphur come up over a region of several hundred acres. Some of the crystallizations of sulphur around the steam vents were exceedingly beautiful, delicate as snow, frosting the rocks with brilliant yellow. We had some grand views of the peak. Although we were higher than our last camp it was not so cold, but it froze some that night.

Fort Crook.
October 5.

SEPTEMBER 26 we made our first ascent of Lassen's Peak—King and I and the three friends who had come with us from their camp. We were up and off early, were on the summit before ten o'clock, and spent five hours there.

We had anticipated a grand view, the finest in the state, and it fully equaled our expectations, but the peak is not so high as we estimated, being only about 11,000 feet.[5] The day was not entirely favorable—a fierce wind, raw and chilly, swept over the summit, making our very bones shiver. Clouds hung over a part of the landscape. Mount Shasta, eighty miles distant, rose clear and sharp against a blue sky, the top for six thousand feet rising above a stratum of clouds that hid the base. It was grand. Most of the clouds lay below us at the north. The great valley was very indistinct in the haze at the south, but the northern part was very clear.

We were back early, and had a hearty dinner of hot coffee, venison and trout, pork and beans—the former for a change, but the latter as a stand-by for fatigues and climbing. All were delighted with their trip.

We had a cold, windy, and cloudy night, and the next day, Sunday, September 27, a snowstorm set in, and our friends left us for the warmer climate of a lower altitude. During the forenoon we had fierce snow squalls, which whitened the ground. Without tent or other shelter than the trees, it was cold and cheerless. But in the afternoon it cleared up, and we had the freshest of air and the bluest of sky. The firs above us were silvered with snow, and the rugged peak whitened. It was too cold to write, so I read *Bleak House*, and finished it by the camp fire at evening.

September 28 the thermometer stood at 17° in the morning; it was cold and nearly clear. I had lain awake half the night to get up early and climb the peak again, but clouds deterred us; so King went down the valley to sketch the mountain, while I took a long tramp around the east side of the peak. We made our preparations for an ascent the next morning, should it be clear.

Tuesday, September 29, we were up at half-past one, had an early breakfast by the light of the bright moon, now two days past its full, and at 2.45 were on our way.

The description that follows I wrote on top of the mountain. It has the merit of rigid truthfulness in every particular.

First up a canyon for a thousand feet, then among rocks and over snow, crisp in the cold air, glittering in the bright moonlight. At four we are on the last slope, a steep ridge, now on loose bowlders and sliding gravel, now on firmer footing. We avoid the snow slopes—they are too steep to climb without cutting our way by steps. We are on the south side of the peak, and the vast region in the southeast lies dim in the soft light of the moon—valleys asleep in beds of vapor, mountains dark and shadowy.

At 4.30 appears the first faint line of red in the east, which gradually widens and becomes a livid arch as we toil up the last steep slope.

We reach the first summit, and the northern scene comes in view. The snows of Mount Shasta are still indistinct in the dusky dawn. We cross a snow field, climb up bowlders, and are soon on the highest pinnacle of rock. It is still, cold, and intensely clear. The temperature rises to 25°—it has been 18°.

The arch of dawn rises and spreads along the distant eastern horizon. Its rosy light gilds the cone of red cinders across the crater from where we are. Mount Shasta comes out clear and well defined; the gray twilight bathing the dark mountains below grows warmer and lighter, the moon and stars fade, the shadowy mountain forms rapidly assume distinct shapes, and day comes on apace.

As we gaze in rapture, the sun comes on the scene, and as it rises, its disk flattened by atmospheric refraction, it gilds the peaks one after another, and at this moment the field of view is wider than at any time later in the day. The Marysville Buttes rise from the vapory plain, islands in a distant ocean of smoke, while far beyond appear the dim outlines of Mount Diablo and Mount Hamilton, the latter 240 miles distant.

North of the Bay of San Francisco the Coast Range is clear and distinct, from Napa north to the Salmon Mountains near the Klamath River. Mount St. Helena, Mount St. John, Yalloballey, Bullet Chup, and all its other prominent peaks are in distinct view, rising in altitude as we look north.

But rising high above all is the conical shadow of the peak we are on, projected in the air, a distinct form of cobalt blue on a ground of lighter haze, its top as sharp and its outlines as well defined as are those of the peak itself—a gigantic spectral mountain, projected so high in the air that it seems far higher than the original mountain itself—but, as the sun rises, the mountain sinks into the valley, and, like a ghost, fades away at the sight of the sun.

The snows of the Salmon Mountains glitter in the morning sun, a hundred miles distant. But *the* great feature is the sublime form of Mount Shasta towering above its neighboring mountains—truly a monarch of the hills. It has received some snow in the late storms, and the "snow line" is as sharply defined and as level as if the surface of an ocean had cut it against the mountain side.

Through the gaps we catch glimpses of the Siskiyou Mountains, and, east of Mount Shasta, the mere summits of some of the higher snow mountains of Oregon.

In the northeast is the beautiful valley of Pit River, with several sharp volcanic cones rising from it; while chain appears beyond chain in the dim distance, whose locality I cannot say, for we have no maps of that region.

In the east, valley and mountain chain alternate until all becomes indistinct in the blue distance. The peaks about Pyramid Lake are plainly seen. Honey Lake glistens in the morning sun—it seems quite near.

In the southeast we look along the line of the Sierra, peak beyond peak, until those near Lake Bigler form the horizon. The mere summit of Pyramid Peak is visible, but the Yuba Buttes, Pilot Peak, and a legion of lesser heights are very distinct. The valleys between these peaks are bathed in smoke.

Nearer, in this direction, are several beautiful valleys—Indian Valley, the Big Meadows, Mountain Meadows, and others—but all are dry and brown.

Like many philanthropists, in looking at the distant view I have almost forgotten that nearer home, just about the peak itself. Great tables of lava form the characteristic features; for Lassen's Peak, like Mount Shasta, is an extinct volcano. The remains of a crater exist, a hollow in the center, with three or four peaks, or cones, rising around it. The one we are on is the highest. The west cone has many red cinders, and looks red and scorched. A few miles north of the peak are four cones, the highest above nine thousand feet high, entirely destitute of all vegetation, scorched and broken. The highest is said to have been active in 1857.

The lava tables beneath are covered with dark pine forests, here and there furrowed into deep canyons or rising into mountains, with pretty valleys hidden between.

Several lower peaks about us are spotted with fields of snow, still clean and white, sometimes of rose color with the red microscopic plant, as in the arctic regions.

Little lakes bask in the sunlight here and there, as blue as the sky above them. Twelve are in sight. And the Boiling Lake is in view, with clouds of white steam rising through the trees in the clear, cold, mountain air.

Here and there from the dark forest of pines that forms the carpet of the hills curls the smoke from some hunter's camp or Indian's fire.

Many volcanic cones rise, sharp and steep, some with craters in their tops, into which we can see—circular hollows, like great nests of fabulous birds.

On the west, the volcanic tables slope to the great central valley. The northern part of this, from Tehama to Shasta City, is very distinct and clear, with its forests and farms and orchards and villages, a line of willows marking the course of the Sacramento River. Farther south, smoke and haze obscure the plain.

But in all this wide view there appear no green pastures or lovely green herbage. Dark green forests, almost black, lie beneath us; desolate slopes, with snow and scattered trees, lie around us, and all the valleys are dry and sere. All is as unlike the mountains of the eastern states, or the Alps, as it is possible for one mountain scene to be unlike another.

As the sun rises it is truly wonderful how distinct Mount Shasta is. Its every ridge and canyon and snow field look so plain that one can scarcely believe that it lies eighty miles distant in air line—a weary way and much farther by any road or trail.

The valleys become more smoky, and the distant Sierra more indistinct, dark and jagged lines rising above the haze.

Until 10 A.M. not a cloud obscures the sky, then graceful cirri creep over from the Pacific, light and feathery.

The day wears on. The sun is warm and the air balmy. Silence broods over the peak—no sound falls on the ear, save occasionally, when a rock, loosed by last night's frost and freed by the day's thaw, rumbles down the steep slope, and all is silent again.

Now and then a butterfly or bird (of arctic species) flits over the summit and among the rocks, but both are silent.

Before 2 P.M. the smoke increases in the valleys, until the great central valley looks like an indistinct ocean, without surface or shores. Mountain valleys become depths of smoke that the sight cannot penetrate. The distant views fade away in haze, and the landscape looks dreamy.

We remained on the top until nearly three—over nine hours—then returned. We enjoyed a slide down a steep slope of snow. I "timed" King on it—he descended a slope four hundred or five hundred feet in fifty-seven seconds. The only mishap of the day was King getting his ears frostbitten.

September 30 King and I took a long walk around to the recent cones on the north of the main peak. It was a tedious walk to reach them, over rocks and ridges and slopes of soft volcanic sand, but more tedious to get over the cone itself. There are four of these cones, the highest over nine thousand feet, with its top red and burned. This is the one said to have been active in 1857. We examined but one of these—all have their bases blended, only their tops are distinct. This one was perhaps 8,500 feet, the top about two miles across in either way, and entirely of broken rocks, mostly loose, but here and there a pinnacle of rock two or three hundred feet above the main mass. These rocks are angular bowlders, of all sizes up to fifty feet or more in diameter, thrown together in the wildest confusion. Lassen's Peak looks sharper from this side than any other, and views seen from among these pinnacles and rocks are some of the most picturesque imaginable. A series of photographs would be treasures indeed.

The place was of great scientific interest. These mountains have been *thrust up* from beneath, and the rocks crushed by the gigantic natural forces. In some places masses of rock two hundred feet high, or more, are all cracked and crushed into fragments, but the fragments are still in place. There were other points that made it especially interesting. Glaciers once streamed from Lassen's Peak, down on every side. The rocks are furrowed and polished by them, canyons show their traces everywhere, but all have passed away.

Camp 153.
October 6.

WE finished our work at Lassen's Peak the last day of September and made our preparations for leaving. I felt anxious enough about the next three days' ride. We were to pass through a desolate region and among Indians reputed to be bad. I could get no definite information about them. Some thought them safe if one were careful and well armed, others thought it dangerous to go through with so small a party. If I could strike a road near Pit River and get to Fort Crook it would save two hundred miles of travel, but we expected that it would take us three days in the "hostile Indian country." Nevertheless, I resolved to try it, although we had no gun and only two working revolvers. We made every preparation, even put on a pistol that would not go off, as a "dummy," put notebooks in pockets, so that if the worst came we might save them, etc.

October 1 we started on our route of supposed peril, and I

will anticipate the trip by saying that of the terrible Hat Creek Indians we saw but two poor fellows gathering grasshoppers for food! Thus vanished the perils and dangers of such a trip.

Well, we started over the ridge, went up among the snow, and struck the headwaters of Hat Creek. We went among rocks, down canyons, along ridges, across treacherous swamps, through chaparral, and at last struck an old emigrant road that looked as if it had not been used for years. We followed along this for some miles and at last struck the Hat Creek road from Battle Creek to Honey Lake, having come through the woods from the Red Bluff road about thirty miles. We struck a band of teamsters passing through to Red Bluff. They told us about the roads, and we had to return some miles to find feed for camp. That night was very cold.

October 2 we were up before four o'clock; the moon was bright and the thermometer down to 20°. We ate our breakfast by moonlight, in the cold morning air, and were off in the twilight. Soon the sun gilded Lassen's Peak. This valley is a surface of lava flow, its top of porous rock, raised in great "blisters"—sort of domes rising from the general surface. In one place there was a cavern where the lava had cooled on the top and the melted interior had run out from beneath the crust, then the top had fallen in.

From Hat Creek (Canoe Creek on your maps) we rose on another lava flow, a table eight hundred feet above Hat Creek, and struck north on this for twenty or twenty-five miles—a dry, rocky slope, no water, and with thin soil. As we sank into the valley of Pit River we had most grand views of Mount Shasta; it looked very sharp and steep and towered up an immense height—it is truly the grandest mountain I have ever seen.

We struck Pit River at the junction of Fall River, and stopped until the next noon. A man named Kaler, a Kentuckian, lives here, and keeps a ferry. He lives alone, and a lonely life he must lead. Bats flitted around his cabin, and spiders crept over the wall.

October 3 we went on to Fort Crook. This lies on Fall River, eight miles north of Pit, on the plain in an open pine forest. We stopped there for two days and three nights. I had to send down the valley and get horses shod—my horse had lost a shoe so long ago that he was getting quite lame. We had a dirty, miserable camp, dusty, without shade, and I felt decidedly rheumatic after my long and cold sojourn at Lassen's Peak. Indians swarmed around our camp, men and women and children, in every style of dress and every state of degradation. We traded old clothes and worn-out blankets for salmon, trout, and bows and arrows. I have a fine bow and a lot of arrows that I shall take home.

Lieutenant Davis, in charge of the post, was very kind and gave us hay for our horses. Except for ten or a dozen men the troops are all away now, fighting Indians. It must be a lazy life, indeed, in such a place.

The valley of Pit River is nearly twenty miles wide at the Fort. It is a lava table, about 3,500 feet above the sea, and has but few settlers. It is pretty, and over portions the soil is fertile; but nights are cold, and it is too far from the rest of the world to be of value yet.

NOTES

1. James Terry Gardner, or Gardiner (1842–1912), was at Sheffield Scientific School for only a brief period in 1862, but was awarded an honorary Ph.B. many years later. Largely for the benefit of his health he accompanied his boyhood friend, Clarence King, across the plains in 1863. Upon his arrival at San Francisco he entered the service of the United States Engineer Corps as a civilian assistant and was assigned to construction of fortifications at Black Point and Angel Island. In the spring of 1864 he joined the Whitney Survey and was a member of Brewer's party that summer. During the next few years he was with King in Arizona, in the Sierra, and on the Survey of the Fortieth Parallel. From 1873 to 1875 he was a member of the Hayden Survey (*U.S. Geological and Geographical Survey of the Territories*). He then returned to New York state, where he became director of the State Survey, 1876–86. Thereafter he practiced

as a civil engineer and engaged in coal mining activities. He had a summer home at Northeast Harbor, Maine. In 1868 he married Josephine Rogers, of Oakland, California, who died in 1872. In 1881 he married Eliza Greene Doane, of Albany, New York. The family name had been spelled Gardiner until James Terry's father dropped the "i." James Terry used the form "Gardner" until mid-life, when he resumed the earlier form.

The meeting with Brewer is described in a letter that James wrote to his mother a few months later. "By stage and cars," he says, "we came to Sacramento and there took the steamboat. It was crowded with people from the mines. Many rough, sunburned men in flannel shirts, high boots, belts, and revolvers were around me, but among them one man attracted my attention. There was nothing peculiar about him, yet his face impressed me. Again and again I walked past him, and at last, seating myself in a chair opposite and pretending to read a paper, I deliberately studied this fascinating individual. An old felt hat, a quick eye, a sunburned face with different lines from the other mountaineers, a long weather-beaten neck protruding from a coarse grey flannel shirt and a rough coat, a heavy revolver belt, and long legs, made up the man; and yet he is an intellectual man—I know it. . . . I went to Clare, told him the case, and showed him the man. He looked at him, and, without any previous knowledge to guide him in the identification, said, from instinct: 'That man must be Professor Brewer, the leader of Professor Whitney's geological field-party.' Clare had never seen a description of Brewer, but had once read a letter written by him [Brewer's letter to Brush about Mount Shasta]. After dinner Clare walked up to this man, the roughest dressed person on the boat, and deliberately asked him if he was Professor Brewer. He was; and Clare introduced himself as a student from Yale Scientific School and was warmly received. He then introduced me and we all spent the evening together. On arriving in this city [San Francisco] Brewer took us to his hotel. The next morning we spent our last money for some decent clothes. Brewer immediately took us around to the State Geological rooms and introduced us to Professor Whitney and the gentlemen connected with the Survey. . . . Through Brewer I was introduced to some civil engineers, who have been valuable acquaintances. In three days Clare was made an Assistant Geologist."

2. Clarence King (1842–1901) became perhaps the most widely known man connected with the Survey. From the moment of his meeting with Brewer he advanced directly and rapidly to the head of geological survey work in America. He served on the Whitney Survey until 1866, organized and directed the United States Geological Survey of the Fortieth Parallel (1867–78), and was largely responsible for the consolidation of various federal surveys into the United States Geological Survey, becoming its first chief (1879–81). His later career as a mining geologist was disappointing. He traveled extensively, was a connoisseur of art and literature, and was an intimate friend of John Hay and Henry Adams. Two of his publications indicate the position he might have attained in literature had he applied himself to writing: *Mountaineering in the Sierra Nevada* (1872); and "The Helmet of Mambrino," in *Century Magazine* (May, 1886). The latter was reprinted in *Clarence King Memoirs—The Helmet of Mambrino,* published for the King Memorial Committee of the Century Association, New York, 1904. The definitive biography is now Thurman Wilkins, *Clarence King. A Biography,* New York, 1958.

3. John Ross Browne (1821–75), native of Ireland, traveler, writer, cartoonist, diplomatist, mining engineer, came to California in 1849. His best known writings are: *Crusoe's Island: With Sketches of Adventure in California and Washoe* (1864); *Adventures in the Apache Country* (1869); *Yusef* (1853); *Report on the Mineral Resources of the States and Territories West of the Rocky Mountains* (1867); *Resources of the Pacific Slope* (1868). Articles by him appeared almost continuously in *Harper's Monthly* from 1860 to 1868. In 1868 he was United States Minister to China (Francis J. Rock, *J. Ross Browne: a Biography* [1929]).

4. In one of Brewer's notebooks the names and addresses are given: R. J. Walsh, Colusi County; Aug. Eastman, Tehama County; Wm. P. Keating, St. Louis, Missouri.

5. The United States Coast and Geodetic Survey figure is 10,466 feet.

CHAPTER VII

SISKIYOU

Frontier Characters—Shasta Valley—Pluto's Cave—Yreka—Indian Styles—Cottonwood—A Sick Man—Deadwood—Scott's Bar—The Klamath—Sciad Ranch—The Siskiyou Mountains—Happy Camp—Sailor Diggings—Arrastras—Lewis' Ranch—Low Divide—A Wager.

Crescent City, California.
November 11, 1863.

It is just a month since I sent my last letter and over a month since I have written anything. I am safely here, alone—King and John have both returned—the rains have set in—and I shall return also the first opportunity—that is, the first steamer—and in the meantime, with this prelude, will continue my journal in the order of events.

October 6 we came on toward Yreka, about fifteen miles, up a volcanic table, through fine forests of pine and spruce.

October 7 we came on twenty-six miles, over tables of lava, the decomposed top forming a rather fertile soil covered with forests, not dense, but of large and beautiful trees—pines, cedars, spruce, and fir—in places the more rocky hills covered with chaparral.

One plain, Elk Valley, three or four miles in width, is without trees, and the views of Mount Shasta rising in its single cone, very sharp, eleven thousand feet above us, its top covered with snow, would delight the painter and enchant the lover of the grand.

There have been some Indian troubles on the road during the summer, and all the people—consisting of but one family, however—have been run off.

We camped by a stream, by a hunter's cabin, known as "Pilgrim Camp." The hunter came in just as we arrived, bringing with him five deer, which he had just shot. We bought the half of a fine, large, fat buck and again luxuriated on venison. We camped in front of his cabin. In the evening an old man came along, a German, from the Moselle, old and gray, who had served ten years under Napoleon. He regaled us with stories of his youthful campaigns. He crossed the plains this summer, from Kansas; the bushwhackers had driven him out and he has taken up a ranch, the previous owners of which were killed by the Indians last year.

We stayed one day at this camp. It is a wild place, no neighbors near; deep snows fall in winter and treacherous Indians infest it in summer. This hunter is a strange character. He is an extraordinary shot, as many an Indian has found—I cannot say to his sorrow, for he never wounds—his first shot, whether with revolver or rifle, is sure death. The Indians have long since ceased to molest him. They hold him in superstitious awe, as they have never been able to hit him with an arrow, while the Indian who made the attempt has always lost his life. No band of savages seems a match for his quick observation and unerring rifle. He is not a young man, rather of middle age, and has a bad reputation, but he treated us cordially. His name is More; he was born in Kentucky, but early ran away from home to the frontiers of Texas, where, between fighting Indians and hunting, he led an adventurous life. Thence he went to the frontiers of Missouri, and then came here. He says that he found the life of the "honest miner" too civilized for him, so he again turned hunter. He has made money here, and still makes it, by the sale of venison and skins, and he hoards his gold like a miser, burying it in the earth. He was very talkative, saying that he had to use his visitors for he did not often get them, and many of his sayings were pithy and witty in the extreme. He is one of those erratic characters with which this state abounds.

October 9 we came on to Shasta Valley, over the pass on the east side of Shasta Peak. This pass is about six thousand feet high and very gradual. The views of the peak were the most sublime we have yet had. We were up to within two thousand feet of the lower edge of the snow, in the sparse timber and pure air of this height. The peak rose over seven thousand feet above us, a *very* sharp cone, against the intensely blue sky. At times light feathery clouds condensed and curled around the peak, but soon dissolved in the warmer air beyond.

We descended into Shasta Valley, a plain of lava, generally barren and desolate, but in places with a thick soil where there are ranches. I was delegated with a message to the first one; I stopped at the cabin, found a squaw with painted face and some pretty, half-breed children—half-white but scarcely half-civilized. We camped at Hurd's Ranch, north of Shasta Peak, which loomed up grandly, over twelve thousand feet above us.

October 10 in the morning we went to visit a cave about three-quarters of a mile distant, just discovered, and of which extraordinary stories were told. It was, indeed, quite a curiosity. It is called Pluto's Cave. The surface of the country is a gentle lava slope, very rocky, with but little soil and with stunted cedars and bushes, the lava rising into innumerable hummocks a few feet high. Under this the cave extends. It looks as if the surface of the great lava flow had cooled, but that the crust had broken somewhere lower down and a long stream of the fluid had run out, leaving a long, empty channel or gallery. The roof of this gallery is beautifully arched—in places it is at least fifty feet high and as many broad. The bottom is of broken blocks of lava, and the sides are occasionally ornamented with fantastic shapes of stone, where the melted or viscous fluid has oozed through cracks, sometimes in a thick, black stream, like tar, then cooled, in others like froth on the surface of the molten mass—but all now cool enough, hard, rough, black rock. We went in near a mile, to the end, or at least to where the fallen fragments blocked up the way. Multitudes of bats lived in it, even to the very end. Near the entrance the roof had broken in in several places, and there were many skulls of mountain sheep that had got in and perished. These are the chamois of the Rocky Mountains and Sierra. They are nearer a goat than sheep, and have enormous horns, hence some hunters call them the "big horn." On one of these skulls the horns were 14½ inches in circumference at the base and 33 inches between the tips.[1]

In the afternoon we rode to Yreka and camped in a field near town. Our way led across the volcanic plain. On the north side rose an innumerable number of small, sharp volcanic cones, the highest but a few hundred feet above the plain. On the south was the majestic form of Mount Shasta, the grand feature in the landscape, the one I never tire in writing about, although you must be tired of the repetition.

At Yreka I found a tremendous pile of letters, no less than fifteen, besides some for King. Some were from friends not heard from for a long time. These letters were read, and on the next day, a quiet Sunday, were again read over and over again, and some few answered. You at home little know the blessed charm that letters can have, their true value to the person that wanders, homeless and desolate, especially when his bed is the ground and his canopy the sky, and when all he holds dear is so far away.

We remained at Yreka three days, camped in a quiet field about a mile and a half from town. We were often visited by Indians—there was a large encampment near us. Some of them were the best looking I had yet seen in the state, far superior to the miserable Diggers of the central part of the state. Some of the squaws were quite pretty, but they had their faces painted in strange ways, often looking absolutely disgusting. Some had streaks of black, others streaks of black and bright red, others red with a red streak running over the top of the head; some appeared as if spattered with black or red or both —mere daubs of color, without any apparent design. I was told that all these styles mean something—married and single women paint differently—but what they mean I did not study out. These Indians are the remains of several tribes, the Klamath, Shasta, Siskiyou, and another tribe—now all united into one which numbers about two hundred warriors.

Yreka is a rather pretty little place, surrounded by low hills, with mining in all the gulches. While here, the temperature sank to freezing every night, but the days were warm.

Wednesday, October 14, we left Yreka and went to Cottonwood, about twenty or twenty-two miles—first across the valley-plain northeast, with grand views of Mount Shasta, then among low hills to Klamath River, which we crossed by a ferry. We camped about a mile from the town of Cottonwood, and as far from the river, in a field, a rather dirty place, without the shelter of trees from the sun by day or the cold of night, and remained there five days.

We stopped at the ranch of a poor man who had a little house, some fields, hay for our animals, a pretty wife and five little children, the oldest apparently not over seven or eight years old. I pitied the poor woman, for he was sick—I did not know how sick until she sent for me to see if I was a doctor or could help him. There was no doctor within twenty-two miles, and to get one costs fifty to seventy-five dollars *per visit*, which the man could ill afford to pay, so he had to put it off until he was not only in intense agony, but his life, I thought, in danger. So I set to work to doctor him, went into town late in the evening for remedies, and happily effected speedy and complete relief. He slept well the rest of the night, the first sleep for some days, and felt well in the morning, so he got up, ran around some, and was, of course, taken down again as a consequence. I will sum up by saying that I doctored him again before I left, to his great relief and the unbounded gratitude of his poor overworked wife. We were supplied with an abundance of good butter and milk while we were there, great luxuries to us, without pay, for they would take nothing. If he had not been relieved that night he would in all probability have died within a week. So much for my medical practice and its happy success.

Cottonwood is a little mining town, once busy and hustling, now mostly "played out," two-thirds of its houses empty, its business dull, the whole place looking as if stricken with a curse. You have no idea of the dilapidation of a mining town in its decline, before it is entirely dead.

We found much of geological interest and were busy enough during our short stay. One day we rode up the valley, crossed the line into Oregon, and climbed the Siskiyou Mountains. The state line was about eight miles from our camp. The view we had was fine, extending south to Mount Shasta and north far into Oregon. Mount Pitt[2] is a grand object, a perfect cone, about nine thousand feet high, rising far above all the surrounding mountains.

The hills have all taken on the colors of autumnal foliage, not so brilliant as we have it in the East, but more so than we have it farther south in this state. Fall weather is coming on. The Sunday we spent there storms played around the peaks of the Siskiyou chain north of us and whitened them with snow.

Here was our last camp. I resolved to send John back to San Francisco Bay with our pack animals and take King and strike across to the ocean. The season is getting too far advanced to live much longer in camp, the clear nights are intensely cold, and without shelter it is impossible to sleep warm. The six nights we were there the temperature sank as follows: 25°, 23°, 15°, 19°, 22°, and the last night 10°—entirely too cold to lie out on the ground under the open sky.

You have no idea what appetites such cold air and such a life promote. In Genesee Valley during six days we three ate thirty-six pounds of beefsteak, besides other food—and on the Shasta trail we three ate forty-four pounds of venison in seven days, besides the other food.

Crescent City.
November 15.

OCTOBER 20 was clear and cold. We ate our last camp breakfast, gave our tin plates, frying pan, coffeepot, etc., to the family where we stopped, for we had our mules packed with specimens, and took our way back to Yreka, where we stopped at a hotel for the night.

October 21 I started John back to San Francisco with one horse and two mules, a long and weary ride of four hundred miles to make alone. I packed up all our baggage except a change of clothes which I carried in my saddlebags, and with King started and rode west over a range of mountains and stopped at Deadwood. Our horses had not been accustomed to grain for some time and were slightly foundered by a mess last night.

Deadwood is a busy little mining town, lively and noisy enough today, for there has been a judicial election, a very important one, all over the state. One of the Democratic candidates for Supreme Judge, a Mr. Todd Robinson, is an out-and-out Secessionist, refuses to take the oath of allegiance, proclaims his Secession proclivities, etc. He had friends here, and the barroom of our house was lively and noisy enough. A motley crowd were drinking, talking politics, and some playing poker.

October 22 we came on to Scott's Bar, near the mouth of Scott River. We rode down Cherry Creek a few miles to Fort Jones, which lies in a beautiful valley near Scott River. Mining is going on along all these streams and many clusters of miners' cabins occur along the route.

From Fort Jones we followed down the lovely Scott Valley for some miles—a rich bottom, with fertile ranches, surrounded with high and very steep mountains, rough and rugged, and furrowed into very deep canyons. Scott River at last flows for some miles in one of these canyons, so we had to go over a high mountain to avoid it. Our road was a mere trail, and the hill was tremendous. We descended the steep slope nearly three thousand feet and struck the river again and followed it down that to Scott's Bar. This was once quite an important town. Placers, rich and abundant, called together a busy and thriving population. Several hotels and stores and many saloons did a thriving business. But the placers are mostly worked out, the population has started after new mines and fresh excitements, over half the houses are empty, four-fifths of the population gone, business has decayed, and the town is dilapidated. We stopped at a rather large hotel, now desolate—its few boarders look lonely in it. It is kept by a rather pretty grass widow, whose husband has left the sinking town and his unfaithful and too frail wife for the northern mines.

The mines are not all exhausted, the deeper bars still pay. Deep excavations are dug below the river bed, large water wheels, turned by the swift current, pump the water out of these claims, and some are paying well. One piece was found this day weighing some two or three pounds, an uncommon good strike. The big wheels creaked dolefully all night long, and seemed to bewail the decline of the decaying town.

From Scott's Bar we followed down the river three miles to the Klamath River, and then followed down that. High mountains rise on both sides, perhaps five thousand feet above the river, and in many places the canyon proper is at least three thousand feet deep. In the bottom of this gorge flows the river, swift and muddy, and precipitous canyons come down from either side. The trail is at times over rocks close by the river, at others it winds over spurs and ridges and abounds in picturesque views.

There was no tillable land as we passed along, but formerly there were rich placers, and ten years ago a large population lived in this canyon, and you will see some places noted on your maps; but all this has passed away, the miner leaves only desolation in his track, and everywhere here he has left his traces.

We passed what was once the town of Hamburg, two years ago a bustling village—a large cluster of miners' cabins, three hotels, three stores, two billiard saloons, and all the other accompaniments of a mining town—now all is gone. The placers were worked out, the cabins became deserted, and the floods of two years ago finished its history by carrying off all the houses, or nearly all—the boards of the rest are now built into a cluster of a dozen huts. A camp of Klamath Indians on the river bank is the only population at present! Their faces were daubed with paint, their huts were squalid. Just below were some Indian graves. A little inclosure of sticks surrounded them. Each grave is a conical mound, and lying on them, or hanging on poles over them, are the worldly goods of the deceased—the baskets in which they gathered their acorns, their clothing and moccasins, arms and implements, strings of beads, and other ornaments—decaying along with their owners.

In contrast with this was a sadder sight—a cluster of graves of the miners who had died while the town remained. Boards had once been set up at their graves, but most had rotted off and fallen—the rest will soon follow. Bushes have grown over the graves, and soon they, as well as the old town, will be forgotten.

Friends in distant lands, mothers in far off homes, may still be wondering, often with a sigh, what has become of loved sons who years ago sought their fortunes in the land of gold, but who laid their bones on the banks of the Klamath and left no tidings behind. Alas, how many a sad history is hidden in the neglected and forgotten graves that are scattered among the wild mountains that face the Pacific!

The population has not entirely left this portion of the river, however. Here and there may be seen a white man, and industrious Chinamen patiently ply with rockers for the yellow dust.

About midway between Scott's Bar and Happy Camp a side stream of considerable size comes in from the northeast, called

Sciad Creek, and here is a fertile little flat of about a hundred acres, the best ranch perhaps in the entire county of Siskiyou. It is known as the Sciad Ranch. We crossed the river by a ferry to it, and stopped two days. It is a delightful spot—it seems an oasis in a desert. Here lives a thriving New York farmer, from Ulster County or Orange County, named Reeves, and he is making money faster than if he were mining for gold. He treated us very kindly indeed and we luxuriated on delicious apples, pears, and plums. His table groaned under the weight of well-cooked food, in pleasing contrast with the miserable taverns of the last few days, and we did ample justice to his good fare.

He came here in 1854, and says that the first year he raised twenty thousand pounds of potatoes per acre, which he sold for *fifteen cents per pound!* But times and prices have changed. His potatoes yield this year about fifteen thousand pounds per acre, and he complains that he gets no price for them—he sells them at four cents per pound, only about $2.40 per bushel—his fruit goes at 12½ cents per pound. The place is a pretty one, picturesque, and fertile. But he wants to get away. He has some pretty little girls growing; who are here caged up from the world, from society, from schools, and all means of improvement—no wonder he wants to sell out.

Just north of this ranch are several high peaks of the Siskiyou Mountains. Three conspicuous points are known as the Three Devils. I climbed one of these. It was a steep slope about four thousand feet above the valley, but several higher peaks lay back of us. Two men from the ranch went up with me, merely for the pleasure of the trip. One was a German who plays the key bugle. He carried it up with him and every little while awakened the echoes of the silent mountains with its notes.

The day was very smoky, and the landscape spread out around us rough in the extreme—the whole region a mountainous one—the peaks five thousand to seven thousand feet high, some indeed much higher—and all furrowed into deep canyons and sharp ridges, many of the former over two thousand feet deep. The hills are covered with scattered timber, not dense enough to be called forests, or in places with shrubby chaparral. With the exception of the ranch below us there is no tillable land; there is nothing to make the region ever a desirable home for any considerable population.

The whole of this wide landscape was bathed in smoky vapor, and the mountains faded in it at no great distance. On a clear day Mount Shasta is in view in the southeast, and the ocean in the southwest, but then both were invisible. It would be difficult to say where the smoky earth ceased and the smoky sky began.

October 26 we left this place and came on thirty miles, over a good but rough trail.

Here let me say that our way is a mere trail—no wagon road enters this part of the state. The region is too rough to admit them, except at an enormous cost of construction, so the county builds a trail, just as wagon roads are built in other places, suitable for riding or packing.

We followed down the Klamath River eighteen miles, the trail abounding in the most picturesque views to be imagined, the mountains rising three or four thousand feet on both sides from the swift river. Once we crossed a spur and rose perhaps two thousand feet or more above the river, commanding a grand view of the canyon beneath, into which we descended again at another turn of the river.

Here and there a poor Chinaman plies his rocker, gleaning gold from sand, once worked over with more profit, but there are few white inhabitants left until we reach Happy Camp.

We passed some huts of Indians and some Indian graves. Over a squaw's grave I noticed a calico dress, such as white women wear, once doubtless a prized article, now fluttering in tatters from a pole stuck in the grave. We passed many deserted cabins and houses during the day—some were once quite neat.

Happy Camp is a group or village of miners, with hotel, saloon, etc., but the place looks on its decline. We merely passed through it, left the river here, and struck north up Indian Creek for twelve miles. There were no houses until we reached Indiantown, where we spent the night. There is some mining here, but not what there once was, the place like all the rest is falling into piteous dilapidation. We stopped at a miserable hole, once a "hotel." Our horses had no hay so they gnawed their ropes and the wooden posts. We fared a little better—we got some salt pork and biscuit.

We found that the dirty, blear-eyed, old, broken-down landlord showed traces of once having had some intelligence. He told us his history, and King chanced to have corroborative evidence of its truth. He was once wealthy, one of the "solid men" of Brooklyn, president of the Northern Transportation Company, a rich and powerful business corporation. He broke and came to California in charge of government stores, in 1847, before the discovery of gold here, and has been here since. Here he grew rich twice, but lost all both times in reckless speculation. Now he is poor enough, looks miserable, broken-down, and sad. There seems no probability of Fortune ever again taking him by the hand as of yore. These histories I so often run against here sadden me and make me pity the poor wretch who makes his grand end and aim of life the acquisition of gold, and who is under the influence of the insane desire to grow suddenly rich.

October 27 we crossed the Siskiyou Mountains to Sailor Diggings, just over the line in Oregon. Our trail led up a hill some three thousand feet or more above our starting point in the morning and stopping place at night—a tremendous hill. The views were very fine. The stupendous canyons—those of Klamath River and its tributaries—the snowy form of Mount

Shasta in the distance, the rugged peaks of the Siskiyou, some of them spotted with snow, the view far into Oregon—all were beautiful. We struck the head of Illinois River and followed it down to Sailor Diggings.

San Francisco.
December, 1863.

SAILOR DIGGINGS, or Waldo, as it is officially called, is a mining town in Josephine County, Oregon, but so near the state line that for a time it voted in California. Here were formerly very rich diggings, and some pay yet. A single piece of gold once found near here weighed over fifteen pounds and was worth over $3,100.

I spent the next day here and rode out ten miles to visit a quartz mine, the only one worked in this region. The principal owner and manager is a very intelligent young German, who treated us very kindly. The quartz is crushed and gold extracted by *arrastra*, the old Spanish method. The machinery consists of a sort of large shallow tub, about twelve feet in diameter and two feet high, the staves of thick plank, the bottom of stones firmly laid in a solid formation. In the center there is an upright shaft with four arms, like the arms of an old-fashioned cider mill. These are stout and short, and to each one several rocks, each weighing several hundred pounds, are fastened by chains. The whole is driven by water power. The quartz is broken with hammers into pieces as large as apples, and several hundred pounds are thrown in this tub with water. The heavy bowlders are dragged over them, grinding and crushing, finally reducing the material to a pulp like thin mush. There is some quicksilver always in the bottom, which runs into the hollows between the stones of the bottom or bed and dissolves most of the gold. After it is thoroughly pulverized it is run through a trough about a foot wide with water in a shallow stream. On the bottom of these troughs are coarse woolen blankets, in the hairs of which the fine particles of gold are caught. Every few minutes these blankets are washed in a large tub of water, which removes this gold.

In this way about twenty-five dollars of gold is extracted from each ton of rock crushed, which is probably scarcely half that it contains. The process is a crude one, and its only recommendation is its cheapness, as a mill of this kind can be built for $2,000 to $5,000, while an improved mill, with stamps, such as I described in a previous letter, costs from $15,000 to $150,000, according to the size and the locality, cost of freight, etc.

October 29 we left Sailor Diggings and went west about twenty-eight miles and stopped at Lewis' Ranch. Sailor Diggings lies in a basin or flat covered with open forests—in places oaks, grand old trees; in others forests of pines.

We rose some three thousand to four thousand feet and commanded a grand view—the distant Cascade Range, in Oregon, the fine cone of Mount Pitt, the rugged mountains of the Siskiyou in the south and southeast; while west, stretching to the distant horizon lay the broad Pacific, blue and quiet. We could see its waters for at least two hundred miles north and south and far out to the west.

We followed along on the crest of ridges for several miles, with deep canyons on every side, the soil barren but supporting a growth of low bushes, scarcely dense enough to be called chaparral, with here and there a small pine or cedar. One species of pine bears cones when but two feet high, and little trees ten feet high were fruitful with them.

The road at last sank into a very deep canyon—perhaps near three thousand feet deep—and steep, but the road is not bad. It is the great artery of supplies for southern Oregon. In this canyon, on the North Fork of Smith River, again in California, we struck Lewis' Ranch, the first house for many miles, and here we found the neatest place and the best supper we had seen for many a day.

This is the center of a new mining district, known as the Rockland District. Copper mines have been discovered this summer, and all is excitement; here we stopped three days. The first two days we spent in exploring the copper leads. Croppings of copper occur along a line of about eight miles, and it is all taken up in claims. Drifts are being run, shafts being sunk, prospecting being done, and many hope for great riches soon—alas, many to be disappointed. Some of the ore looks very fine, and some few of the large number of claims may eventually pay, but more money will be made by shrewd speculation in "stocks" of the infant mines than from the ore derived from them. The region is a very rough one, the canyons deep and steep, the hills rising two or three thousand feet.

Sunday, November 1, was a dull, foggy, drizzly, rainy day, the first we had seen for many a long month—since last February or March. But we were in a very cozy place. We sat in the parlor and chatted with the ladies and had a pleasant time.

Mr. Lewis was born in Georgia but moved to Pennsylvania, lived among the Quakers, and married there. His wife was a Quaker, and uses the "thee" and "thou" in the good old-fashioned style. She has a rather pretty daughter, lately married, and still living at home. They live in a quaint house in a deep canyon, no neighbors near. Although it is a lonely place so far as neighbors are concerned, many teams pass, and the teamsters stop and make the house ring with their noisy mirth. An immense amount of teaming goes over this road into southern Oregon.

We sold our two poor, jaded, worn-out horses here—both, with their saddles and bridles, for fifty dollars. They were two that we bought this summer, and although I had ridden them but two or three hundred miles, yet they had been in the party and I felt like parting with old friends when I bade them goodbye. You cannot imagine how one gets attached to the poor brutes, when you travel with them by day and almost eat and sleep with them, when they have carried you over long and laborious trails, when they are your continual care and anxiety, as if of the family; and when from this long intimacy they have an affection for you almost human—it is not strange that we come to regard them not merely as beasts of burden, but as trusty companions and tried friends. There are two animals especially, a mule and a horse, which have been with us from the start, that I have ridden nearly five thousand miles, to which I feel more attached than I ever imagined I could be to any of the brute creation. We have them yet, and we may yet continue our companionship.

November 2 we footed it to Low Divide, or Altaville, about eighteen miles. The road was very crooked, running over high ridges and sometimes commanding grand views of the wide Pacific and of the surrounding rough landscape. The hills are covered with low bushes, and here and there in the canyons heavy timber, as we approach the sea.

Low Divide is a little town on a sharp ridge—a "low divide," in truth, between higher hills. It is a regular mining town, of miners' cabins, a few stores, saloons, and a "hotel." At this last I stopped the entire week, and a filthier, dirtier, nastier, noisier place I have not struck in the state. The whole scene was truly Californian—everyone noisy. We found that the landlord had killed a pig that afternoon, and over sixty dollars had been lost or won in betting on its weight! You cannot differ with a Californian in the slightest matter without his backing his opinion with a bet.

I will digress and relate an incident that happened at Crescent City a few days later, which illustrates the thing perfectly. I was in the sitting room of the boarding house where I stopped, playing with two little girls on my lap—one of them a puny, slender little thing of three years, the daughter of the landlord. They had a tape measure of mine, playing with it. The father and mother were standing by as I said to the little thing, "What a big girl—half as tall as your papa." This led to a laugh, when I said in earnest to him, "Certainly, she is *half* as tall as you are." "Oh, no," was the reply. "Certainly, she is," I again remarked; "such little children are taller than they seem—she is half as tall as you." "I will bet you fifty dollars on it," was the quick response, and he slid his hand into his pocket and drew out gold to twice that amount. I carefully pulled out fifty dollars, his wife looking on and offering no comment. "Now," says I, "I never bet—you see I have the money, but I will not stake it, for I take no man's money in that way—but now let me show you how easy it is for you to lose, betting on a thing you have never given thought to before." He still was willing to stake his money. I stood his little girl down, she measured 2 feet 11½ inches; he, although a stout man, was but 5 feet 9 inches. He was astonished, of course, and *then* his wife lectured him on betting. Such is human nature.

Well, to go back to Low Divide. Copper ore occurs here, scattered over quite an extent of country. A great number of claims are taken up and much work is being done on them. Only one of the mines, out of over thirty, has paid expenses, and this has produced as yet scarcely over five hundred tons of ore, sold at perhaps $50,000. But all *hope* to get rich. I trudged over the hills by day and sat in the dirty barroom or saloon during the evenings, and watched men lose their earnings at poker.

King went to Crescent City, and, finding letters for him calling him back to San Francisco, he started immediately for Jacksonville, 120 miles, where he could take stage. He took the barometer and our beautiful Indian bows that we had carried all the way from Pit River. He got safely back with the barometer, but the bows were stolen from him in Marysville, after we had carried them some seven or eight hundred miles. It was too bad.

NOTES

1. The Bighorn or Mountain Sheep (*Ovis canadensis californiana*). John Muir tells of these animals in a chapter, "The Wild Sheep," in *The Mountains of California*, and a chapter, "Wild Wool," in *Steep Trails*.

2. Now known as Mount McLoughlin, a name used as early as 1838. The name Pitt is a corruption of Pit, derived from Pit River, so called because of pits dug by the Indians for trapping game (Lewis A. McArthur, *Oregon Geographic Names* [Portland, 1928]).

CHAPTER VIII

CRESCENT CITY AND SAN FRANCISCO

Redwoods—Crescent City—Indian Wars—A November Storm—Driftwood—To San Francisco by Steamer—Population—Advantages of San Francisco—The Vigilance Committee—Adieu to 1863.

San Francisco, California.
December, 1863.

My last letter left me at Low Divide, the region of the principal copper mines of Del Norte County. We had rainy weather there a part of the time, which increased my discomforts—standing at night in a crowded barroom, with seats for half a dozen, while twenty or thirty wet, dirty men from the mines steamed around the hot stove. To go to bed was no relief. We slept on the floor upstairs, some twenty or twenty-five of us—they kept running in and out all night. The noise from below prevented sleep until late, and the last of the card players would be getting to bed in the morning after the first risers were up.

But all things must come to an end, and so must this. Saturday, November 7, I footed it four miles down the road to a little Dutch tavern, where I got the luxury of a clean bed and a clean table, yet had to sit with wet feet and cross feelings all the evening and listen to a drunken miner who was determined to enlighten me on the subject of geology.

I got to talking with our fat, clever little Dutch landlady, who was perfectly delighted to find that I had passed her father's house, and must doubtless have seen it, in the Kingdom of Württemberg, and that, too, since she had left it. It made us good friends. I stopped there two days and looked at some copper mines near, which were unfortunately pure, unadulterated "wild cat." It was a rather pretty place—I could hear the roar of the distant ocean all night long, and the sunsets over the hazy Pacific were beautiful.

From there I walked to Crescent City, sixteen miles. I descended from these barren hills, and soon a marvelous change in the vegetation came on the scene. I passed through a forest, called everywhere "The Redwoods," a forest of redwood timber —and such forests the world probably does not show elsewhere.

As I have before told you, the redwood is a sort of gigantic cedar. It has foliage like the hemlock, and wood like coarse, poor red cedar. The forest is narrow, and mostly made up of gigantic trees—large groups of trees, each ten to fifteen feet in diameter, and over two hundred feet in height, the straight trunks rising a hundred feet without a limb. The bark is very thick and lies in great ridges, so that the trunks seem like gigantic fluted columns supporting the dense canopy of foliage overhead. They generally swell out at the bottom, so that a tree but ten feet in diameter at thirty feet high, will be fifteen or more at the ground. They grow so abundant that the sun cannot penetrate through the dense and deep mass of foliage above. A damp shady atmosphere pervades the forests, and luxuriant ferns and thick underwood often clothe the ground. Large trees fall, mosses and ferns grow over the prostrate trunks, trees spring up among them on the thick decaying bark. The wood is so durable that a century may elapse before the fallen giant decays and mingles with its mother earth. In the meantime, trees a century old have grown on it, their bases twelve to fifteen feet from the ground, sustained on great arches of roots that once encircled the prostrate log upon which they germinated. A man may ride on horseback under some of these great arches.

The wood is soft and brittle, like pine. Fire sometimes in the summer spreads through the woods. The thick bark protects the large trees, but the fire often gets into the wood and burns out the dry center, leaving a cavity as large as a small house. A reliable gentleman, Doctor Mason, told me that once caught in a storm he with his four companions sought shelter in one of these hollows. They built a fire, and the five men with their five horses passed the night in this novel shelter! I fully believe the story.

At times a number of trees start from the same base—an immense woody mass thirty to forty feet in diameter, with half a dozen huge trunks rising from this great gnarled base. Mosses accumulate in the hollows and nooks, bushes and ferns take root and grow parasitic among the trees to a great height, trailing lichens festoon from the branches and give a somber look to the dense shade. I saw in one place a tree six to eight inches in diameter that had taken root thirty or forty feet from the ground on the trunk of a larger tree, its roots twining like great serpents over the bark until they reached the ground.

The largest tree that I measured was fifty-eight feet in circumference, and looked three hundred feet high. It was perfectly symmetrical. Much larger trees are reported at various places along the coast. These trees belong to the same genus with the Big Tree of the Sierra. There are two species of the genus: one only in the Sierra, the other only near the sea—both of them grand wonders of the vegetable world. The amount of timber in one of these trees is almost incredible. A man will build a house and barn from one of them, fence a field, probably, in addition, and leave an immense mass of brush and logs as useless.

These forests have almost an oppressive effect upon the mind. A deep silence reigns; almost the only sound is that of some torrent coming down from the mountains, or the distant roar of the surf breaking upon the shore of the Pacific.

A restless genius who has seen much of the world and of adventure, but is yet poor, has conceived the idea of a speculation in one of these trees, which is rather novel. There is a large tree near Trinidad so near the water that he proposes to build a large boat of it and launch it—cut from a single tree, like a canoe, but modeled like a schooner. He says he can have a boat of twenty-four or twenty-five foot beam, and eighty to one hundred feet long, and have the sides, the bottom, and floor of the grand saloon of the natural wood, the interior being dug out of the solid tree. A great saloon fifteen to eighteen feet wide in the widest part, but tapering aft, and sixty or seventy feet long, would be fitted up in it. A vessel of over three hundred tons register, or near five hundred tons burden could thus be built! He proposes to take his novel craft to the larger seaports of the world and make a fortune by its exhibition. However, he will probably never have the means to carry it out.

Crescent City lies on a little plain of a few miles in extent that juts out from the hills into the sea. A little cove, but no harbor exists here. A town was built up here ten years ago, which grew rapidly, but like too many Californian towns it has passed its zenith of prosperity and is on its decline. Scarcely half of its houses are occupied; the rest are deserted, their windows broken, their looks dilapidated, giving a sad note to the place. A dense forest of tall firs and spruce once grew just back of the place, but this was killed by fire, and now the dead and blackened trunks, many of them over two hundred feet high, hundreds in number, stand like specters haunting the city. The main business street lies on the sandy beach, so close to the water that it has houses on but one side, and the water of extreme high tides comes up across the streets to the very stores, which look out on the lovely cove.

I met an old friend, a Mr. Pomeroy, at this place, and spent some very pleasant evenings with him and his wife. They are from Massachusetts. I stopped in the place twelve days, making excursions in every direction and looking at various coal and copper mines, but we had much rain, and often I was not out at all. One of these excursions carried me through the redwoods in two other places; I wish it had been oftener.

Quite a number of Indians live in the city, and not a few white men have squaws for their wives—a sad feature of the civilization of many of these back places. One sees as many half-breed children as he does pure bloods of either race. What is to become of these half-breeds, and what their situation is to be in the future society of various parts of this country, is a serious problem. It is a good American doctrine that a man not entirely white has few rights or privileges that a pure white is bound to respect, and as abuse and wrong has thus far failed to civilize and raise the Indian, it is, indeed, a serious problem.

The Indian wars now going on, and those which have been for the last three years in the counties of Klamath, Humboldt, and Mendocino, have most of their origin in this. It has for years been a regular business to steal Indian children and bring them down to the civilized parts of the state, even to San Francisco, and sell them—not as slaves, but as servants to be kept as long as possible. Mendocino County has been the scene of many of these stealings, and it is said that some of the kidnapers would often get the consent of the parents by shooting them to prevent opposition. This was one cause. Some feeling arose between the races, and doubtless the Indians stole cattle —at least the whites accused them of it, and retaliated fearfully.

About three years ago fragments of two or three tribes were at Trinity Bay, or Humboldt Bay. The warriors were out hunting and fishing farther north, while the women, children, and the old and infirm were left on an island near Eureka. Some "bold" whites saw a chance for an easy victory. They went in the night to this island and *murdered the whole of these people!* Women, children, infants at their mothers' breasts, decrepit, infirm, and aged people were killed in cold blood and with the most revolting cruelty. Some of these squaws had white husbands, some of the children were half-breeds. *Over a hundred were slain!* The husbands, sons, and brothers of these victims swore eternal revenge, and fearfully have they gratified it. Men, women, and children have been alike murdered. They take no prisoners—their white foes took none. Desolated farms, the ashes of dwellings, and mutilated dead mask their track. They have nearly depopulated Klamath County and made life unsafe for near two hundred miles in the coast ranges.[1]

Of course, the innocent people suffer. And yet these hostile Indians are but very few—not two hundred are left. They are the desperadoes and outlaws from several tribes, with whom the friendly tribes have no dealings. Nothing short of their absolute extermination can bring peace, and it is a costly matter. They are well acquainted with all the intricacies of the mountains, they are brave to desperation, and live only to wreak their vengeance on the race that has wronged them. The three counties they infest are the only ones in the state we have not visited.

But in talking about these Indians I have again wandered from Crescent City. On Saturday night, November 14, a tremendous storm raged on this coast. It did immense damage to the shipping at San Francisco, and sank the ship that had on board the iron "monitor" sent out to protect the harbor. At Crescent City it was heaviest in the night, with the heaviest southeast wind the place has seen for ten years. There were no vessels to injure, but it did other damage. The wind blew down some unoccupied houses. It blew down a shed attached to the house where I was stopping and did damage there to the amount of about $200. The surf broke up so high that it brought driftwood and heavy logs up to the doors of the front buildings. I could not sleep all night—the breaking of the surf seemed almost like the booming of artillery. The wind partially ceased, but the surf was heavy all day Sunday, and the wind returned, but with less violence on Sunday night.

As before remarked, a flat stretches out into the sea here, about five or six miles wide and ten to fifteen miles long. It is mostly covered with very heavy timber and a tangled undergrowth of ferns and bushes, but here and there are openings where pretty farms abound. There are some lakes in here, beautiful sheets of water. I went out to one—with grassy swamps around it and rushes and reeds growing up in the shallow margin. Dense dark forests surrounded it. There were a few canoes tied up to the shore, and by two cabins that I passed were some really beautiful half-breed children. Myriads of ducks and geese and other waterfowl swarmed, and some white swans and pelicans enlightened the scene. These waterfowl, especially ducks, are very abundant. I saw a hunter, an Indian, coming in town with a horse loaded with them. He must have had a hundred. They cost only $1.50 per dozen, and I luxuriated on wild ducks all the time I stayed there.

The floods of two years ago brought down an immense amount of driftwood from all the rivers along the coast, and it was cast up along this part of the coast in quantities that stagger belief. It looked to me as if I saw enough in ten miles along the shore to make a million cords of wood. It is thrown up in great piles, often half a mile long, and the size of some of these logs is tremendous. I had the curiosity to measure over twenty. They were worn by the water and their bark gone, but it is not uncommon to see logs 150 feet long and 4 feet in diameter at the little end where the top is broken off. One I measured was 210 feet long and $3\frac{1}{2}$ feet at the little end, without the bark.

In the afternoon of Friday, November 20, unexpectedly, the steamship *Oregon* arrived from San Francisco, and suddenly the town was all astir. It did not leave until the afternoon of the next day, when we went aboard about 3 P.M. There is no dock—the steamer anchors two or three miles from the shore. Surfboats come up as close as possible. We got into a cart, which was drawn out by a horse to the boat. We clambered in and were rowed to the ship. This cost us two dollars. We were soon under way, and that evening I sat and listened to the purser tell stories of Utah. He spent eighteen months among the Mormons. We had a stiff breeze all night, and it was clear, but the ship rolled heavily. I slept well, however, although it has been three years since I have been on shipboard.

All day Sunday we ran down the coast. The wind was fair and we made fine headway. The coast ranges were in sight, some of the higher peaks covered with snow. A beautiful night set in, but the breeze was heavy. The moon was light and we ran so close to Punta de los Reyes that we could hear the barking of the innumerable seals and sea lions that thronged the rocks there. A Russian man-of-war had been wrecked there but a few days before.

We ran into the Golden Gate about midnight. It is a most beautiful entrance to a more beautiful harbor. The whole scene lay so lovely in the soft moonlight that I stayed on deck until we anchored in front of the city. Monday, November 23, we had a lovely morning. Another steamer lay at our wharf, so that we did not get ashore until nearly noon.

I found the men all back at the office except Gabb, who came a week later. Ashburner and King were just starting for the Mariposa Estate—they went that afternoon. I once more donned the habiliments of civilization and went and took rooms at a boarding house with the same landlady that I had last winter, who has moved to new and more comfortable, as well as more fashionable, quarters. I met many old friends, among them Mrs. Ashburner, had a most pleasant dinner, and thus again began civilized life, calling on Mrs. Whitney in the evening.

All hands are now very busy in the office, hurried the worst kind—the old monotonous life has set in and it is irksome enough. To leave the free open air for the confined office and bedroom, and the laborious outdoors work for writing, is a great change and is irksome.

San Francisco, California.
December 27.

SAN FRANCISCO is not only the metropolis of the state, but in reality the most prosperous portion, growing the fastest, and the growth being healthy. Most of the interior towns of the state are at best growing but very slowly, and a large majority are actually decreasing in population. In fact, the state is. This will surprise you, but it is true, and arises from several causes.

First, the newly discovered mineral regions in Arizona, Idaho, Nevada, and Colorado get most of their roving, adventurous population from this state, where there are tens of thousands who have long hoped in vain to get suddenly rich, who have no ties to bind them to any spot, who love adventure and pursue it to each new region. This class, ever on the move, has furnished not less than 150,000 men to these other new regions, and the state is the loser in point of population. These all come from the mines, and the mining towns, as a consequence, decline as the placers become poorer and the population leaves.

Next, there is an enormous preponderance of males in the population—in some counties there are, on the average, eight men to one woman! Even in this city there are 20,000 more men than women. As a consequence, the natural increase of the population is far less than in a population normally constituted. The women, what there are of them, are prolific and fruitful to a satisfactory degree—there is no complaint on that score—it is their lack of numbers from which the population suffers.

These are the two principal causes that check the increase of population. Another, but smaller cause is found in the men who have made some money and return East to enjoy it.

But these causes, which have been at work at large in the state, have not checked San Francisco. *Its* growth has been rapid, it has grown as if by magic. Fifteen years ago two or three ranch houses and barren sand hills marked the spot; today it is a city of over 100,000 inhabitants, and growing fast.

Since I arrived here three years ago building has been going on at an almost incredible rate. I live now in a fine, large boarding house, with stores under it, on a growing and fashionable street. When I arrived streets were laid out there, through barren sand hills, with here and there a sort of shell of a house standing.

The first day of last January the first street railroad car started. There was, indeed, a sort of street railroad to the Mission Dolores, three miles distant, but not regular street cars. Today they run through all the streets, some of them running over three miles—there must be over a dozen miles of street railroads in active operation.

Here is a healthy climate. When the interior is scorching with intense heat in summer, this is cool with sea breezes. When, in winter, fathomless mud abounds in the interior, here are more pleasant days than elsewhere. Everyone who can lives here, at least a part of the year, and miners, when out of work or full of dust, come here to spend their money and enjoy themselves.

This is not only the great seaport of the state, but of the western coast of America—there is not another *good* harbor between Cape Horn and the Bering Straits, and this is not only a good one, but one of the very finest in the world—so this place must ever be of necessity the commercial metropolis of the Pacific—the New York of an immense region, not only of this state, but the center of commerce for the whole coast—all parts must pay tribute to it. Capitalists seeing this, invest their money here. They make it elsewhere in the state, perhaps—in mines or trade—but invest it here. Huge buildings have gone up this year, built with money made in Washoe, but invested here.

The place is in such easy communication by bay, rivers, and coast, to most of the rest of the state, and is so easy of access, that it is gradually absorbing the trade of the smaller interior towns, and it fattens on their decay. All these and other causes make the city what it is, and lead to such bright hopes of the future.

A part of the city is scattered over steep hills, but most of it is built on sand flats that stretch along the bay or are built out into it. The location is lovely. A range of hills six or eight miles wide separates us from the ocean. The city fronts east, and across the bay, which is here about six or seven miles wide, little villages are growing up. Oakland is the largest, and grows as Brooklyn does, only it is farther off and grows slower. A new railroad has just been opened along the west side of the bay to San Jose, fifty-six miles distant.

The city abounds in fine mansions, substantial buildings, palatial hotels, and all the accompaniments of a large city—the only thing strange is that it has grown in fifteen years.

It is the best-governed city in the United States—there is less rowdyism than in any other city I know of in America. This will surprise you. Previous to 1856 it was terrible—its fame for murder and robbery and violence spread over the world. It was even vastly worse governed than New York, by the vilest of all politicians. They held the elections, and by election frauds, double ballot boxes, etc., legally kept the power. Robbers were policemen and murderers were judges. The life of any respectable man who dared raise his voice against the iniquities of officers was endangered, and from the corruptions of the courts there was no redress. The most prominent citizens were shot in the streets.

At last the people rose in their might and formed the celebrated Vigilance Committee. This was composed of the best and most prominent citizens, who usurped the government, chose leaders, made courts, tried and executed or banished criminals, and enforced decrees with the bayonet and revolver. At the tap of the alarm bell all stores were closed, and ten thousand armed men were in the ranks to enforce justice, though not law. They publicly hung a few of the worst offenders and banished many of the less prominent ones. They held control of the city until election, when decent officers were elected. They appointed a committee to nominate officers for the government of the city—the ticket called the Citizens' Ticket or People's Ticket, the nominees being chosen from *both* political parties. No man of this committee could hold office. This goes on still. The committee is changed yearly, the old one nominating a new committee, all of business men, and they cannot nominate one of their own number to any office. How unlike the caucuses of the roughs in eastern cities.

Well, from that time the city has been well governed; roughs have tried to get the upper hand once or twice but have been most overwhelmingly defeated. Once, indeed, three or four men were nominated by the committee itself who were not good; an independent meeting was called, a new ticket was made out on which the regular nominees were retained if they were decent men, but rejected if not, and it carried the city. So much for the city government—it is not perfect, but compared with New York City it is as far ahead of that, as that is ahead of the Fiji Islands.

We will bid adieu to the year 1863, thankful for its mercies and penitent for its sins—and look with hope toward the new year which approaches. I have traveled 4,243 miles this year, making a total in the state, since I came, of: horseback (or mule) 6,560 miles; on foot 2,772; public conveyance 4,175— a total of 13,507 miles, or enough to reach more than halfway around the earth.

NOTE

1. The massacre on Indian Island in Humboldt Bay took place on the night of February 25, 1860. There is an account of it in A. J. Bledsoe, *Indian Wars of the Northwest* (San Francisco, 1885), pp. 302–309. The *Humboldt Times* (Eureka, California) of March 3 also describes it and mentions attacks on other groups of Indians the same night. "The whole number killed at the different places on Saturday night," says the *Times*, "cannot fall far below a hundred and fifty, including bucks, squaws and children." The event acquires importance because of an indirect effect upon American literature. Bret Harte was at that time living in Humboldt and was employed on the Uniontown paper, the *Northern Californian*. "Harte

was temporarily in charge of the paper, and he denounced the outrage in unmeasured terms. The better part of the community sustained him, but a violent minority resented his strictures and he was seriously threatened and in no little danger. Happily he escaped, but the incident resulted in his return to San Francisco." (Charles A. Murdock, *A Backward Glance at Eighty* [San Francisco, 1921], p. 79.)

BOOK V
1864

CHAPTER I

SAN JOAQUIN VALLEY—GIANT SEQUOIAS

Off for the Field Again—Pacheco's Pass—A Reunion —San Luis de Gonzaga Ranch—Drought—The San Joaquin Valley—Mirage and Whirlwind—The Passing of Jim—Visalia—Foothills—Thomas' Mill —Sequoias—Two Giants.

Pacheco Pass.
May 29, 1864.[1]

We left San Francisco on the afternoon of May 24. All were delighted to get off again. Before leaving we had a pleasant little dinner at a French restaurant and in the evening three of us called on J. Ross Browne. We had a very jolly time; he drew an amusing caricature of us, with our big boots, woolen shirts, and closely cropped hair.

The party consists of Hoffmann, King, Gardner, and Dick. The first two have been with me much before. Gardner is a very nice young man, an engineer. He and King give their time, the Survey paying only their traveling expenses in return for their time and labor. Dick is a young man from Pike,[2] who goes as packer, etc.—a very good fellow, but most unfortunately knows nothing about packing, so we have to teach him and I think that he will do well. Packing is an intricate art. To put a load of baggage on a mule and make it stay there, and at the same time not hurt the mule, is a great art. We have but two pack-mules, scarcely enough, to tell the truth.

We came on eighteen miles and stopped at Haywards that night, and were nearly devoured by mosquitoes. The next day we went on to San Jose, thirty-one miles. We stopped for noon at Warm Springs and, for the first time in this state, because of our camp shirts did not dine in the public dining room, but when it was ascertained who we were there were very funny apologies.

At San Jose I called on Mr. Hamilton. His congregation is building a fine church, but the earthquake a few days ago frightened them much. I met there Mrs. Mead, Mrs. Hamilton's mother. She looked scarcely older than when I had seen her last, nine years ago, in Maine.

May 27 we came on up the San Jose Valley, twenty-one miles. The day was intensely hot, 97°, the air scorching and dusty. The drought is terrible. In this fertile valley there will not be over a quarter crop, and during the past four days' ride we have seen dead cattle by the hundreds. The hot air trembled over the plain, and occasionally a mirage seemed to promise cool weather ahead, only to vanish as we approached. The mountains on either side were bathed in a haze and seemed all tremulous in the heated air.

We got to the 21-Mile House and camped under the old oak trees. We had camped there before, once in '61, and again in '62. The spot seemed familiar and awoke pleasing memories, and that night, on the ground under the trees, sweeter sleep came than had for many a long night before.

Yesterday we came on here, thirty miles farther. We followed south to Gilroy, then struck east and entered the pass, and are now camped under the same tree that we camped under two years ago when we came through the same pass. The hills are terribly dry, totally bare of forage, parched and brown. The scattered oaks, evergreen, seem dark—they are so tenacious of life that the drought fails to kill them. It must be a terrible year for the thousands of sheep that are kept here. A hot wind sweeps up the canyon in which we are camped. It is 92°, and the air intensely dry. It always draws through this gap in an almost ceaseless current—the trees and bushes are all bent by it. There is a little tavern here, a stage and telegraph station.

A little event occurred this morning, so characteristic of California that I must relate it. A traveler stopped at the house last night, tall, swarthy, heavily bearded. As I stood by this morning, along came another, with gun on his shoulder, knife in his belt. A heavy beard covered his face; his dress, his whole look told of a rough life. He eyed the stranger a minute. Then ensued the following conversation, and much more of the same sort—only the names and strong oaths I leave out:

B. "Is your name Smith?"

S. "Yes."

B. "Why, how do you flourish? My name is Brown—don't you know me?"

S. "No."

B. "Why, d—n it, we were together at —— in Mexico, in '47."

S. "By G—d, that's so, but I didn't know you—you've changed."

B. "Certainly—I was a boy then but seventeen years old."

S. "And I but eighteen. How did you escape?"

B. "By G—d, the d—n greasers couldn't keep me."

S. "Where do you live?"

B. "Up here in the mountains."

S. "How long you been here?"

B. "Since '49. Where do you live?"

S. "By G—d, nowhere, and I have been almost everywhere since I saw you."

B. "To the States?"

S. "Yes, half a dozen times."

B. "Where's the old man?"

S. "Father?"

B. "Yes."

S. "In old Arkansaw, fighting Yankees I suppose. Have you seen any of the other boys?"

B. "Yes, lots of 'em. You remember Bob?"

S. "G—d, yes, don't I—where's he?"

B. "Gone to Idaho, saw him last winter. They call him 'Cherokee Bob' here—you know he is very dark."

S. "He ain't no Cherokee, he's a regular white man."

B. "Yes, but since he killed that feller in Stockton he has gone by the name of Cherokee Bob."

S. "Has he been back?" (Meaning East.)

B. "Yes, but he killed a man in Missouri and had to leave again. He's killed four in this state but always got off. He's a d—d lucky fellow."

S. "Where's Bill?"

B. "Dead. Killed about four year ago. There's Dick, too, he got shot in Utah."

S. "Where's ——, that tall feller, you know?"

B. "Oh, he's in this state, a shiftless cuss. Been mining, but ain't worth a cent."

S. "D—n me, it does me good to see an old friend. I hain't seen one of those old boys afore since I came here fifteen year ago. I left here soon. Let's drink."

The two imbibe and recall old reminiscences, and each goes on his way. Much of the conversation, of course, I don't report, only a part of it. But the whole thing was so Californian that I had to tell it. They had been fellow prisoners, taken by the Mexicans in the Mexican War.

> Under a Tree, Tulare Plain.
> Sunday, June 5.

MONDAY, May 30, we came on to San Luis de Gonzaga Ranch, at the eastern entrance of the pass. Our road lay over the mountains. They are perfectly dry and barren, no grass—here and there a poor gaunt cow is seen, but what she gets to eat is very mysterious.

As we cross the summit the Sierra Nevada should be in view,

with its sharp outline and cool snows; but not so—we look out on the dry plain, which becomes more indistinct and finally fades away into the hazy air, shutting out like a veil all that lies beyond. The wind blows heavily over the pass, and we descend to the San Luis Ranch. The wind is so high that we can build no fire, so we cook in the dirty kitchen. Dust fills the air —often we cannot see fifty yards in any direction—it covers everything. We cook our dinner, but before it can be eaten we cannot tell its color because of the dirt that settles on it. Our food is gritty between our teeth, and as we drink out our cups of tea we find a deposit of fine sand in the bottom. Dirt, dirt, dirt—eyes full, face dirty, whole person feeling dirty and gritty.

All around the house it looks desolate. Where there were green pastures when we camped here two years ago, now all is dry, dusty, bare ground. Three hundred cattle have died by the miserable water hole back of the house, where we get water to drink, and their stench pollutes the air.

This ranch contains eleven square leagues, or over seventy-six square miles. In its better days it had ten thousand head of cattle, besides the horses needed to manage them. Later it became a sheep ranch, and two years ago, when we camped here, it fed sixteen thousand sheep besides some few thousand cattle. Now, owing to the drought, there is no feed for cattle, and not over one thousand sheep, if that, can be kept through the summer. The last of the cattle, about one thousand head, were lately sold for $1,500, or only $1.50 each! Such is the effect of the drought on one ranch.

We spent a miserable night there, the wind and dust almost preventing sleep, and paid fourteen dollars in gold for the hay that our seven animals ate.

May 31 we came on to Lone Willow, a stage station out on the plain, where there has been a sheep ranch until the present year. The ride was over the plain, which is utterly bare of herbage. No green thing greets the eye, and clouds of dust fill the air. Here and there are carcasses of cattle, but we see few living ones—not twenty during the day, where nearly as many thousands could have been seen two years ago. There is a sink hole of alkaline water, by which stands the "lone willow," the only tree for many a weary mile. Our camp here is as dirty, dusty, and miserable as the last. There is a well that supplies water for drinking that is poorer than any you ever tasted, yet quite good for the region.

June 1 we came on to Firebaugh's Ferry, on the San Joaquin, twenty-five miles. Portions of this day's ride, for miles together, not a vestige of herbage of any kind covered the ground; in other places there was a limited growth of wire grass or alkali grass, but not enough to make it green. Yet cattle live here—we passed numbers during the day, and countless carcasses of dead animals. We camped at Firebaugh's, where we got hay for our animals and took a grateful bath in the cold San Joaquin. The bad water, dust, alkali, and our change of diet begin to tell on the boys, but all are cheerful.

June 2, to Fresno City. For the first ten miles the ground was entirely bare, but then we came on green plains, green with fine rushes, called wire grass, and some alkali grass. The ground is wetter and cattle can live on the rushes and grass. We now came on thousands of them that have retreated to this feed and have gnawed it almost into the earth.

The air is very clear this day—on the one side the Coast Range loomed up, barren and desolate, its scorched sides furrowed into canyons, every one of which was marvelously distinct; on the other side the distant Sierra, its cool snows glistening in the sun and mocking us on our scorching trail. We camped by a slough of stinking, alkaline water, which had the color of weak coffee. It smelt bad and tasted worse, and our poor animals drank it protesting. We drank well water which looked better and tasted better, but I think it smelt worse. But in this dry, hot, and dusty air we must drink, and drink much and often.

At Fresno City we got barley but no hay. I cannot conceive of a much worse place to live, unless it be the next place where we stopped; yet here a *city* was laid out in early speculative times, streets and public squares figure on paper and on the map, imaginary bridges cross the stinking sloughs, and pure water gushes from artesian wells that have never been sunk.

June 3 we came on to Elkhorn Station, an old Overland station. We came southeast across the plain. The day was hot, as usual, but not so clear. The mountains were invisible through the dusty air; the perfectly level plain stretched away on every side to the horizon, and seemed as boundless and as level as the ocean. It is, in fact, sixty miles wide at this place, and neither tree, nor bush, nor house breaks the monotony. Thus we slowly plodded our weary way over it, league after league, day after day. During the entire day we saw beyond us, behind us, sometimes all around, the deceitful mirage. I never cease to wonder at this phenomenon, although it has been so long a familiar thing. It looks so like water, its surface gently rippled by the wind, clear and sparkling, trees and mountains as vividly reflected in it as in genuine lakes! But it always vanishes as you approach it—heated air, and not cool water, we find in its place.

During these days whirlwinds stalked over the plain. The high winds I spoke of as occurring near Pacheco's Pass ceased. Fitful, often hot, puffs blew first this way and then that, giving rise to little whirlwinds that looked like waterspouts at sea, moving for a time over the plain, then breaking and vanishing. They were continually about us during the heat of the day. Sometimes they were slender columns of dust but a few feet in diameter and several hundred feet high; at others the columns were larger. Sometimes they were like cones with bases upward; then again they would break and throw out branches which fell down on all sides with beautiful effect—all the time moving over the plain, some slowly, some swiftly. It is not uncommon to see a dozen of these at once, and I have counted twenty-seven at one time.

We camped at Elkhorn Station, nearly in the center of the plain. There is some feed here, and a well supplies the cattle with water, poor though it is. Again we got barley, but no hay.

Here a calamity befell us. I was awakened at about midnight by our mule, Jim. He was sick and in a terrible agony. Poor feed, change of diet, bad water, alkali, dust, heat—all had probably combined to produce the result. We watched with him all night, bled him, gave him such remedies as we thought best under the circumstances, but at six o'clock in the morning he died. He was our most valuable animal, a most excellent mule, worth $150 or $200. He had been a faithful beast, was very sagacious and very true, and had been with us since we started at Los Angeles, nearly four years ago. I did not think that I could feel so sad over the death of any animal as I did over that faithful old mule, who has been our companion for so long a time and under such varied circumstances. He died near the house. I hired a man to drag him away. We left him out on the plain to the vultures and coyotes, both of which species are fat this year, for the starving cattle have been their harvest. Luckily a wagon from this place was going into Visalia and I sent in his saddle and pack.

June 4, yesterday, we came on here to Kings River. Here we struck good water again, good hay for our animals, and fine oak trees for shade. And what a relief! We are again in good spirits. Last night, again, we had a camp fire, and the boys sang songs.

Camp 164, Thomas' Mill.
June 14.

I sent you a long letter from Visalia a few days ago, which brought up the history of our wanderings to Kings River, where we were camped June 5. The next day we came on to Visalia, twenty-four miles. In a few miles we passed the belt of oaks that skirts the river for a couple of miles on each side; then across the barren, treeless plain, still perfectly level—in places entirely bare, in others with some alkali grass. The surface of the soil was so alkaline that it was crisp under the horses' feet, as if covered with a thin sheet of frozen ground.

Before reaching Visalia we again struck timber. The region about Visalia is irrigated from the Kaweah River, and is covered with a growth of scattered oaks—fine, noble, old trees. The town is a small place on the plain, but very prettily situated among the fine trees. That night was intensely hot, and we roasted in the hot beds of the hotel where we stopped. We stopped there the next day, getting provisions and a horse to supply the place of old Jim who died on the plain. There is a camp of cavalry here and I had authority to call out an escort, should it be deemed necessary on account of hostile Indians, so I called on the commandant. The soldiers were anxious to get into the mountains and begged me to make the requisition, but I shall wait until I see some need of it, which I do not anticipate.

June 8 we packed up a part of our provisions and came on. We left most of them to be brought up to our present camp by ox teams that are hauling lumber from this mill. We came on twenty-nine miles that day, through a most intense heat, at times above 100°. We struck east until we reached the hills, then up a canyon northeast. Soon after we started we passed the belt of trees, then over the barren plain, so dry and so hot. The heat was so bad that it nearly made us sick. The low foothills are as dry as the plain—no grass, but covered with a growth of scrubby oaks and bushes scattered over them. Our poor animals got but little to eat that night, but we had a glorious night's sleep in the cool open air.

We were up at dawn the next day and off early, and at noon came to camp by a stream of delicious water, where our hungry animals had good grass. Our course had been up all the time since striking the mountains. We had now got up some three thousand feet, had passed lower, dry foothills, and had just struck the region of pines. Grand old trees grew in the valley where we camped and over the neighboring ridges, large, but scattered, hardly forming forests. And how delicious the cool, pure mountain water tasted—our first real good water for many a long day! In the afternoon I climbed a high point above camp, commanding a fine view of the surrounding region.

Friday, June 10, we came on but four miles to this camp. Up, up, up, over a high ridge, and at last into a dense forest of spruces, pines, firs, and cedars. We then sank into a little depression where there is a beautiful grassy meadow of perhaps two hundred acres, surrounded by dense, dark forests. Here there is a steam sawmill, where two or three families live.[3]

And here let me describe this delightful camp, so refreshing after the monotony, heat, dust, alkali, discomfort, and tedium of the great plain. The level, grassy meadow lies in front, with a rill of pure, cold water. Ridges are all around, clothed with dark pines and firs, with here and there the majestic form of some scattered Big Trees, the giant *sequoias* that abound here, although so rare elsewhere. We are at an altitude of over five thousand feet, or just about one mile above the sea. We are far above the heat and dust of the plain. It has been cold every night—from 23° to 32°—the days cloudless, the sky of the clearest blue, the air balmy and so cool that it is just comfortable without our coats. You cannot imagine the relief we feel both by day and night after the discomfort of the previous two weeks.

As I have said, the Big Trees are abundant here, scattered all along between the Kings and Kaweah rivers. We are on the south branches of Kings River. Saturday we all went up on the ridges about a thousand feet above to see the largest trees. I will describe but two. The largest one standing is 106 feet in circumference at the ground and 276 feet high. But it swells out at the base, so that at twelve feet from the ground it is only seventy-five feet in circumference. It is finely formed, and you can but imperfectly imagine its majesty. It has been burned on one side, and were it entire its circumference at the base would be 116 to 120 feet!

Now for the other tree. It is prostrate and no larger, but the story seems bigger. It has been burned out so that it is hollow, and we *rode* into it seventy-six feet and turned around easily. For forty feet *three* horsemen could ride in abreast, but we had but one horse along, which we took up on purpose to take this wonderful ride. Nor is it difficult, for most of the cavity is nine feet high and as wide. The greatest width is 11 feet 6 inches, and the greatest height is 11 feet 8 inches in the clear. Our horse was very gentle, and in this part I *stood* erect in the saddle and could just fairly reach the top! The tree is broken in two places, and fire has widened the fracture. At 120 feet from the base the tree is still 13 feet 2 inches in diameter inside of bark, and at 169 feet it is still 9 feet in diameter inside of the bark! All of these measurements were carefully made with an accurate tape line, except the height of the largest tree, which was got by measuring a base and triangulating.

There are trees of this species of every size, many being over twelve feet in diameter. Two of the smaller ones have been cut and split into fence posts—how it takes away the romance of them, using them for fence posts—and in a few years more many of the smaller ones will be sawn into lumber by the mill here!

About six miles east of this is a high bald mountain, about eight thousand feet, which we ascended, and a description of the view will answer for any of the higher points near. It commands a view of the whole western slope of the Sierra, the snowy peaks on one side, the great plain on the other.

Along the crest, twenty-five miles east, are the rugged snow-covered peaks that we hope to explore. The western slope is rough in the extreme, and both the topography and aspect are unlike anything else I have seen. The region is so *very* rough that I am filled with anxiety as to the possibility of reaching it. On the west the great plain stretches away to the horizon, the Coast Range being shut out by the hazy hot air that hovers over it.

NOTES

1. There are no letters in 1864 prior to this date. Brewer's notebooks indicate that he was in San Francisco from January to May.

2. Meaning Pike County, Missouri. Richard D. Cotter went to Alaska after his season with the Whitney Survey, suffered a severe illness, and later settled in Montana. He wrote to Professor Brewer from time to time between 1870 and 1898. In a letter from York, Lewis and Clark County, Montana, September 18, 1898, he says: "I have been a J.P., Post Master, & the last office I was honored with was Sabbath School Supt. I imagine I can see you laughing as I have often seen in Camp. Of course I had other work to do or I should have starved to death long ago." The correspondence indicates that he was miner, rancher, carpenter, and Jack-of-all-trades.

3. The mill was built by Joseph H. Thomas. It was situated a short distance west of the present General Grant National Park. The big trees described are in or near the park.

CHAPTER II

THE HIGH SIERRA OF KINGS RIVER

Forests and Meadows—A View of the Sierra—Penetrating the Mountains—Granite—Venison and "Biled Owl"—Mount Silliman—Grizzlies—Sugarloaf Creek—Roaring River—Ascent of Mount Brewer—Clarence King and Dick Cotter—Independence Day at Brewer Lake—Adventures of King and Dick—To Visalia with a Toothache—Return by Moonlight—The Soldier Escort—Kings River Canyon—Search for a Trail.

Camp 166, near the Big Meadows.
June 22, 1864.

I AM in camp today and will begin another letter, although the air is cold and my fingers stiff. I sent my last from Thomas' Sawmill about a week ago.

We had some provisions brought up to that camp by lumber teams, and on their arrival we started. June 17 we left that camp and came on about eight miles. We left all trails behind at the mill, but we had looked out a way beforehand. We struck back on the divide between the Kaweah and Kings rivers, where an old Indian trail formerly ran. We divided our baggage between all the animals and walked, for the way was terrible. At times it lay over and along the ridge, in forests of fir and pines, and then over rocky hills and up steep slopes—so steep that our animals could hardly cling to them. We passed hundreds of the Big Trees, which are everywhere scattered through the forests here.

At last we struck a little meadow surrounded by forests,

where we camped. It was at an altitude of about 7,400 feet. I went beyond, on a granite knob, where I had a grand view over this rough region, with the snowy peaks ahead, which gave me a lively sense of the difficulties we would have to surmount. That night the thermometer stood at 19°.

June 18 we came on here, about eight miles farther, over a region fully as rough, sometimes through forests, and at others over and among rocks. We at last struck a trail that has recently been cut for the purpose of bringing in cattle. We came to camp here, by a little meadow, where our animals have good grass and we plenty of wood and water. It is at an altitude of about 7,800 feet. Here is a succession of grassy meadows—one called the Big Meadows is several miles in extent—and some men have cut a trail in and have driven up a few hundred cattle that were starving on the plains. Back of these lie the sharp and snow-covered peaks of the crest. We have now been here four days. The thermometer has not been up to 50° by day, and at night it has sunk to 18°, 16°, 17°, and 20°—surely not comfortable nights for sleeping without shelter.

Monday, June 20, three of us went on a peak about five miles east to spy out the country.[1] The view was grand—on the west the whole slope of the Sierra and the great plain, ending in haze—around us the roughest region imaginable—along in front the crest of the Sierra, its more prominent points not less than twelve thousand feet high, with rocks, precipices, pinnacles, canyons, and all the elements to make a sublime landscape. We were up about 9,700 feet, and it was very cold —only three degrees above freezing—and yet there were trees three to four feet in diameter. We had a weary and rough walk back by missing the way; but a hearty supper awaited us and we slept well that night regardless of the cold.

The nineteenth was Sunday and we stayed in camp. I shall not work Sundays this year as I did last; the state can afford to do without it. I will not use myself up as I did last summer.

Yesterday, June 21, the longest day in the year, ought to

have been one of the hottest. I stayed in camp alone, while Dick and King went out to shoot a bear, if they could. Hoffmann and Gardner went on a ridge for bearings. At about noon it began to snow violently, and it continued all the afternoon. You cannot imagine how uncomfortable it is without any shelter. The boys got back wet and numb with the cold, but I had made a big kettle of soup, which was pronounced an eminent success. We made our preparations for an uncomfortable night, but at about sundown it cleared up very suddenly and today is again sunny.

Before we started on this trip I heard that there were hostile Indians somewhere in here, driven out of Owens Valley, so I wrote the Governor, and he to the Commandant of the Pacific, who in turn issued orders to the various military posts within two hundred miles of here to furnish me with a military escort if I should demand it. At Visalia there is a company of cavalry. The order had been received and the soldiers were very anxious to get away and begged me to make requisition. I did not, hoping that I would not need them. This morning two men from Visalia were here, an underofficer and a private, who wanted to go. As it will please the men, cost no one anything, more than it would to have the soldiers stay in camp, and as I thought they might be useful even if we found no Indians, I sent in a requisition with one of them for an escort to join us in two weeks, when we will strike north into the region possibly hostile, but most probably not.

Camp 168.
Sunday, June 26.

WHEN I wrote the other day my fingers were cold and another snowstorm threatened; in fact, it did snow a little, but the storm passed east of us.

June 23 the weather still looked unsettled. I went to a hunter's camp and got forty pounds of dried venison and bear meat, and then went on a hunt with him. He shot one deer and had left three others where he had shot them a few days before.

He gave me the meat of two and packed them out where we could get them the next day.

There is a succession of granite flats and meadows and low granite knolls about that camp, the whole covered with forests. On my way back to camp I got bewildered and lost—for the first time in the state. I walked about two miles to a granite ridge where I could see the country, saw where I was, and then went to camp. It could hardly be called "lost in the Sierra," as I was not detained an hour, yet it was uncomfortable for the time.

The weather looked better and the next day, June 24, we left that camp. We cached all the baggage that was not absolutely necessary, and came on about eight miles, over hills, through brush and forests, among rocks, and finally came to camp in a little grassy meadow in a canyon. We were surrounded by forests of firs and pines. We were tired and the amount of venison we ate for dinner might seem fabulous were it stated.

June 25, yesterday, we came on about eight miles farther, and so rough was the way that we found this distance a good day's work. Our route lay along the divide between the head branches of the Kings and Kaweah rivers, over steep ridges, some of them nearly ten thousand feet high, and then along ridges covered with forests of subalpine pines and firs. There are two species of pine and one of fir. All grow to a rather large size, say four to five feet in diameter, but are not high. All are beautiful, the fir especially so, but there is difference enough in the color of the foliage and habit of the trees to give picturesque effect to these forests, which are not dense. All have a very dark green foliage, in harmony with the rugged landscape they clothe. The ground under the trees is generally nearly bare. There is but little grass or undergrowth of either herbs or bushes.

The rocks are granite, very light colored, the soil light-gray granite sand. Here and there are granite knobs or domes, their sides covered with loose angular bowlders, among which grow bushes, or here and there a tree. Sometimes there are great slopes of granite, almost destitute of soil, with only an occasional bush or tree that gets a rooting in some crevice. Behind all this rise the sharp peaks of the crest, bare and desolate, streaked with snow; and, since the storms, often great banks of clouds curl around their summits.

The whole aspect of this region is peculiar; the impression is one of grandeur, but at the same time of desolation—the dark pines, the light granite, the sharp cones behind, the absence of all sounds except the sighing of the wind through the pines or the rippling of streams. There is an occasional bird heard, but for most of the time silence reigns. At night the wind dies down, the clouds disappear, if any have occurred during the day, and everything is still. During the night there is no sound. The sky is very clear and almost black; the stars scarcely twinkle, but shine with a calm, steady, silvery light from this black dome above.

Our present camp is by a little meadow, at an altitude of about 9,500 feet.[2] The barometer stands at less than 21½ inches, water boils at 193.5 degrees. Yet when one is still and not climbing he does not perceive the lightness of the air. It is a calm Sunday. The sky is intensely blue, a few white clouds float above, but it is cold in the shade, only 43° to 44°, and my fingers are cold enough.

Since writing the above we have got dinner and the weather has become warmer, a few degrees. We had a most glorious venison soup. We eat venison three times a day. A few days ago, before we got the deer, the boys shot a large arctic owl, an enormous fellow. They dressed and cooked him. I have often heard of "biled owl," but this is the first time that I have practically tested it, and it is nothing to brag of—strong, tough, and with a rather mousy taste.

We have a book of sermons in camp, and thus far we have had one read aloud each Sunday. A laborious week lies ahead, and when I next write I hope to have seen the high peaks.

Sunday, July 3.

WHEN I wrote a week ago we were camped at a little meadow on the divide between the Kings and Kaweah rivers. Not over a mile to the east an entire change of country begins. We exchange the granite hills that have the form of domes for those rising into pinnacles and sharp peaks.

Just east of that camp we climbed a steep hill, and came suddenly to a precipice. Beyond was a great basin, or valley, the head of which is an immense rocky amphitheater, the rocky sides very steep, in places tremendous perpendicular precipices. There is a little lake in this basin, about 1,600 feet below the brink of the cliffs. From this there widens a valley, which runs directly back toward the crest of the Sierra a few miles and then turns, the waters finding their way to Kings River by a deep canyon.[3]

We went on only about three miles and camped in a canyon where there was a little grass.[4] We explored the amphitheater that day, but in the afternoon it began to rain. All the evening it rained. We turned into our damp blankets and prepared for a miserable night's sleep, but it stopped raining about nine o'clock and then cleared up, although the trees dripped water all night, and it was cold—the thermometer sank to 25°.

I find it too cold to write more. I must defer until a warmer day or warmer camp.

In camp on the south fork of Kings River.
July 7.

IT is a pleasant, clear day. For three days the sky has been of the intensest blue, not a cloud in sight day or night. I am alone in a beautiful camp and I will write.

We have come down into a deep valley, where it is warmer and there is good grass. We are still camped high, however—about 7,500 feet. A fine breeze plays up the valley, very pleasant, but it makes it hard to write—it flutters the paper and gives much trouble. The desolate granite peaks lie in sight—bare granite and glistening snow. It freezes every night.

Tuesday, June 28, we had a fine clear morning, and four of us started to visit a peak a few miles distant. We had a rough trail, over sharp ridges, and finally up a very steep pile of granite rocks, perhaps a thousand feet high, to the peak, which is over eleven thousand feet high, and which we called Mount Silliman, in honor of Professor Silliman, Junior.

In crossing a ridge we came on fresh bear tracks, and soon saw the animal himself, a fine black bear. We all shouted, and he went galloping away over the rocks and into a canyon. We had gone but a short distance farther when we saw a very large female grizzly with two cubs. She was enormous—would weigh as much as a small ox. After we looked at her a few minutes we all set up a shout. She rose on her hind legs, but did not see us, as we sat perfectly still. We continued to shout. She became frightened at the unseen noise, which echoed from the cliffs so that she could not tell where it came from, so she galloped away with the cubs. These would weigh perhaps 150 pounds each; she would weigh perhaps 900 pounds or more. We also saw a fine buck during the trip.

We reached the summit after a hard climb, and had a grand view of the rough landscape. Great rocky amphitheaters surrounded by rocky ridges, very sharp, their upper parts bare or streaked with snow, constituted a wild, rough, and desolate landscape. Clouds suddenly came on, and a snowstorm, which was a heavy rain in camp. We got back tired enough.

The next day King, Gardner, and I took some sketches. Dick went after a deer, but saw only a bear. The animals strolled off, and several hours were consumed in getting them, during which the boys saw more bear and deer.

June 30 we were up early and left. We changed our route and came on about ten miles, by such a terrible way that it was a hard day's work—over rocks, through canyons and brush.

We sank into a canyon and camped about two thousand feet below our last camp. We had some trouble with our fire—it got into some dead logs and we feared a general burn. We fought it, and Gardner came near being bitten by a rattlesnake that was driven out. He was an enormous fellow, but had lost most of his rattles.

July 1 we came on by a still rougher way, about eleven miles. We crossed the south fork of Kings River, down over tremendous rocks and up again by as rough a way. We struck a ridge which is a gigantic moraine left by a former glacier, the largest I have ever seen or heard of. It is several miles long and a thousand feet high.[5]

We were working back toward high peaks, where we hoped to discover the sources of Kings, Kaweah, and Kern rivers, geographical problems of some considerable interest and importance. We got back as far as we could and camped at an altitude of 9,750 feet, by a rushing stream, but with poor feed. Wood was plenty, dry, from trees broken by avalanches in winter. A beautiful little lake was near us.[6] About five miles east lay the high granite cone we hoped to reach—high and sharp, its sides bristling with sharp pinnacles.

Saturday, July 2, we were up at dawn, and Hoffmann and I climbed this cone, which I had believed to be the highest of this part of the Sierra. We had a rough time, made two unsuccessful attempts to reach the summit, climbing up terribly steep rocks, and at last, after eight hours of very hard climbing, reached the top.[7] The view was yet wilder than we have ever seen before. We were not on the highest peak, although we were a thousand feet higher than we anticipated any peaks were. We had not supposed there were any over 12,000 or 12,500 feet, while we were actually up over 13,600, and there were a dozen peaks in sight beyond as high or higher!

Such a landscape! A hundred peaks in sight over thirteen thousand feet—many very sharp—deep canyons, cliffs in every direction almost rivaling Yosemite, sharp ridges almost inaccessible to man, on which human foot has never trod—all combined to produce a view the sublimity of which is rarely equaled, one which few are privileged to behold.

There is not so much snow as in the mountains farther north, not so much falls in winter, the whole region is drier, but all the higher points, above 12,000 feet are streaked with it, and patches occur as low as 10,500 feet. The last trees disappear at 11,500 feet—above this desolate bare rocks and snow. Several small lakes were in sight, some of them frozen over.

The view extended north eighty to ninety miles, south nearly as far—east we caught glimpses of the desert mountains east of Owens Valley—west to the Coast Range, 130 or more miles distant.

On our return we slid down a slope of snow perhaps eight hundred feet. We came down in two minutes the height that we had been over three hours in climbing. We got back very tired, but a cup of good tea and a fine venison soup restored us.

Sunday, July 3, we lay until late. On calculating the height of the peak, finding it so much higher than we expected, and knowing there were still higher peaks back, we were, of course, excited. Here there is the highest and grandest group of the Sierra—in fact, the grandest in the United States—not so high as Mount Shasta, but a great assemblage of high peaks.

King is enthusiastic, is wonderfully tough, has the greatest endurance I have ever seen, and is withal very muscular. He is a most perfect specimen of health. He begged me to let him and Dick try to reach them on foot. I feared them inaccessible, but at last gave in to their importunities and gave my consent. They made their preparations that day, anxious for a trip fraught with so much interest, hardship, and danger.

July 4 all were up at dawn. We got breakfast, and King and Dick packed their packs—six days' provisions, blankets, and instruments made packs of thirty-five or forty pounds each, to be packed into such a region! Gardner and I resolved to climb the cone again, as I had left instruments on the top, expecting someone would go up. Our way lay together for five miles, and up to thirteen thousand feet. I packed Dick's heavy pack to that point to give him a good start. I could never pack it as far as they hope to. Here we left them, and as we scaled the peak they disappeared over a steep granite ridge, the last seen of them.

Gardner and I reached the summit much easier than Hoffmann and I had two days before. The sky was cloudy and the air cold, 25°. We were on top about two hours. We planted the American flag on the top, and left a paper in a bottle with our names, the height, etc. It is not at all probable that any man was ever on the top before, or that any one will be again—for a long time at least. There is nothing but love of adventure to prompt it after we have the geography of the region described.[8]

We were back before sundown; a hearty dinner and pleasant camp fire closed the day. We sang "Old John Brown" around the camp fire that night—we three, alone in these solitudes. Thus was spent Independence Day. The last was with Hoffmann alone, in the Sierra farther north. We heard not a gun. Would that we might know the war news—we are over a month behind.

The next morning we lay in our blankets very late, after the fatigue of the previous day—in fact were in bed eleven hours. We stayed in camp and took latitude observations. It was a most lovely day.

That camp was in a valley that runs back to the cones. High granite ridges rose to above thirteen thousand feet on both sides; that on the south rose in great precipices, nearly perpendicular, over two thousand feet high. Patches of snow lay in the nooks and corners. This granite is of a uniform light ash-gray color, inclining to pearly, and by the lights of sunset showed the most beautiful rosy tints. Scraggy pines grew in the crevices up to eleven thousand feet, gnarled and twisted by the winter storms of these desolate regions. One new species I found here, not known to botanists. By day the sky is generally of a deep blue-black, by night almost black; the stars shine with a mild silvery luster almost without twinkling.

Camp 180, on the ridge north of Kings River.
July 21.

KING and Dick got back in five days, and had a tremendous trip. They got on a peak nearly as high as Mount Shasta, or some 14,360 feet, and saw five more peaks still higher. They slept among the rocks and snow one night at an altitude of twelve thousand feet, crossed canyons, and climbed tremendous precipices, where they had to let each other down with a rope that they carried along. It was by far the greatest feat of strength and endurance that has yet been performed on the Survey. The climbing of Mount Shasta was not equal to it. Dick got his boots torn off and came back with an old flour sack tied around his feet.[9]

Upon their return we went back to the Big Meadows, as we were out of flour, salt, bacon, and sugar—in fact had nothing but venison and beans to eat. An escort of seven soldiers had been there several days waiting for us. They were having a good time and were eating venison at a heavy rate. A severe toothache had set in two days before—I spent sleepless nights and was incapacitated for a week. I had ridden along with the party, but was in intense agony all the time. It ulcerated badly, and Tuesday, July 12, I started for Visalia, sixty miles distant, to have it out. King was going down in order to take another trail and reach the high peaks and the region which had been inaccessible to us.[10]

We started early in the morning and rode twenty-five miles before noon. The trail went down, down, down all the time—we sank five thousand feet in that twenty-five miles. Most of the way the trail led through magnificent forests. The giant *sequoias*, or Big Trees, were abundant; they occurred for several miles along the trail—hundreds of them from fifteen to twenty-five feet in diameter.

At twenty-five miles we struck a pleasant ranch, in a little valley, where we stopped all the afternoon. It was a nice place, and we got two very nice meals—the first square meals for some time and we did them ample justice. It was hot, we dared not ride farther by day, but just at sundown we were off again.

It was moonlight until midnight, but we rode all night and got into Visalia just after sunrise. Once we missed the way, and for two hours plodded over the plain. Just before daylight the ulceration in my jaw broke, and what a relief it was. My face was badly swollen, and for over fifty hours it had been terrible—by far the worst toothache I had ever had. We got our breakfast, then went to bed, and slept until noon. It was intensely hot, and we felt it, coming from the cool mountains. I had the tooth pulled that day, and stopped there over one night and two days. I got an escort of two men from the camp to assist King.

Thursday evening, July 14, I started back alone. I rode about twenty-five miles before the moon went down. I had got into the foothills and could not see my way, so I pulled off the saddle and lay down under a tree and slept two hours, when day dawned and I went on. I got my breakfast at a miserable cabin, where I had a vivid idea of what stuff some people can live on. I got to Lewis' Ranch, the nice place I spoke of, about nine or ten o'clock, and I stayed there all day and night, as it was too far to go on that day. I got back to the Big Meadows Saturday afternoon.

Hoffmann and Gardner had been to Thomas' Mill and had obtained the provisions we had there. We were now ready, and the next day, Sunday, July 17, we started and came on about seven miles. There are seven soldiers with us, fine fellows, who are right glad to get out of the hot camp at Visalia.[11] They are mounted, armed with Sharp's carbines and revolvers, and have a month's rations on three pack mules. We made quite a cavalcade—eleven men and sixteen animals—and left quite a trail. We followed back on an old Indian foot trail, a hard trail. Once, one of their pack-mules upset and tumbled down some rocks. He was bruised and cut, but not seriously injured. We camped at a fine meadow. The boys saw a bear, but he got away.

Monday, July 18, we continued, and in about four miles came on a camp of half a dozen men, prospectors, who had crossed the mountains from Owens Valley and had worked their way thus far.[12] Never before were so many white men in this solitude. Three of them were going back, and luckily for us, showed us the way into the canyon of Kings River.

It was a horrible trail. Once, while we were working along the steep, rocky side of a hill, where it was very steep and very rough, old Nell, our pack-mule, fell and rolled over and over down the bank *upward of a hundred and fifty feet*. Of course, we thought her killed. She rolled against a log which stopped her, but a part of her pack went farther. Strangely, she was not seriously hurt. We got her back to the trail, put on her pack, and she has packed it since. A bag of flour went rolling down the hill, burst, and we lost a part of it.

We sank into the canyon of the main South Fork of Kings River, a *tremendous* canyon. We wound down the steep side of the hill, for *over three thousand feet*, often just as steep as animals could get down.

It begins to rain. I must quit.

July 23.

I AM spending today in camp, the first for two weeks and I will go on with my story.

We got into the canyon of the South Fork of Kings River, and forded the stream, which is quite a river where we crossed, and camped at a fine meadow in the valley. It was a very picturesque camp, granite precipices rising on both sides to immense height. The river swarmed with trout; I never saw them thicker. The boys went to fishing and soon caught about forty, while the soldiers caught about as many more.

We left there the next morning and worked up the valley about ten miles. Next to Yosemite this is the grandest canyon I have ever seen. It much resembles Yosemite and almost rivals it.[13] A pretty valley or flat half a mile wide lies along the river, in places rough and strewn with bowlders, and in others level and covered with trees. On both sides rise tremendous granite precipices, of every shape, often nearly perpendicular, rising from 2,500 feet to above 4,000 feet. They did not form a continuous wall, but rose in high points, with canyons coming down here and there, and with fissures, gashes, and gorges. The whole scene was sublime—the valley below, the swift river roaring by, the stupendous cliffs standing against a sky of intensest blue, the forests through which we rode. We would look up through the branches and see the clear sky and grand rocks, or occasionally, as we crossed an open space, we would get more comprehensive views.

We camped at the head of this valley by a fine grassy meadow where the stream forked. On both sides rose grand walls of granite about three thousand feet high, while between the forks was a stupendous rock, bare and rugged, over four thousand feet high. We luxuriated on trout for the next two meals. The rattlesnakes were thick—four were killed this day.

The next day, July 20, we started in different directions. Hoffmann and Gardner climbed the cliffs on the south side. They got up two thousand feet by hard climbing, only to find walls a thousand feet above them which they could not scale. I explored a side canyon, to the south, where the Indian foot trail ran, to see if we could get out that way with our animals. I had a grand climb, but found the way entirely inaccessible for horses. I followed up a canyon, the sides grand precipices, with here and there a fine waterfall or series of cascades, making a line of foam down the cliffs. I climbed over bowlders and through brush, got up above two very fine waterfalls, one of which is the finest that I have seen in this state outside of Yosemite. I had a hard day's work.

In the meantime a soldier had explored another canyon, and reported that we could get out to the north that way,[14] so the next day we started, and came to this camp. It was worse than any of our other trails. We are not over 4½ or 5 miles from our last camp, and have come up over four thousand feet!

It was heavy for our animals. Twice we had very steep slopes for a thousand feet together, where it seemed at first that no animal could get up with a pack. Once our pack horse fell, turned a complete somersault over a bowlder, and landed below squarely on his feet, when he kept on his way as if nothing had happened. His pack remained firm and he was not hurt in the least. Fortunately it was not so steep there. There were places where if an animal had once started he would have rolled several hundred feet, but all went safely over. We camped at a little over nine thousand feet where we are now, by a meadow on the hillside where we have a grand view of the peaks in front and the canyon beneath us.

Yesterday Gardner and Hoffmann went on a peak about twelve thousand feet, which commands a comprehensive view of all the ground we have been over lately; while two soldiers, Dick, and I explored ahead for a trail. We were unsuccessful, but we got on a ridge over eleven thousand feet high that commands a stupendous view. The deep canyons on all sides, the barren granite slopes, clear little lakes that occupy the beds of ancient glaciers, the sharp ridges, the high peaks, some of them rising to above fourteen thousand feet, like huge granite spires—all lay around, forming a scene of indescribable sublimity.

We killed a rattlesnake at ten thousand feet. I have never before seen them so high in the mountains. Dick also killed a grouse, a fine bird nearly as large as a big hen, and splendid eating. We had it for breakfast this morning.

We thought the region north impassable for our animals,

but Hoffmann and a soldier both saw another way they thought practicable, and three have gone today to explore it. If that too should prove impracticable we shall be in a hard fix and will have to make our way to Owens Valley and cross the Sierra again at some point north.

NOTES

1. Probably Shell Mountain.
2. Near J. O. Pass.
3. Sugarloaf Creek, flowing into the canyon of Roaring River.
4. Probably at the head of Clover Creek.
5. The route was down Sugarloaf Creek, across Roaring River near Scaffold Meadow. Brewer, King, and the Whitney Survey reports call Roaring River the south fork of Kings River.
6. At the head of Brewer Creek.
7. This was the first ascent of Mount Brewer. Brewer and Hoffmann must have attempted to climb the nearest face; had they gone around to the south or a little farther to the north they would have found it easier.
8. "Thirty-one years later, C. L. Cory, Harvey Corbett, and I [Joseph N. LeConte] ascended the peak from the west, but did not find the record left by Professor Brewer, as deep snow covered the summit. But in 1896, after ascending from the east, one of the members of the party found the bottle containing the record, which was carefully removed, and which I succeeded in photographing. It bore the record of but one other climber during the period of thirty-one years. The record remained on the summit for a number of years after that, but the fragile paper was broken by continual handling, and it was finally removed to the Sierra Club rooms for preservation, where it was unfortunately destroyed in the fire of 1906" (Joseph N. LeConte, in *Sierra Club Bulletin*, XI, No. 3 [1922], 252).
9. The experiences of King and Cotter are set forth with considerable spirit in King's *Mountaineering in the Sierra Nevada*. A somewhat more restrained account appears in the official report of the Survey (*Geology*, I, 384–387). They reached the summit of Mount Tyndall (14,025 feet) and from that point beheld and named Mount Williamson (14,384 feet) and Mount Whitney (14,496 feet). Lack of provisions prevented an attempt to reach the latter.
10. On this occasion King reached the mountain, but made a bad choice of routes and failed to attain the summit. The first ascent was not made until 1873. For a further account of Clarence King's efforts, see Francis P. Farquhar, "The Story of Mount Whitney," in *Sierra Club Bulletin*, XIV, No. 1 (February, 1929).
11. The names of the men are given in Brewer's notebook: Judd, Webster, Cole, Orr, Bump, Spratt, Heisley.
12. A statement by one of these prospectors, Thomas Keough, is given in *Sierra Club Bulletin*, X, No. 3 (1918), 340–342.
13. John Muir expressed the same opinion in an article, "A Rival of the Yosemite," in the *Century Magazine* (November, 1891).
14. Copper Creek.

CHAPTER III

OWENS VALLEY AND THE SAN JOAQUIN SIERRA

The Prospectors' Trail—Kearsarge Pass—Owens Valley—Desert Characteristics—Incidents of Inyo—Indian Signals—A High Pass—The Upper San Joaquin—Attempt to Reach Mount Goddard—Newspapers in Camp—Canyons of the San Joaquin—Hoffmann Seriously Ill—Pushing On to Clark's Ranch—To Yosemite with Mr. Olmsted—Notice of Professorship at Yale.

Camp 189, on the middle fork of the San Joaquin.
August 5, 1864.

I WROTE last on July 23, while the boys were exploring for a way to get north. They were unsuccessful, and I decided to cross the summit to Owens Valley.

Sunday, July 24, we remained in camp for latitude observations, but the forenoon was cloudy and the afternoon rainy—heavy showers. In keeping instruments dry I got my blankets wet, and although it cleared up at evening, I had a rheumatic night in my wet blankets. There were slight showers during the night, but no heavy rain, and the next morning was clear. We packed up, got back into the canyon by our steep trail, killed a tremendous rattlesnake on the way, and camped again at the head of the valley, where we had been a week before. One of the soldiers caught a fine mess of trout. We have seen deer in abundance, but have not succeeded in getting any lately.

Some prospectors had come over the summit to this place, as I told you, and we resolved to follow their trail, assuming that where they went we could go. Tuesday, July 26, we started and got about eleven miles, a hard day's work, for we rose 4,300 feet. First we went up a steep, rocky slope of 1,000 to 1,500 feet, so steep and rough that we would never have attempted it had not the prospectors already been over it and made a trail in the worst places—it was terrible. In places the mules could scarcely get a foothold where a canyon yawned hundreds of feet below; in places it was so steep that we had to pull the pack animals up by main strength. They show an amount of sagacity in such places almost incredible. Once Nell fell on a smooth rock, but Dick caught her rope and held her—she might have gone into the canyon below and, with her pack, been irretrievably lost. We then followed up the canyon three or four miles and then out by a side canyon still steeper. We camped by a little meadow, at over nine thousand feet. Near camp a grand smooth granite rock rose about three thousand feet, smooth and bare.[1]

July 27 we went over the summit, about twelve miles. The summit is a very sharp granite ridge, with loose bowlders on both sides as steep as they will lie. It is slow, hard work getting animals over such a sliding mass. It is 11,600 feet high, far above trees, barren granite mountains all around, with patches of snow, some of which were some distance below us—the whole scene was one of sublime desolation. Before us, and far beneath us lay Owens Valley, the desert Inyo Mountains beyond, dry and forbidding. Around us on both sides were mountains fourteen thousand feet high, beneath us deep canyons.[2]

We descended down the canyon of Little Pine Creek[3] and camped at a little meadow, in full view of the valley below and the ridges beyond, which were peculiarly illumined by the setting sun. On both sides of us were great rocky precipices. During the day's progress we passed a number of beautiful little lakes.

Thursday, July 28, we were up at dawn and went to Owens River, sixteen miles. Six miles brought us out of the canyon on the desert—then ten miles across the plain in the intense heat, and we camped on the river bank, without shade or shelter, the thermometer 96° in the shade, 156° in the sun. Yesterday in the snow and ice—today in this heat! It nearly used us up.

Owens Valley is over a hundred miles long and from ten to fifteen wide. It lies four thousand to five thousand feet above the sea and is entirely closed in by mountains. On the west the Sierra Nevada rises to over fourteen thousand feet; on the east the Inyo Mountains to twelve thousand or thirteen thousand feet. The Owens River is fed by streams from the Sierra Nevada, runs through a crooked channel through this valley, and empties into Owens Lake twenty-five miles below our camp. This lake is of the color of coffee, has no outlet, and is a nearly saturated solution of salt and alkali.

The Sierra Nevada catches all the rains and clouds from the west—to the east are deserts—so, of course, this valley sees but little rain, but where streams come down from the Sierra they spread out and great meadows of green grass occur. Tens of thousands of the starving cattle of the state have been driven in here this year, and there is feed for twice as many more. Yet these meadows comprise not over one-tenth of the valley—the rest is desert. At the base of the mountains, on either side, the land slopes gradually up as if to meet them. This slope is desert, sand, covered with bowlders, and supporting a growth of desert shrubs.

Here is a fact that you cannot realize. The Californian deserts are clothed in vegetation—peculiar shrubs, which grow one to five feet high, belonging to several genera, but known under the common names of "sagebrush" and "greasewood." They have but little foliage, and that of a yellowish gray; the wood is brittle, thorny, and so destitute of sap that it burns as readily as other wood does when dry. Every few years there is a wet winter, when the land of even these deserts gets soaked. Then these bushes grow. When it dries they cease to put forth much fresh foliage or add much new wood, but they do not die —their vitality seems suspended. A drought of several years may elapse, and when, at last, the rains come, they revive into life again! Marvels of vegetation, some of these species will stand a tropical heat and a winter's frosts; the drought of years does not kill them, and yet the land may be flooded and be for two months a swamp, and still they do not die. Such for instance is the common sagebrush of the deserts, *Artemisia tridentata.*

The aspect of these deserts is peculiar. In the distance, when individual bushes cannot be distinguished, they look like a gray plain of uncovered soil; near by they are still gray, or yellowish gray, but covered with bushes—no grass, few herbs, and no trees, only half a dozen species of bushes.

The Inyo Mountains skirt this valley on the east. They, too, are desert. A little rain falls on them in winter, but too little to support much vegetation or to give birth to springs or streams. They look utterly bare and desolate, but they are covered with scattered trees of the little scrubby nut pine, *Pinus fremontiana,*[4] and some other desert shrubs, but no timber, nor meadows, nor green herbage. There are a few springs, however. These mountains were the strongholds of the Indians during the hostilities of a year ago. They are destitute of feed, and the water is so scarce and in such obscure places that the soldiers could not penetrate them without great suffering for want of water. Camp Independence was located in the valley, and for a year fighting went on, when at last the Indians were conquered—more were starved out than killed. They came in, made treaties, and became peaceful. One chief, however, Joaquin Jim, never gave up. He retreated into the Sierra with a small band, but he has attempted no hostilities since last fall. These Indians are in the region where we are now, and it was against them that we took the escort of soldiers as a guard. There are a number yet, however, in the valley, living as they can—a miserable, cruel, treacherous set.

Mines of silver and gold were discovered in the Inyo Mountains some two or three years ago. They made some excitement, a few mills were erected, and three villages started—Owensville, San Carlos, and Bend City. The last two are rivals, being only 2½ miles apart; the first is 50 miles up the river. We camped on the river near Bend City and went into town for fresh meat and to get horses shod. It is a miserable hole, of perhaps twenty or twenty-five *adobe* houses, built on the sand in the midst of the sagebrush, but there is a large city laid out—on paper. It was intensely hot, there appeared to be nothing done, times dull, and everybody talking about the probable uprising of the Indians—some thought that mischief was brewing, others not.

Friday, July 29, we were up at dawn and started up the valley, and traveled twenty-two miles, about three-fourths of the way over deserts, the rest over grassy meadows. Many settlers have come in with cattle this year. Most of them are Secessionists from the southern part of the state. We passed the old Camp Independence, so lively a year ago—now a ranch, and the *adobe* buildings falling into ruin.

It was a terrible day. The thermometer ranged from 102° to 106°, often the latter, and most of the time 104°. It almost made us sick. There was some wind, but with that temperature it felt as if it came from a furnace. It came from behind us and blew the fine alkaline dust into our nostrils, making it still worse. We camped by the river and took a cooling bath. Our camp was the scene of a fearful tragedy a year ago. The Indians attacked a party of one man, a nigger, two women, and a child. The nigger was on horseback and fought well, killing several. In attempting to cross the river the team horses were drowned. He gave his horse to the women—both of whom got on it, and they and the white man escaped to Camp Independence. The Indians caught the negro and afterward said that he was tortured for three days.[5]

During this day's ride the Sierra loomed up grandly. The crest is at the extreme eastern part of the chain—grand rocky peaks, some of them over fourteen thousand feet, their cool snow, often apparently not over ten miles from us, mocking our heat.

Saturday, July 30, we kept on our way up the valley. It was not so hot—100° to 104°, most of the time 101°. We had the same features—deserts most of the way, grassy meadows where streams came down from the Sierra and spread on the plain, the barren Sierra on the west, the more barren Inyo Mountains on the east. The eastern slope of the Sierra is here almost destitute of trees, save in the canyons and along the streams.

We camped by a creek, where also a stream of warm water comes from some copious hot springs about a mile distant. We took a luxurious bath. Although the stream of cold water was abundant, it dried up in the intense heat before night and ceased to run, but flowed again the next morning. We have seen Indians along up the valley, but they shunned us, afraid of the soldiers. One only came into camp. There was a sweat house and other Indian "fixens" near camp, but the Indians all kept away.

We wanted some fresh meat, so two of the boys went out and shot a fine heifer and brought in the beef. They assumed that she belonged to a Secessionist and confiscated her. It is very common here for men traveling to supply themselves with beef from the large herds, but this is the first time that we have ever done it—in fact, it was the soldiers who did it, but we helped eat the beef, and it answered us a very good turn indeed.

At our camps in the valley our only fuel was sagebrush, which burns like tinder, but is little better than straw to cook by. No trees grow in the valley.

Sunday, July 31, we were up very early, but not refreshed, for the mosquitoes had allowed us but little sleep. We went up the valley sixteen or seventeen miles where a lava table crosses the entire valley, barren in the extreme. Here a stream comes down, and there is a sort of basin where there are nine or ten square miles of the best grass I have seen in the state. Three or four settlers have come in this year with cattle and horses, but there is feed for ten times as many. One has started a garden, to sell vegetables in Owensville, ten or fifteen miles distant, and Aurora, sixty-five miles distant. He came into camp and wanted to sell vegetables. I bought some, also four pounds of butter—all luxuries. Perhaps you would be interested to know what prices are asked for vegetables so far from any market. They were: green peas in the pod, ten cents a pound; turnips, eight cents a pound; cucumbers, twenty-five cents a dozen; radishes, thirteen cents a dozen; butter, seventy-five cents a pound, in gold—now worth three times its face in greenbacks. But they were very acceptable notwithstanding their price.

Monday morning, August 1, we were up early. Some of the horses strayed away and two hours were spent in finding them, so we got a late start. We struck into the mountains by a blind Indian trail. Last year some soldiers were piloted over the summit by friendly Indians, in pursuit of hostile ones, by this trail. One of them was with us and knew the way. Our way was continually up and we had grand views of the desert plain below and the desert mountains beyond. The Sierra lay ahead, with high peaks and cool snows.

We camped in the canyon of the southwest fork of Owens River at over eight thousand feet. As soon as we halted signal smokes rose from the hills around, and after dark signal fires blazed, the Indians telegraphing our progress. The fires were bright, but short. Ice formed that night, a pleasant change from the heat of the last few days.

August 2 we were off in good season. As soon as we started the signal smokes again showed that we were watched. I will not repeat, but let it suffice to say that all of our movements up to the present time have been thus telegraphed. As soon as we stop, smokes rise, when we start they appear, and at night their blaze is seen on the heights—so the Indians know all of our movements.

We crossed the summit that day. As we approached it, it

seemed impassable—great banks of snow, above which rose great walls and precipices of granite—but a little side canyon, invisible from the front, let us through. The pass is very high, nearly or quite twelve thousand feet on the summit. The horses cross over the snow. So far as I know it is the highest pass crossed by horses in North America. There is no regular trail, but Indians had taken horses over it before the soldiers did. The region about the pass is desolate in the extreme—snow and rock, or granite sand, constitute the landscape.[6]

We came down a very steep slope 2,500 feet and camped on the middle fork of the San Joaquin, near its head. The next day we climbed a ridge south, to see the country, and had a grand view, but as it is so similar to those already described I will give no description. We met eleven Indians, all well armed with rifles, but they avoided us. Their signals blazed, however, until near midnight.

August 4, yesterday, we came down the valley to this camp, eighteen miles. In places the canyon widens into a broad valley. There are many beautiful spots, but they have been rarely seen by white men before.[7] It is the stronghold of Indians; they are seldom molested here, and here they come when hunted out of the valleys. We saw their signs everywhere; their fires and smokes on the cliffs near showed their presence, but we saw not a man, woman, or child. Once we came on a camp fire still burning, but the Indians were out of sight.

In this valley hundreds of pine trees have the earth dug around them to protect them from fire, for pine seeds or nuts form an important article of food with the Indians. One species has *very* large cones, with large seeds—hundreds of bushels of seeds are gathered for food.

This morning four soldiers left for Fort Miller for supplies. They took an unknown way, following the foot trail, but hope to be back in seven days, while we will work south into the region we could not penetrate from Kings River. I have stopped here today to wash clothes, mend, write up my notes, etc., for we have been on the go incessantly for the last eleven days.

Middle fork of the San Joaquin.
August 14.

WE are again back here, stopping over a day, and I will go on with my story. And yet I have a mind to pass it over, it is so like the rest. Rides over almost impassable ways, cold nights, clear skies, rocks, high summits, grand views, laborious days, and finally, short provisions—the same old story.

Yet one item is worth relating. A very high peak, over thirteen thousand feet, rises between the San Joaquin and Kings rivers, which we call Mount Goddard, after an old surveyor in this state. It was very desirable to get on this, as it commands a wide view, and from it we could get the topography of a large region. Toward it we worked, over rocks and ridges, through canyons, and by hard ways. We got as far as horses could go on the ninth, and thought that we were within about seven miles of the peak.

We camped at about ten thousand feet, and the next day four of us started for it—Hoffmann, Dick, Spratt (a soldier), and I. We anticipated a very heavy day's work, so we started at dawn. We crossed six high granite ridges, all rough, sharp, and rocky, and rising to over eleven thousand feet. We surmounted the seventh, a ridge very sharp and about twelve thousand feet, only to find the mountain still at least six miles farther, and two more deep canyons to cross. We had walked and climbed hard for nine hours incessantly, and had come perhaps twelve or fourteen miles. It was two o'clock in the afternoon. Hoffmann and I resigned the intention of reaching it, for it was too far and we were too tired.

Dick and Spratt resolved to try it. We did not think they could accomplish it. Dick took the barometer and I took their baggage, a field glass, canteen, and Spratt's carbine, which he had brought along for bears. Hoffmann and I got on a higher ridge for bearings, and then started back and walked until long after sunset—but the moon was light—over rocks. We got down to about eleven thousand feet, where stunted pines begin to grow in the scanty soil and crevices of the rocks. We found a dry stump that had been moved by some avalanche on a smooth slope of naked granite. We stopped there and fired it, and camped for the night.

We had brought along a lunch, expecting to be gone but the day. It consisted of dry bread and drier jerked beef, the latter as dry and hard as a chip, literally. This we had divided into three portions, for three meals, a mere morsel for each. This scanty supper we ate, then went to sleep. The stump burned all night and kept us partially warm, yet the night was not a comfortable one. Excessive fatigue, the hard naked rock to lie on—not a luxurious bed—hunger, no blankets—although it froze all about us—the anxieties for the others who had gone on and were now out, formed the hard side of the picture.

But it was a picturesque scene after all. Around us, in the immediate vicinity, were rough bowlders and naked rock, with here and there a stunted bushy pine. A few rods below us lay two clear, placid lakes, reflecting the stars. The intensely clear sky, dark blue, *very* dark at this height; the light stars that lose part of their twinkle at this height; the deep stillness that reigned; the barren granite cliffs that rose sharp against the night sky, far above us, rugged, ill-defined; the brilliant shooting stars, of which we saw many; the solitude of the scene—all joined to produce a deep impression on the mind, which rose above the discomforts.

Early in the evening, at times, I shouted with all my strength, that Dick and Spratt might hear us and not get lost. The echoes were grand, from the cliffs on either side, softening and coming back fainter as well as softer from the distance and finally dying away after a comparatively long time. At length, even here, sleep, "tired nature's sweet restorer," came on. Notwithstanding the hard conditions, we were more refreshed than you would believe. After months of this rough life, sleeping only on the ground, in the open air, the rocky bed is not so hard in reality as it sounds when told. We actually lay "in bed" until after sunrise, waiting for Dick. They did not come; so, after our meager breakfast we started and reached camp in about nine hours. This was the hardest part. Still tired from yesterday's exertions, weak for want of food, in this light air, it was a hard walk.

At three in the afternoon we reached camp, tired, footsore, weak, hungry. Dick had been back already over an hour, but Spratt had given out. Gardner and two soldiers, supposing that Hoffmann and I also had given out, had started with some bread to look for us. We shot off guns, and near night they came in, and at the same time Spratt straggled into camp, looking as if he had had a hard time. Dick and he did not reach the top, but got within three hundred feet of it. They traveled all night and had no food—they had eaten their lunch all up at once. Dick is *very* tough. He had walked thirty-two hours and had been twenty-six entirely without food; yet, on the return, he had walked in four hours what had taken Hoffmann and me eight to do.

Supper and sleep partially restored us, but we came back here the next day, a long, hard day's work. This was necessary, for we were out of flour, rice, beans—in fact, had only tea, sugar, and bacon. This was to be our rendezvous with the soldiers, and they had got in only an hour ahead of us, with an abundance of provisions. Not a great variety, but an abundance of salt pork, flour, coffee, and sugar, so we are all right again.

I have stopped here two days, Saturday and Sunday, to rest, wash, and mend clothes, and write up notes.

The soldiers brought back a lot of newspapers from the camp at Fort Miller—papers from the East, from various parts of this state—old many of them, but very acceptable.

Yesterday, after washing my clothes, I spent the rest of the day in reading. There is a sort of fascination in reading about what is going on in the busy world without, in the noisy marts of trade and commerce, in society and politics, in the busy strife of war, of brilliant parties and gay festivities, and sad battles, and tumultuous debate, while we are here in these distant mountain solitudes, alike away from the society and the strife of the world.

Camp 195, near north fork of the San Joaquin.
August 18.

MONDAY, August 15, we packed up and started and made a big day's march toward the north fork.[8] We had rather rough going and finished by going down a steep hill about two thousand feet. The north fork runs in a very deep, rocky canyon still a thousand feet deeper. We camped at good grass, but poor water—some pools that remained in a little swamp. Near our camp the canyon is very grand—a notch in the naked granite rocks. During the day we crossed a number of streams bordered by dense thickets of alders through which we had to cut our way with our knives. In one of these, Buckskin, our pack horse, caught a leg between two rocks and bruised it badly.

The next morning we were off as usual, but soon found that we were "in a fix"—great granite precipices descended ahead of us. We turned back, and after much trouble found a very steep, rocky place where we could get down about one thousand feet to the river. Buckskin was so lame that I camped before noon, at the river, after going only about four miles. It was a lovely spot—a little flat of a few acres, with grand old trees, and high, naked granite cliffs around. The river runs through this, entering and leaving it by an impassable canyon. We found the river full of trout, and the boys caught a fine mess.

San Francisco.
October 12.

Two months have passed since I have written anything about our trip, and I must now resume the story to make it complete. I was just telling about getting into the canyon of the north fork of the San Joaquin River. One of the soldiers, Win Orr, was ahead. We tried to stop him but could not. The upshot of the matter was that he got lost, and we heard nothing of him for many days. He gave us much trouble. During the day a deaf and dumb Indian came into camp. It was remarkable to see how he would express what he had to say by his actions and pantomime.

August 17 we started again, putting the pack on another animal. We had a tremendous climb to begin with. First a very heavy hill, three thousand feet high, and just as steep as animals can climb. We struck the trail of Orr, the lost soldier, but we soon came to a region where there were cattle, and their abundant tracks prevented us from seeing which way he went, so we camped, and I sent out two soldiers to look for him. The search was continued for the next day. They found his tracks in places, but there were Indian tracks also and we feared that they had got him.[9]

We had a grand view from the hill we crossed. The peaks along the summit are very black and desolate, and with much snow, and the canyons very deep and abrupt. We found cattle in the woods, and the soldiers shot a beef and we had fresh meat again, a true luxury. We then worked west one day, and then another was spent in getting the topography of the region. I was not at all well and lay in camp one day, miserable enough.

For a week or ten days Hoffmann had been complaining of a sore spot on one of his legs, but nothing could be seen to indicate any hurt or injury. He came back from a trip away from the camp so lame that he could scarcely walk. I resolved to push out. I was getting run down, felt nearly sick, our provisions were low, and Hoffmann was getting rapidly worse. So the next morning, August 21, Sunday, we started.

We had a terribly rough way at first, but finally struck some cattle trails, and, at length, the first dawn of returning civilization. We found two men camped under a tree, watching cattle which they had driven up from the plains. We rode twenty-three miles and then camped at the altitude of about eight thousand feet. Hoffmann had grown so much worse that by night he had to be lifted from his horse and could not walk a step. It had been cloudy all day, and soon after dark a cold rain set in and it rained hard all night. There is no need of again describing the discomforts and miseries of sleeping on the ground in a cold, rainy night, when the rheumatism creeps into every nook and joint of one's frame. It was hard on poor Hoffmann. It is bad enough for a well man, but for a sick man, racked with pain, tired from the day's riding, feeble, weak, and used up, it is hard, indeed.

The next day it rained but little, but he got steadily worse. We had to lift him on and off his horse and travel on the slowest walk. We had trails most of the day, however. We passed several cabins, in some of which white men were living with squaws, and lots of half-breed "pledges of affection" were seen. We descended rapidly and camped in the valley of the Fresno River, five thousand feet lower than the place we started from in the morning. We stopped at a ranch and got some potatoes and green corn. Ah, what real luxuries they seemed! The proprietor had a squaw wife and a growing family. It is a noteworthy fact that nearly all the "squaw men" (men living with squaws) in this state are rank Secessionists—in fact, I have never met a Union man living in that way. They are generally the "poor white trash" from the frontier slave states, Missouri, Arkansas, and Texas.

The night was very hot, the air steamy, for it rained some, and Hoffmann grew worse—his leg was very painful. The next day, August 23, we got to Clark's Ranch, on the Yosemite trail.[10] Hoffmann was in a bad state; indeed, we could scarcely get him so far.

This looked like civilization. We met Mr. and Mrs. Ashburner, Mr. Olmsted,[11] and other friends, who had been at Yosemite. You can imagine the joyous meeting. We were a hard-looking set—ragged, clothes patched with old flour bags, poor—I had lost over thirty pounds—horses poor. We camped there and hoped for Hoffmann to recruit. There he remained three weeks from that time, some of us with him all the time, but he grew no better, in fact, rather worse. After stopping there three days I went to Yosemite Valley with Mr. Olmsted. He is the manager of the Mariposa Estate. His family were in the valley, where they spend several weeks during the heat of the summer.

The great Yosemite is ever grand, but it was less beautiful than when I saw it before. The season has been so *very* dry that the grassy meadows at the bottom were brown and sere, the air hazy, the water low. Over the great Yosemite Fall but a mere rill trickled, so small that it was entirely dispersed by the wind long before it reached the bottom of its great leap of half a mile.

We took a week and went back to the summit on the Mono trail, where we had such a pleasant time last year. It was a charming trip. I threw off the anxiety that had weighed me down of late, and enjoyed the scenery of that grand region. Mr. Olmsted is a very genial companion and I enjoyed it.[12]

I got back to Clark's Ranch September 6 and found Hoffmann no better. He was anxious to get where he could get medical advice. We tried to get Chinamen to carry him to Mariposa, for we were twelve miles from a wagon road. At last, September 10, we four started to carry him ourselves. We made a litter and put a bed on it and started. The trail was so narrow that only two could carry at once. The trail led over a hill over six thousand feet high, and he grew heavier and heavier mile by mile, but it was successful. Gardner and I returned for our animals, while King and Dick went on, and we all met again at Mariposa. There we got a carriage, put a bed in it, and King and Dick got him to Stockton, a hundred miles distant, and thence by steamer to San Francisco. He is still very sick and may never recover.

Gardner and I spent a few days on the Mariposa Estate, looking at its mines and mills, but I will give no description. It is a tremendous estate. I rode to Stockton with Mr. Olmsted in his private carriage, carrying $28,000 in gold bullion—quite a load.

I have received unofficial notice of my election as professor at Yale, and shall be on the road in a week if I can. I am now working hard to get off early, but will not close my journal yet, for a long trip still lies before me. I have counted up my traveling in the state. It amounts to: horseback, 7,564 miles; on foot, 3,101 miles; public conveyance, 4,440 miles—total, 15,105 miles. Surely a long trail!

NOTES

1. The route was up Bubbs Creek (named for John Bubbs, one of the prospectors) and up Charlotte Creek.
2. Kearsarge Pass.
3. Now called Independence Creek.
4. Usually classified as *Pinus monophylla* Torrey; also called one-leaf piñon.
5. A circumstantial account of this episode is told by W. A. Chalfant, *The Story of Inyo* (Bishop, California, 1922), pp. 136–138.
6. This was Mono Pass, leading from the head of Rock Creek (the southwest fork of Owens River) to Mono Creek (called by Brewer the middle fork of the San Joaquin). The name Mono is that of a widely scattered division of Shoshonean Indians and is found in many localities throughout the Sierra. This pass is not to be confused with the one farther north at the head of the Tuolumne.
7. For thirty years after Brewer's visit this region continued to be practically unknown save to Indians and sheepherders. In 1894 Theodore S. Solomons explored it and named some of its features, including the broad valley (Vermilion Valley) (*Sierra Club Bulletin*, Vol. I, No. 6 [May, 1895]).
8. This is now known as Middle Fork; there is a branch coming into it farther up now called North Fork.
9. Brewer's notebook states that on August 21 they learned that Orr had passed through the cattle camp safely several days before.

10. Wawona.

11. Frederick Law Olmsted, famous landscape architect, was in California in 1863–64 as superintendent of the Mariposa Estate, the huge mining property of which Frémont, once the sole owner, was at this time losing control.

12. In the notebook there is an account of a climb of Mount Gibbs with Mr. Olmsted, of which the following is an abridgment: "August 31, started [from camp at Mono Pass] for the summit, but took the next peak south of Mount Dana, fearing Olmsted could not reach the other. On this peak I managed to get his horse up, so we rode to the top. He named the peak Mount Gibbs, after O. W. Gibbs [Oliver Wolcott Gibbs, Professor of Science at Harvard]. Strange enough we saw a group of persons on Mount Dana, clear against the sky. These turned out to be the party of a Mr. St. Johns, including a little girl six years old and a man sixty-two years old and lame. We met them at the Soda Springs next day."

CHAPTER IV

THE WASHOE MINES

Nevada—Placerville—Staging across the Sierra—Discovery of the Washoe Mines—Speculation, Boom, and Collapse—The Comstock—Gould & Curry Mine—Sandy Bowers—Story of the Plato Mine—Mount Davidson—Salaries in Nevada.

Virginia City, Nevada.
November 4, 1864.

I sit down this lovely morning to write a letter from a *new state*, emphatically, in every sense of the word. I am in a city of twelve thousand to fourteen thousand inhabitants, where four short years ago the desert bushes grew unmolested, the desert mountains sent back no echoes of the sounds of human industry, and the desolate hills showed no signs of human inhabitants, or at least of the homes of civilized men. New in another sense—a state constitution has just been adopted, has been telegraphed entire to Washington, and the wires just bring the news that the President has issued his proclamation making Nevada a state.

I have all along intended to visit this city and the famous silver mines of the region, expecting to take it in on my trip overland. Five days ago I had to give up that idea, relinquishing long-cherished hopes and plans. The season is late, my health is not entirely good, there are fresh Indian outbreaks, there is no certainty of my getting through, and even if I should, the expense would be enormous, for I have failed to get a through ticket; so I gave up the plan and came up here for a short visit.

November 1, I took the steamer for Sacramento, took the cars at midnight and passed over the hot plains in the cool night; then the stage. Daylight found us at Placerville, where we took our breakfast. It was my first visit to Placerville—a pretty mining town of considerable size and importance, but lacking the bustle and life that it had when surrounded by the rich placers that gave it a name. The old name of Hangtown, so well bestowed in early days, is about forgotten.

Here we took the Overland stage across the Sierra. For some miles the stage road runs through the dry, barren foothills, with signs of worked-out placers on all sides. But the road rises and the scene changes as it enters the granite region and the great forest belt of the western flank of the grand old Sierra Nevada. This has been the great road to Washoe—sometimes three thousand teams with twenty-five thousand animals at work at one time. We passed hundreds of freight wagons, with from six to ten horses or mules each. Three or four rival roads now divide the patronage—all, like this, toll roads and built at an enormous expense. Hotels and stables abound, even on the very summit, where the snow falls to an immense depth every winter. An aggregate of *fifty feet* fell there the "wet winter." But what of that? Where money is to be made, be it in the cold and snows of the Sierra or the intense heat of the deserts, no matter, there will be a man to get it. "For wheresoever the carcase is, there will the eagles be gathered together."

The day was lovely, the sky clear, and the air balmy. The heat grew less, until near the summit the ice lay thick around the springs, and snow lay in patches among the trees. Six inches of snow fell in a recent storm, but it is now mostly gone. We had all the features of this grand Sierra scenery—high, barren peaks streaked with snow, rising above the dense, dark forests of the lower slopes, and deep canyons and steep precipices, around which at times the road passes at such a dizzy height that an upset would send the persons of the passengers several hundreds of feet below and their souls to the other world. But such *drivers* as there are! They manage the big coaches and their six-horse teams as if it were but play, and go dashing along steep grades and around short curves at a rate to make the hair stand on end.

We crossed the summit at last, over 7,200 feet high, then descended a steep grade to Lake Valley, with the lovely Lake Tahoe (once Bigler), finer even than Lake Thun or Lake Geneva. Night came on, clear and cold, long before we crossed the eastern summit. We descended into the Carson Valley, part fertile and part desert, then crossed the desert hills to this city, where we arrived about midnight. We passed through Silver City and Gold Hill, lively mining towns, where the clatter of nearly a thousand stamps greets the ear for miles. At the former place a political procession was in progress, indicative of the enthusiasm of the "honest miners."

I find here various acquaintances and much to see. I had heard of all of this before, time and time again, yet had no just appreciation of it. One can hear of Niagara for many a year, yet really have no just conception of it until he sees it.

It is a popular idea on this coast that a country utterly barren, uncomfortable, with bad climate, worse water, nothing inviting about it, must of necessity contain the precious metals. "What else is it good for?" is their triumphant shout if you dispute it. Following this idea, this region was "prospected" by various parties and at various times from 1850 on. Gold was found quite early, and at last, about 1858 or 1859, silver ore. It attracted no attention, however, until 1860, when suddenly the tide set in here—such an immigration as only gold and silver can cause. Thousands poured in from California, Oregon, the Colorado, the States—adventurers, gamblers, speculators, miners, prospectors, lawyers—in short, all of that numerous class that "makes haste to grow rich" who could get here. Mines were located and opened, mills built, towns were laid out and grew like mushrooms. Capital was active.

Some of the mines proved immensely rich, and as a consequence "wildcats" flourished in all their glory. A few mines turned out bullion at such a rate that statesmen and financiers feared that silver must fall in value. Last year it culminated—when nearly a *ton a day* of gold and silver bullion was shipped from here; when more than five thousand teams were employed in bringing freights here; when stock in the paying mines was held at perfectly wild rates and values, even to *over twenty thousand dollars per foot;* when probably a dozen mines were paying high dividends, and millions of dollars were being expended in working the thousand "wildcat" mines whose stock was in the market; when a population variously estimated from sixteen thousand to twenty-four thousand was congregated here in a few miles, all drawing their "supplies and forage" from over the mountains; when speculation ruled the day and in the streets were met men poor three years ago, now worth their hundreds of thousands—then, I say, the excitement raised and culminated—and this summer the vast bubble broke.

All at once the stocks of the good mines went down from far above their real value to as much below, and "wildcat" collapsed. Ophir stock, once held at over $4,000 per foot, is now a drug at as many hundreds. Gould & Curry, once over $6,000 per foot, now sells for less than $1,500, and everything in the same way. As a consequence, thousands have been ruined, work has stopped on much "wildcat," and good mines must in future be worked in a more healthy manner. Of course, a portion of the floating population has left, but still perhaps sixteen thousand are in Virginia City, Gold Hill, Silver City, and the environs.[1]

This is the center of a region on which many hundreds of gold and silver mines have been located, but only a few of these have ever paid expenses, certainly less than twenty of them. All of these (unless it be one or two) are located on one line or vein—the famous Comstock Lead as it is called—all within a distance of 1½ miles. The mines are reckoned by the *foot*, measured along the lead, and the stock sales are quoted at so much per foot, rather than at so much per share of the capital stock.

I was introduced to the superintendent of the Gould & Curry Mine, Mr. Bonner. He took me through the mine, and a description of that will answer for all.

The vein is several hundred feet wide, but not all rich ore. It is distributed in great irregular masses, called chambers, because when the ore is taken out a great room or chamber is left. The miners learn by experience what ore is good and what to reject as poor. The vein runs down slanting, so that men who commenced on one side at the surface are now at work over four hundred feet under the city. It requires great skill to explore the mine and keep the sides from caving in. The timbering is unlike any I have ever seen. Stout timbers, a foot square, in short pieces from five to seven feet long, placed in a peculiar manner, run entirely across the mine in every direction, further strengthened by braces where the pressure is greatest. Yet, at times, these timbers are crushed as if they were but straws. Long tunnels run into the mine from low down the hill, and out of these the ore is taken.

We descend into the mine and get out by a cage, a small platform running between guides down a shaft. The sensation is peculiar. Three or four men stand on this, the engine starts, and down we go rapidly in this shaft, like a deep well. It grows dark, as down, down, down we go. We hear the rope rattle over the pulley overhead and the rattle of the cage in its guides. We know that a chasm hundreds of feet deep is beneath. We are hung on this rope at a height twice as great as that of the tallest spire in the United States.

At last we strike the "lower level" and step out into the mine, where we find great timbers running in every direction. Here and there a candle sheds a feeble light. Long, dark galleries or "drifts" run from the landing place, with narrow railroad tracks in them, and little cars in which the ore is moved.

We hear the rumble of the cars on these tracks, and the distant clatter of picks and drills as the miners are loosening the ore and rock. Perhaps a heavy pump is groaning, worked by a powerful engine, to get the water out of the lowest parts.

The miners load the good ore in the cars and reject the poor, which is sometimes left in the mine and sometimes is carried outside and dumped out of the way. These "dump piles" are among the most conspicuous features of the place—hundreds of thousands of tons of this useless material are piled up. They often contain silver and gold, for unless the ore runs twenty-five dollars of the precious metals to the ton of ore, it is rejected. As a matter of fact, it must contain more, for twenty-five dollars must be got out of it to make it pay, and no mill gets it all out, so generally one-fourth, sometimes even one-half is left in the tailings. Some of the ore will yield from one to two thousand dollars per ton, but if it yields a hundred dollars it is thought very rich.

We ascend in the mine by ladders, stage beyond stage. Men are at work everywhere, digging the ore, getting it away, putting in the heavy timbers. Here galleries branch off; there great chambers have been opened. We walk along planks, through galleries, sometimes dry, sometimes dripping with muddy water. Look out! That little black hole by your feet may be deep! Only two days ago a miner slipped into such a hole and fell 115 feet; he was buried the day we were in the mine.

The Savage Mine joins this, and you can pass from one to the other far underground. It is now the richest of the large mines. It is like the Gould & Curry in all its looks. They take out about two hundreds tons of ore per day, and last month sent $300,000 worth of bullion to market.

After the ore is taken out it is crushed in mills and goes through long processes to get out the gold and silver, but I will not attempt to describe them, for it would take a hundred pages to make all the various processes intelligible.

The Gould & Curry mill is the finest. It cost over a million dollars. A friend of mine has charge of it, and he showed me all through it. First is the ore room, where the ore is stored; then eighty ponderous "stamps," which crush it with a noise and clatter that is almost deafening. Then it goes through the "pans," until the rock is washed away and the precious metal is held, alloyed with quicksilver. The quicksilver is expelled by heat and the bullion is then cast into bars or "bricks," which weigh generally about eighty or one hundred pounds. This bullion is worth twenty-five to twenty-six dollars per pound and in value consists of about one part of gold to three of silver.

You can get some idea of the magnitude of the works when you learn that the Gould & Curry Company employs five hundred men in its mine and mills, not counting carpenters, blacksmiths, and so forth. Last year this mine produced nearly six millions of dollars. The bullion alone weighed over one hundred tons! But it will produce far less this year.

The yield of the mines of this vicinity last year was about nineteen or twenty million dollars—nearly a ton a day of bullion went to San Francisco. Such are the resources of this new state.

Gold Hill lies 1½ miles south on the same "lead." Its mines are even richer, but are smaller, so make less noise. They look like these. I visited several, only two of which I will notice, and that because of their owners.

The Bowers Mine is only twenty feet. Sandy Bowers, the lucky owner, four years ago drove an ox team in the territory, his wife an industrious, hard-working Irishwoman. He located a claim and held onto it, letting other men work it. It rose in value enormously. In three years it netted him an income of some *ten thousand dollars per month,* and he refused $400,000 for it, or at the enormous rate of $20,000 per foot. Last year he built a house costing him $240,000, adorned with $3,000 mirrors, and all the window curtains cost $1,200 each. The wife, who three years before thought $3 per day good wages for her man, told a friend of mine that she "didn't see how one can get along on this coast with less than $100,000 per year." Poor Sandy—he lived too fast—his $10,000 per month was too little, and he ran into debt $100,000, on which he pays Nevada interest, so I suppose he is a poor man again—if not, he soon will be. Do not say that "shoddy" is peculiar to any state or place.

One more—the Plato Mine is only ten feet. Five years ago, in San Francisco, an Irishwoman sold apples by day, and by night "received calls" for compensation. The latter business proved the more profitable and required the less capital, so she gave up her apple stand and came up here to the new mines, when they first began to attract attention, to ply her trade. She was successful, as is usually the case in the new mining towns, and an admiring lover gave her ten feet of his claim—either as a token of esteem, as he says—or "for value received" in her line, as she says. No matter which—she got the feet, which rose to be worth some $150,000 to $200,000. He finally married her to get his "feet" back, but died soon after, and she owns the mine yet, which, of course, yields her a handsome income.

Last night I dined with Mr. Bonner, superintendent of the Gould & Curry, and had a very pleasant time, indeed. There was a grand Union procession, with banners and fireworks. Rockets were fired over the town from the summit of Mount Davidson.

This Mount Davidson is one of the grand features of the place. It rises very steep two thousand feet above the city, or to a height of about eight thousand feet above the sea. It commands a sublime view, like a grand panorama, which extends more than a hundred miles in every direction. Of course I climbed it as soon as I got here. The afternoon was very clear. On the west and south were the grand old Sierras, clouds hanging over them and snow streaking the peaks. But everywhere nearer, and to the north and east, was a dry, desolate, desert region—sharp mountains, of no particular chain or name, very steep, entirely without trees or other green thing, of a uniform dry grayish-brown color—the canyons are deep and sharp—all is dry and desolate. There are desert valleys, as dry as the mountains, and often apparently as level as the sea, with here and there an alkaline lake, or a dry lake bed, now a shining sheet of white salt and alkali. No painting could convey any idea of the scene. The color is nearly uniform, the mountains much furrowed, so their details are endless—a photograph would show it, a painting could not.

Nevada Territory embraces over eighty thousand square miles, but it is nearly all desert. It has just been made a state, but I see no elements here to make a state. It has mines of some marvelous richness, but it has nothing else, nothing to call people here to live and found homes. Every man of any culture hopes to make his fortune here, but to enjoy it in more favored lands. The climate is bad, water bad, land a desert, and the population floating.

Enormous salaries are paid, of course, for it takes great skill to run a mine producing half a million per month. In fact, I think the highest salaries in the United States are paid here. Last year, Palmer, superintendent of the Gould & Curry got a salary of *forty thousand dollars per year!* I would have no serious trouble in getting a place here with a salary of three hundred to five hundred dollars per month—men of less knowledge in my line get twice that.

NOTE

1. Virginia City had its revival a few years later, however. For a popular account of the Washoe mines see Charles Howard Shinn, *The Story of the Mine* (New York, 1896).

CHAPTER V

HOMEWARD BOUND—NICARAGUA

Steamboat Springs—The Patio Process—Election Day at Virginia City—Truckee River and Donner Pass—A Moonlight Ride—Goodbye to San Francisco—Voyage to Central America—Volcano Pecaya — Nicaragua — The Lake — San Juan River — Steamer Golden Rule—*New York—Home.*

Enfield Center, New York.
December 22, 1864.

I LEFT my journal incomplete. The last number was written in Virginia City, Nevada, and I intended to write more on the steamer on the way home, but neglected it. Now, although home, I will finish my Californian journal that you may know the rest of the story and to make the journal more complete.

In my last letter I told something about the mines at and near Virginia City. One day I rode out to see the Steamboat Springs with Professor Silliman and Baron von Richthofen. These are very extensive hot springs about ten miles northwest of Virginia City. They lie in a valley about two thousand feet below the city and extend along for about a mile and a quarter. Near the Steamboat Meadow there is a wide valley, and at the foot of one side hill are the springs. Many are of water, of course, but the majority are merely steam jets. Nature's chemistry has been going on here on a grand scale. The hot water contains silica dissolved, which is precipitated as the water cools. This has made a hill over a mile long, and in places two or three hundred feet high. Steam issues from thousands of vents, hissing, puffing, spouting—it can be seen for miles. Some of the springs are intermittent in their activity. One I timed and found that the water would sink in the basin and then rise and overflow again at intervals of about five minutes. They are very extensive, and there are signs that many others have existed in the region at some former times. Much of the country is volcanic.

Monday, November 7, I went home with a Mr. Hill, who lives near Dayton, on the Carson River about ten miles from Virginia City. He works tailings from the mills by the patio process. Many of these tailings are very rich in silver still, for the mills fail to get it all out. Often over a hundred dollars per ton is left. The patio process is Mexican, and this is probably the only place in the United States where it is used. Great wooden floors are laid on the ground, tight, with a rim around, like a dish or tub, forty-five feet in diameter and fifteen inches deep. Into this are put the tailings—a mixture of sand and sticky mud—to the depth of ten or twelve inches. Some salt, sulphate of copper, and quicksilver are sprinkled in, and after the mass is moistened it is tramped with horses. Every day for a short time horses are driven round and round in this mud, for a month or more, when the process is finished. More of the chemicals are added from time to time as the process goes on. When finished, the mass is washed in sluices, the mud carried away by water and the amalgam left. This process, which looks so rude, extracts the silver cleaner than any of the so-called refined and perfected processes. This man is getting rich working over tailings, which he must buy at the mills. The horses are poor old "plugs," picked up wherever he can get them cheapest. They get very poor and look comical enough, stained by the sulphate of copper—white horses, with green legs and tails, and blue and yellow spots over them, streaked and speckled and miserably poor.

The next day we returned to Virginia City, arriving about noon. It was election day. I had no vote, of course. One should see the elective franchise exercised in such a new place to have a realizing sense of the free and enlightened voters exercising their rights. Men argued but little, but opinions were freely given, rendered more forcible by bowie knife and revolver. I don't know how many I saw drawn during the day, yet I saw no one hurt. Some were shot in the evening, however, as the excitement rose and the whiskey began to work. I was up until very late that night around town.

I intended to leave the next day, but found the stages filled, so I had to wait over. I took a stroll over the hills a few miles, looking at the rocks and geology of the country.

On November 10 I bade goodbye to my friends at Virginia City and got upon the stage at noon for California again. I rode on the box with the driver, so I had a good chance to see the country. We crossed the Truckee River and then ascended the grand Sierra. We went by the Truckee Pass, where the Pacific Railroad is to cross, according to present hopes. We passed by Donner Lake, a beautiful sheet of water three miles long, the scene of the fearful tragedy that overtook the Donner party at "Starvation Camp" a few years ago. Here we took supper; then again on the stage and over the pass.

It is the grandest of the wagon road passes that I have yet seen. The moon was at its full, and as a little snow lay around and as the naked granite rocks are so very light colored, we could see almost as well as by day. The road winds up the rocky height through a scene of unmitigated rocky desolation —great bare crags, with here and there a tree or bush struggling for existence in the corners and crevices—all else naked rock and snow. How a railroad is to be got through and over this pass is not easy to see. It will take much money.

The night was cold as well as clear. After crossing the summit we were soon in the grand forests of the western slope. We had a jolly set on the stage, that is on the top, and we awoke the echoes of the mountains with our songs. I slept but very little that night. We passed through Dutch Flat, now so familiar to every Californian because of its name in connection with the railroad, and finally reached the railroad soon after daylight, near Auburn, and were in Sacramento in time for breakfast. I stopped there until the afternoon and then went to San Francisco. During my stop I called on the comptroller and got the cheering news that we would probably be paid up in January next.

It was a mild lovely afternoon as we sailed down the muddy Sacramento. The hazy air shut out all views of the Sierra. Nothing but the level banks of the river was in sight, yet at times even these were picturesque, with their festoons of vines and the gray willows hanging over the lazy water.

The next two days I was busy enough. I finished packing up and called on friends to bid them goodbye. I had made many warm friends in that city, and I parted with many, probably never more to see them in life. I spent the last evening, Sunday, at my boarding house, where I felt almost at home.

Monday morning, November 14, I was up early and off for the steamer. I bade goodbye to dear Hoffmann. He was out of bed, but could not walk yet. He wept like a child when I left, and I felt like parting from a brother. For over three years we had been together almost all the time—in winter in the office, and during summer in the field.

Several of my friends came down to see me off. We were off at half-past ten. It was a dull, foggy morning, and the hills of Oakland were dim. We stopped out in the stream while the passengers were marshaled and a boatload sent ashore—those who had forgotten to provide themselves with tickets. It was "opposition day" and we were on the opposition steamer, *America*. The steerage was full—over 100 had been turned away, and 650 were crowded in there. The second cabin was crowded, but the first cabin was not full. I had a stateroom to myself, on deck, one of the pleasantest on the boat, thanks to the agent at San Francisco, whom I had once met in Yosemite Valley and who had thus kindly remembered our meeting.

We soon settled down to the monotony of ocean life. We got acquainted with genial or congenial friends, we gazed off on the ocean, we ate, we read, we smoked, we slept. Occasional glimpses of the rugged coast of California, a school of porpoises, a whale, or the miseries of some seasick passenger as he hung over the gunwale of the ship and contemplated the mighty deep, were our only excitements. We were not yet acquainted and did not feel at home.

As we ran down along the coast a change came over the spirit of the scene. We left fogs behind and entered the sunny climate and balmy air of the Mexican coast. Bright skies and blue sea came on, the days waxed longer and the nights shorter. We passed Cape San Lucas on the nineteenth, scarcely half a mile from the forbidding shore, and the next day crossed the Gulf of California. We had an unusually cool passage—the captain says the coolest he has ever made on this route. On the twenty-first we passed a steamer of the French blockading fleet off Acapulco, which some thought a pirate of the Rebel persuasion until we were fairly up to her.

Along the coast of Central America the grand volcanic cones rose against the sky, some in sharp outline, others with clouds curling over them, but all very grand. One morning a fellow passenger awoke me before five o'clock to see a grand sight, the volcano of Pecaya in eruption. It was forty or fifty miles distant, but the night was clear. A pillar of clear, white smoke or steam curled up into the blue sky from the sharp cone, and a great stream of lava ran down, forming a broad, glowing belt down the side of the mountain. At times great volumes of steam and smoke would roll up, while along the line of the lava stream steam curled up here and there, perhaps where the hot lava found moist ground or water. It was less distinct after day broke, but for nearly a hundred miles we could see the smoke rising from it. This volcano is in the southern part of Guatemala.

Sunday, November 27, we ran down close to the land all the morning and before noon anchored in the little harbor of San Juan del Sur, where we landed in launches.

The transit company was on hand with wagons, horses, and

mules to transport the passengers to Virgin Bay, on Lake Nicaragua, twelve miles distant. Our little group of six or eight, called by envious passengers "The Committee," had taken means to get good animals, ordered for half-past one. Captain Merry of the *America* was to go with us. For five tedious hours we waited in the heat for those animals. We could not go out for fear they would come, for they were often promised in a few minutes. Well, the animals did not come until night, and we found out that we could leave the next morning and reach the lake steamer in time, so we went back on board the *America* and spent the night.

First, however, we looked around the village. Houses were scattered here and there among the trees, without any order, and of very primitive construction. The furniture is very scanty, and the people live very simply. They are vastly superior to the natives of the Isthmus of Panama in looks, and have much less negro blood. They appear to be the genuine descendants of the aboriginal natives, with but little foreign blood intermingled. The women appear finer than the men, and many of them struck me as decidedly beautiful—too dark for brunettes, yet not black—complexions rich and soft, lips well set, magnificent teeth, large and liquid black or hazel eyes—hair long and flowing, fine as silk and gently wavy or straight, which they had put up in the neatest manner—fine forms, which their very thin dresses showed off to good advantage. The climate is such that dress is worn merely as a matter of taste and decency rather than for warmth. That of the men consists of pants and shirt, that of the women consists of a very thin skirt extending to the ankles, and a jacket or sack which hangs over the skirt. They go bareheaded, or with a gay, light shawl thrown over the head. Such are the natives of Nicaragua as I saw them in three towns—San Juan del Sur, Virgin City, and Castillo Viejo.

We slept on the ship that night and were off at dawn for

Virgin City. Our party was about a dozen—all the rest had crossed the day before or in the night. We presented an imposing appearance. Ahead was your humble servant on a poor mule, who went better than he looked, his nose (the mule's not the writer's) begirt with a hackamore which answered as a bridle, and with curious native saddle in which it was hard to see which predominated, wood, rawhide, or straw—a small American flag fastened on his umbrella floated to the morning breeze. Next came Higby, M. C. from California, a capital fellow; next Captain Roberts, a millionaire from California, also a capital fellow; then Captain Merry of the *America*. Not the least of the party was Ossian E. Dodge, the concert singer, on a poor horse, a great linen coat fluttering, and a bunch of bananas hung to his saddle, which every few rods left one of their number along the road, so he had scarcely any when he got through.

As we crossed the summit Lake Nicaragua came in view, with its deep, blue waters, its rich shores, and its sharp volcanic cones. We arrived at Virgin City in due time. All the passengers were there, and soon the whole of the baggage had arrived. Baggage is hauled over on carts of the most primitive character—the wheels cut from some big trees, no iron used in their construction, only wood and rawhide—four or six oxen hitched to each by yokes that are strapped before the horns.

At noon we were on the lake steamer *City of Leon* and the minute the last baggage was on board the boat left the wharf, leaving behind thirteen passengers who had been tardy without excuse—the brutal captain would not even send a boat for them. They shouted wildly on the wharf, scarcely thirty feet from us, but all in vain—they were left at that miserable hole for a whole month, all because of the bullheaded stubbornness of Captain Hart of that steamer. The passengers will have no redress, of course, as it was owing mostly to their own carelessness that they were left. Nearly a thousand of us were crowded on that boat, with bad food and but little of it, and berths for about one hundred.

The lake is about a hundred miles long by twenty or twenty-five wide. In the middle, on an island, are two volcanic cones rising from the water to 4,200 and 5,100 feet, respectively—perfectly sharp and regular. We ran down the lake about fifty miles before we entered the River San Juan. Toward the foot of the lake are many small rocky islands, all covered with rich vegetation. We continued down the river some miles after dark until we got to where the water was too shallow for our boat, when we came to a little river steamer. The boats remained here until morning, and the passengers distributed themselves over the two boats and thus diminished the crowd. There were no berths. Luckily I had bought a hammock of a virgin at Virgin City, and more luckily Captain Merry had had a nice lunch put up for us. We opened it at midnight after a fast of twenty-one hours.

At daylight the rest of the passengers were transferred and we went down the river two miles to some rapids. Here we got off and walked around the rapids, some two or three hundred yards—a railroad with rude horse cars carries the baggage. This is at Castillo and Castillo Viejo, two little towns, with a curious mud fort on the hill just above. This was Tuesday, November 29. The rainy season had not finished, and it rained part of the day—hot, steamy showers. The shallow river, the rich and varied shades of green in the forests, the strange forms and species, the gorgeous colors of flowers, the great masses of rich foliage and festoons of vines, screaming parrots and paroquets and other birds of brilliant plumage, monkeys jumping about on the trees and chattering at the steamer as it passed, lazy alligators lying alongshore and tumbling into the water at shots from pistols, scaly iguanas almost as brilliantly colored as the flowers themselves—all these sights and sounds, not to mention the smells, told us we were in another clime from that of our homes. It brought back all the old stories I had ever read of tropical countries, of Spanish adventure, and the romance of Central America.

We got into the harbor at Greytown, or San Juan del Norte, after dark. Some of the passengers went ashore, but I stayed on board. I bought some bananas and bread for supper and got a good night's rest in my hammock. The bar has filled up so of late that the steamer cannot get into the harbor—it must anchor outside. We were transferred to a tug the next morning and were carried out. The tug rolled heavily, so did the steamship, and it began to rain hard. It was found impossible to transfer passengers and baggage directly to the ship, so we anchored and were put aboard with small boats.

We were now on the steamer *Golden Rule*, Captain Babcock—a fine steamer and fine captain. We lay there two days, then started. We were getting on in very good season, when the engine broke—something about the eccentric—so the engine had to be worked by hand. This delayed us. Then the last three days we had heavy gales from the north—very heavy the last day, Saturday, December 10. The ship rolled terribly—the wheelhouses would go under on both sides. We got up to Sandy Hook after dark, and failing to get a pilot, ran in without one. The moon was bright and the hills on all sides were white with snow. Three days before thin shirts and linen coats were in demand, now overcoats and furs.

Some of the passengers went ashore that night, others the next morning—I among the rest. Monday I ran about the city, delivered some packages, saw some men on business, and left for home that night, arriving at Ithaca the next morning at eight o'clock. I footed it home and arrived at 11 A.M. on Tuesday, December 13, 1864.

All looks familiar—some few changes, but Providence has been kind. We are all together again, all in good health. It seems as if I had not been away as we gather about the fire of a cold, wintry evening. All of us look a little older in the face, but hearts and affections are as young as when I left to ramble so far from home. We have snow and cold weather, a winter's landscape and a winter fireside, where we mutually have many things to tell.

And here let my long letters cease. It is by no means probable that I shall ever write so long a series again, or at least under such exciting circumstance or amid such interesting scenery. I trust you have had as much pleasure in reading as I have in writing them.

INDEX

Academy of Natural Sciences (California Academy of Sciences), 120, 366
Acapulco, Mexico, 7, 565, *5*
Adams, James Capen ("Grizzly"), 208, 212, 279
Agriculture and agricultural products, 22, 120, 178, 182, 184, 185, 188, 300, 480
Alameda County, 351; Creek, 352
Alamo Pintado Ranch, 76
Alcatraz Island, 355, 356
Alexander, Mr. ——, 385, 386
Allison Ranch, 454
Altaville, *see* Low Divide
Alviso, 247, 248, 371
Amador Road, 440
Amador Valley, 184, 202, 211
America, S.S., 564, 566, 567
American River, 295, 447, 453
American Valley (Feather River), 456
Antelope, 138, 185, 280, 286, 289, 382
Antelope Creek, 336
Antisell, Dr. Thomas, 75, 89
Arizona, 196
Arrastras, 483
Ashburner, William, 18, 26, 46, 246, 248, 352, 360, 496, 547; Mrs., 26, 352, 359, 496, 547
Asphaltum, 59, 63, 118, 126
Atascadero, 92, 93
Auburn, 451, 563
Aurora, Nevada, 420, 421, 422, 427, 539
Aurora Mine (New Idria), 138, 141
Averill, Chester, 18, 27, 157, 162, 164, 261, 269, 303, 321

Babcock, Captain ——, 569

Bache, Mount, 155, 161, 162, 163, 164, 166, 264
Baker, Mrs., formerly Miss Mills, 454
Balley, *see* Shasta Balley and Weaver Balley
Banks, Mr. ——, 267
Bass's Ranch, 303, 304, 322, 323
Bats, 474
Battle Creek, 334, 335, 336, *319*
Beale, Edward F., 387, 396
Bear Creek (upper Sacramento Valley), 335
Bear River, 454
Beardsley, Mr. ——, 218, 219
Bears, 67, 93, 94, 185, 279, 335, 523, 529; grizzly, 35, 38, 47, 66, 94, 95, 96, 107, 108, 109, 112, 130, 131, 178, 208, 282, 287, 523
Belden, Josiah, 175, 189, 365, 366
Bend City, 537
Benicia, 214, 215, 235, 289, 291, 292, 347, *213, 291*
Bidwell, John, 337, 344, 345
Big Meadows (Feather River), 458, 463
Big Meadows (near Kings River), 518, 527, *517*
Big Oak Flat, 401, 402
Big Trees (*Sequoia gigantea*), 154, 166, 398, 399, 400, 403, 413, 491, 514, 515, 517, 527, 528
Bigler, Lake, *see* Tahoe
Birds, 96, 117, 130, 278, 348, 418, 465, 495; *see also* Condors, Ducks, Eagles, Geese, Grouse, Herons, Humming birds, Owls, Pelicans, Swans
Bishop, Mr. ——, 384
Black Mountain, 178, 189
Blake, Gorham, 263, 266, 273, 454
Blake, William Phipps, 194, 212, 236
Bloody Canyon, 415, 416

Boiling Lake, 459, 464
Bonner, Mr. ——, 555, 558
Booker's Ranch, 135, 145
Boschulty, Mr. ——, 386
Botany and botanical collections, xxii, 12, 156, 305, 307, 319, 430, 433
Boundary Commission, 12, 26, 30
Bower Cave, 402, 413
Bowers Mine, 557
Bowers, Sandy, 557, 558
Branch, Francis Zida, 81, 89
Brandy City, 455
Brewer Creek, Lake, 524, 532
Brewer, Mount, 524, 525, 526, 532
Brewer, William H., vii, viii, xvi, xvii, xviii, xix, xxii, xxiii, xxiv, xxv, xxvi, 10, 172, 259, 261, 355, 365, 366, 368, 370, 373, 429, 430, 469, 548
Bridal Veil Fall (Pohono), 403, 404
Bridges, Thomas, 432, 449
Briggsville, 299
Browne, J. Ross, 132, 427, 452, 470, 505
Brown's Mountain, 326
Brush, Professor George J., xvi, xvii, 10, 307, 308, 429
Bubbs Creek, 534, 548
Buckeye, 303
Buckhorn Station, 326
Buena Vista Lake, 385
Butte Creek, 340

Cahuenga Pass, 45
Calaveras County, *387*
Calaveras Grove (Big Trees), 398, 399, 400, 429, 430, 435
Calaveras Valley (Alameda County), 202
California, Gulf of, 565
California, the name, 124, 132
Calistoga, 225, 238
Camels, 26, 30, 41; fossil, 344
Camp scenes and camp life, 12, 15, 16, 41, 64, 85, 86, 91, 92, 112, 121, 170, 171, 172, 174, 181, 186, 193, 194, 195, 198, 205, 218, 263, 265, 267, 276, 282, 284, 285, 287, 291, 313, 317, 409, 411, 509, 514, 519, 526, 542, 543
Campbell, Mr. ——, 311
Campbell, "Mrs.," 321, 322
Camptonville, 455
Cañon Agua Fria, 35
Caribbean Sea, 4, 5, *4*
Carmel Mission, 106, 107, 114
Carmel River, Valley, 106, 108, 110
Carmelo, see Carmel
Carne Humana Ranch, 225
Carpinteria, 51, *48*
Carquinez, Straits of, 262, 291, 333
Carson Canyon, River, Valley, 431, 433, 438, 553, 562
Carson Pass, Spur, 437, 438
Castillo Viejo, Nicaragua, 566, 568
Castle Mountains, 306
Castle Peak, 265, 403
Castle Rocks (Crags), 306
Cathedral Peak, 412
Catholic School, at Benicia, 215, 216
Cattle, 20, 40, 74, 78, 87, 93, 95, 99, 106, 178, 187, 203, 283, 285, 286, 351, 509, 510, 518, 535, 545, 546
Cave City, 400
Caves, 117, 118, 400, 402, 413, 473, 474
Cayeguas Ranch, 49
Cedars, 312, 324, 398, 409
Centerville, 444
Charlotte Creek, 534, 548
Cherry Creek, 477
Chickering, J. W., 3, 10
Chico, 337, 338
Chinese, 243, 250, 251, 252, 330, 366, 367, 369, 370, 481
Chino, 32
Choual, Mount, 155, 163
Christmas festivities, 20, 359, 360
Church services, 12, 21, 50, 69, 76, 101, 103, 104, 120, 127, 158, 173, 179, 214, 247, 258, 344, 359

INDEX 573

Circus, 296
City of Leon, S.S., 567
Civil War, 88, 102, 103, 119, 120, 123, 175, 176, 179, 195, 196, 197, 233, 250, 258, 273, 283, 293, 319, 353, 359, 365, 370, 408, 420, 426, 427, 452
Clark, Miss ——, of Folsom, 160, 167
Clark's Ranch (Wawona), 546, 547, 549
Clayton, 200, 235, 267, 269, 376, *195, 198*
Clayton, Mr. [Joel], 193
Clear Creek, 325
Clear Lake, 226, 230, 264
Cleaveland, Mr. ——, 263
Clover Creek, 522, 532
Coal mines, 64, 67, 183, 192, 198, 199, 204, 206, 207, 210, 211, 276, 332, 333
Coast Range and coast ranges, 82, 174, 236, 298, 380, 382, 510
Coast Survey (U.S. Coast and Geodetic Survey), 10, 191, 238, 256, 273, 323, 344, 470
Cobb, Dr. ——, 162, 173
Cobb, Mount, 226, 230
Colima, volcano, Mexico, 8
Columbia, 397, 398, 400, 425
Condors, 96, 114, 130
Confervae, 233
Conness, John, 177
Contra Costa County, 350
Cooper, Dr. James Graham, 126, 132, 453
Copper Creek, 532
Copper mines, 184, 376, 385, 395, 438, 457, 485, 487, 489
Corral Hollow, 200, 204, 206, 212, 276, *200, 205*
Cotter, Richard D., 505, 516, 525, 527, 532, 541, 542, 543
Cottonwood (Siskiyou), 475, 476
Cottonwood Creek, 301, *297*
Cow Creek, 332

Coyote Springs, 381
Coyotes, 79, 382, 416, 512
Crane Flat, 403
Cranes, *see* Herons, 52
Crescent City, 487, 490, 492, 494, *471, 477*
Crystal Creek, 325
Cypress, Monterey, 107, 114

Dana, Mount, 408, 409, 411, 413, 433, 548
Darlingtonia, 305, 307
Davidson, Mount, 558
Davis, Lt. ——, 468
Day, the Misses, 160, 167
Day, Sherman, 157, 159, 162, 163, 166, 167
Dayton, Nevada, 562
Deadfall (Clayton), *198*
Deadwood (American River), 447
Deadwood (Siskiyou), 477
Deer, 40, 48, 67, 95, 109, 111, 112, 185, 208, 285, 286, 386, 472, 519, 533
Deer Creek (Sacramento Valley), 336, 344
Deer Creek (Tulare Valley), 381
Del Norte County, 489
Diablo, Mount and Range, 173, 174, 184, 185, 187, 191, 193, 195, 199, 200, 201, 203, 212, 260, 262, 263, 264, 265, 266, 267, 268, 273, 275, 294, 377, 462, *195, 261*
Dodge, Ossian E., 567
Dogtown, 321
Dominguez, José María, and Doña Marcellina, 59, 72
Donner Lake, 563
Donner party, 563
Dos Pueblos, 73, 89
Douglas City, 326, 328
Douglas fir (spruce), 166, 409
Downey, Gov. John G., 10, 18, 27, 41, 156, 157, 166, 452
Downey, Mount, 41
Downieville Buttes, 455

DuBois, Henry, 120
Ducks, 40, 348, 418, 495
Dutch Fred's, 377
Dye, Mr. ——, 336

Eagle Creek Ditch, 299
Eagles, 47, 130
Earthquakes, 127, 185, 186, 190, 358, 387
Eastman, Aug., of Tehama County, 460, 470
Ebbetts Pass, 431, 434
Eclipse, lunar, 280
Economic conditions, 119
El Capitan (Tutucanula), 403, 404, 413
El Dorado Canyon, 447
Elizabeth, Lake, 388
Elizabethtown, 444
Elk, 185
Elk Valley, 471
Elkhorn Station, 511, 512
Enfield Center, N.Y., xviii, 569, *561*
Enriquita Mine, 161, 165, 167, 372
Esmeralda, *see* Aurora
Estrada, Don Joaquin de, 93
Eureka (Humboldt County), 493
Eureka (Yuba River), 455

Fair, California State, 187, 188, 189; Contra Costa, 197; World's, at London, 236
Fales Hot Springs, 422, 427
Fall River, 467, 468
Feather River, 456, 458
Field, Justice Stephen J., xv, 10, 26, 244
Financial difficulties, 52, 183, 237, 242, 243, 244, 245, 246, 248, 354, 448, 564
Finch, Charles W., and James, 114
Finch's Ranch, 109, 110, 113
Firebaugh's Ferry, 275, 378, 510
Firs, 153, 155, 312, 324, 398, 409, 520
Floods, 25, 26, 241, 242, 243, 244, 246, 249, 250, 269, 295, 495

Forest Hill, 448, 453, *446*
Fort Crook, 468, *460*
Fort Jones, 477
Fort Miller, 395, 396, 543
Fort Tejon, 383, 384, 387, *381*
Fossils, 18, 46, 47, 59, 74, 75, 78, 87, 113, 185, 186, 199, 200, 206, 207, 212, 299, 300, 302, 326, 333, 336, 338, 339, 343, 344
Fowler's Ranch, 225
Foxen, Benjamin, 77, 89
Frame, Mr. ——, 311
Frear, Walter, 454
Frémont, John C., 10, 11, 309, 323, 449, 549
Frémont Pass, 389
Fresno City, 379, 396, 510, 511
Fresno River, 546

Gabb, William More, 261, 269, 273, 376, 395, 453, 496
Gabilan Peak, Range, 123, 128, 151, 152, 156
Gabriel, John Peter, 19, 27, 125, 126, 171, 193, 213
Galena Hill, 455
Gambling, 20, 104, 189, 198, 248, 330, 421, 486, 487
Gardiner, James T., xxii, 451, 468, 469, 505, 525, 526
Gardner, *see* Gardiner
Gaviota Pass, 75
Geese, wild, 40, 52, 216, 217, 218, 220, 223, 225, 495
Gem (steamer), 295
General Grant National Park, 516
Genesee Valley, 456, 476, *451*
Geological Survey, California State, xv, xvi, xx, xxi, xxix, 9, 10, 122, 245, 354, 370, 373, 375, 452
Geological Survey, United States, xxi, 344, 413, 469
Geysers, 228, 229, 230, 231, 232, 233
Gibbs, Mount, 549
Gifford, Mr. ——, 457
Gilroy, 506

Glaciers, 318, 324, 409, 412, 413, 466, 524
Goddard, Mount, 541, 543
Gold and gold mining, xv, 299, 304, 326, 327, 328, 341, 343, 397, 448, 454, 478, 483, 484, 536, 553, 554, 556, 557
Gold Hill, 553, 554, 557
Golden Age, S.S., 6, 5
Golden Gate, 8, 9, 10, 11, 264, 496
Golden Gate, S.S., 289, 294, 307
Golden Rule, S.S., 569
Gore, John C., 105, 114
Gould & Curry Mine, 446, 555, 556, 557, 558, 559
Grapevine, at Montecito, 59, 60, 72
Grass Valley, 453, 454
Gray, Professor Asa, xxiii, 3
Grayson Ferry, 280
Greenhorn, 394; Mountain, 394
Greytown (San Juan del Norte), Nicaragua, 569
Griswold's, 137, 143
Grizzlies, *see* Bears
Grouse, 531
Guadalupe Mine, 165, 371, 372
Guadalupe Ranch, 101, 97
Guirado, ——, assistant on Survey, 18, 19, 51, 76, 157
Guirado, María Jésus (Mrs. Downey), 27, 166

Haight, Judge Fletcher Mathews, 106, 107, 108, 114, 131
Haight, Henry Huntley, 114
Haight, Sarah (Mrs. Edward Tompkins), 114, 263
Halleck, General [Henry W.], 348
Hamburg, 479
Hamilton, Mount, 173, 189, 289, 365, 462
Hamilton, Mrs. (formerly Miss Mead), 158, 159, 167, 173
Hamilton, Rev. Laurentine, 158, 159, 167, 173, 175, 247, 248, 506
Hangtown, 552

Happy Camp, 481, 482
Hart, Captain ——, 567
Harte, Bret, 500, 501
Hastings, Judge Serranus Clinton, 214, 215, 237
Hat Creek, 467
Haywards, 182, 351, 505
Henry, Mount, 224, 225, 238
Hermit Valley, 431, 434
Herons, 278, 289
Higby, Mr. [William, of Calaveras], 567
Hill, Mr. ——, of Dayton, Nevada, 562
Hillebrand, Dr. William, 430, 448, 449
Hill's Ferry, 378
Hitchcock's Ranch, 111
Hoffmann, Charles F., xxi, xxiv, 128, 132, 163, 164, 171, 172, 234, 261, 303, 313, 323, 325, 332, 347, 398, 429, 453, 505, 524, 545, 546, 547, 548, 564
Hoffmann, Mount, 407, 413
Hollenbeck's tavern, 150, 288, 507
Holy Week in Santa Barbara, 68
Honey Lake, 463
Hooker, Mrs. Katherine Putnam, *see* Putnam, Katie
Hope Valley, 437, 438, 449
Horn, Dr. ——, 392
Hornitos, 395
Horsetown, 299
Hospital Canyon, 278
Hubbard, Mr. ——, 299, 300
Humboldt Bay, County, 226, 493, 500
Humming-birds, 96, 117, 118
Hurd's Ranch, 473
Hydraulic mining, 326, 327

Illinois River, 483
Independence, Camp, 536, 537
Independence Creek, 534
Indian Creek, 482
Indian Valley, 456, 458, 463

Indian Wells, 392, 393
Indians, 39, 51, 69, 70, 77, 128, 195, 196, 215, 222, 223, 296, 300, 301, 302, 304, 320, 321, 322, 335, 338, 339, 380, 381, 387, 390, 391, 392, 393, 395, 417, 419, 438, 443, 466, 467, 468, 471, 472, 474, 479, 493, 494, 500, 513, 519, 536, 537, 538, 539, 540, 545
Indiantown, 482
Insects, 97, 417, 433
Inyo Mountains, 534, 535, 536, 538
Iowa Hill, 453
Ithaca, N.Y., xviii, 569

Jagor, Friedrich, 191, 212
Janin, Louis, 165, 166, 167
Japan, 213, 236
Joaquin Jim, 536
Johnson, Professor ———, of Yale, 429
Jolon, *93*
J. O. Pass, 521, 532
Jordan, Mr. ———, 153

Kaweah River, 513, 514, 517, 520, 522
Kearsarge Pass, 534, 548
Keating, Wm. P., of St. Louis, 460, 470
Keough, Thomas, 529, 532
Kern River, 382, 390, 391, 392
Keysville, 391, 394
King, Clarence, xxi, xxvi, xxix, 324, 413, 451, 453, 461, *465*, 468, 469, 471, 487, 496, 505, 525, 527, 528, 532
King [Starr], Mount, 267, 268, 273
King, Thomas Starr, 120, 187, 258, 263, 273, 359, 366
Kings River, Canyon, 379, 380, 381, 396, 512, 514, 517, 520, 522, 524, 541, 529, 530, 531, 533, *522, 527*
Kirker Pass, 200
Klamath County, 493, 494
Klamath River, 316, 475, 478, 481, 482

Knight's Ferry, 397
Knight's Valley, Creek, 226
Knoxville, 445
Koochahbee, 417, 419

Lake Valley, 439, 441, 553
Lake Vineyard, 21
Lassen County, 344
Lassen, Peter, 336, 344
Lassen's Butte, Peak, 265, 296, 303, 316, 319, 333, 334, 336, 344, 455, 458, 461, 462, 463, 464, 465, 466, 467, 470, *456*
Last Chance, 447
Laurel (Bay tree), 155, 350
Le Conte, Joseph N., 532
Lewis' Ranch (near Visalia), 528
Lewis' Ranch (Oregon), 484, 485
Liebre Ranch, 387, 388
Lincoln's election and inauguration, 10, 48
Lion, California (panther), 96, 109, 285, 289
Little Pine Creek, 534
Livermore Pass, Valley, 202, 204, 211, 272, *267*
Llagas Creek, 169
Lobos, Point, 130
Loma Prieta, *see* Bache, Mount
Lone Tree Canyon, 277
Lone Willow, 509
Los Angeles, 12, 13, 14, 20, 30, 41, 243, 386, 387, *11, 12, 20, 39*
Loveless, Mr. ———, 458
Low Divide, 486, 487, 489
Lower California, 8, 37, 565
Lunatic asylum at Stockton, 259, 260
Lyell Fork (Tuolumne River), 410, 413
Lyell, Mount, 411, 413

Madroña, 110, 114, 225, 226
Mammoth Trees, *see* Big Trees
Marin County, 348, 367
Mariposa, 547, 548

INDEX

577

Mariposa Estate, 547, 548, 549
Marsh, Dr. John, and his son, 202, 269, 270, 271, 272, 273
Martinez, 214, 261, 291, 376, *194, 255*
Marysville, 344, *332*
Marysville Buttes, 264, 462
Maxwell, Mr. ———, 138
Mayhew, Dr. ———, 165, 166, 372
McCloud River, 304
McClure, Mr. and Mrs., of Weaver, 329
McDonald, Mr. ———, 226, 227
McLoughlin, Mount, *see* Pitt, Mount
Mead, Mrs. ———, 506
Mendocino County, 493
Merced River, 401, 402, 403, 405
Merry, Captain ———, of the *America*, 566, 567, 568
Mexico, 7, 368, 565
Middletown, 299
Mike, 19, 70, 71, 121, 122, 151, 170, 171
Millerton, 395, 396, 540
Mines and mining methods, 35, 138, 139, 140, 141, 142, 157, 158, 159, 160, 165, 184, 228, 229, 298, 299, 343, 421, 432, 445, 454, 478, 483, 484, 554, 555, 556, 557, 562
Mirages, 78, 100, 272, 276, 283, 377, 511
Mission San Jose, *see* San Jose Mission
Missions, Spanish-Californian, *see* Carmel, Los Angeles, Monterey, San Buenaventura, San Fernando, San Gabriel, San Jose, San Juan, Santa Barbara, Santa Inez, Santa Margarita
Mokelumne River, 431
Mono Creek, Pass (San Joaquin), 539, 540, 548
Mono Lake, 265, 408, 415, 417, 418, 419, 427, *415*
Mono Pass (Tuolumne), 408, 415, 548

Montecito, 59
Monterey, 103, 104, 129, *101, 103*
Monterey Bay, 100, 105, 110, 155
Monterey, Mission at, 114
Monument Peak, 174, 189
More, Mr. ———, of Kentucky, 472
Mormon Station, 457
Mormons, 185, 449, 496
Morro Rock, 82
Moses, W. S., 323, 324
Mount Shasta City, 324
Mountain Charlie's, 155
Mountain climbs, 19, 22, 23, 33, 36, 37, 60, 61, 83, 84, 109, 123, 150, 151, 161, 162, 163, 164, 173, 174, 178, 179, 191, 217, 218, 223, 224, 256, 263, 264, 265, 266, 267, 268, 278, 287, 288, 311, 312, 313, 314, 315, 316, 317, 328, 329, 331, 332, 335, 407, 408, 409, 411, 415, 416, 423, 431, 433, 437, 441, 448, 460, 461, 462, 463, 464, 465, 476, 480, 518, 523, 524, 525, 526, 527, 531, 532, 541, 543, 549, 558
Mountain View, 175, *173*
Muir, John, 413, 488
Mules, characteristics of, and incidents with, 16, 17, 62, 63, 65, 66, 73, 80, 98, 111, 112, 113, 145, 152, 170, 174, 175, 204, 207, 277, 306, 307, 308, 312, 379, 512, 529, 534
Murphy's, 398, 400, 425, 429, *387, 423*

Nacimiento River, 100, *91*
Nahalawaya, 66
Napa (Town, Valley), 216, 218, 220, 224, 226, 230, 259, 347, *214*
Naples, *see* Dos Pueblos
Negroes, 45
Nelson's Point, 456
Nevada, Territory and State, 422, 441, 443, 551, 559, *551*
Nevada City, 455
Nevada Fall, 405
New Almaden quicksilver mines,

156, 157, 158, 159, 160, 167, 247, 289, 371, 372, *156, 164*
New Idria quicksilver mines, 138, 139, 140, 141, 142, 143, *135, 140*
New Year, 23, 365; Chinese, 243, 369, 370
Nicaragua, 565, 566, 567, 568, 569
Nipomo Ranch, 78, *76*
Nojoqui Ranch, 75
North Star, S.S., 3
Nutmeg tree (*Torreya californica*), 227, 238
Nye, Governor ——, 188, 189

Oak Creek, 389
Oakland, 186, 351, 368, 499
Oakland College, *see* University of California
Oaks, 76, 89, 93, 109, 114, 136, 144, 155, 170, 186, 193, 221, 262, 263, 269, 287, 352, 376, 384, 389, 401, 409, 506
Ojo de Agua de la Coche, 170, 171
O'Keefe, Father Jeremiah, 58, 72
Olmsted, Frederick Law, 547, 548, 549
Oregon, S.S., 495
Oregon, State, 476, 482, 483, 484
Orestimba Canyon, 283, 284, *275*
Oroville, 340, 344, *325*
Orr, Win, 545, 548
Oso, Mount, 277, 278, 279
Overland stage, road, stations, and mail, 33, 34, 45, 56, 74, 79, 122, 157, 243, 288, 439, 511, 552; telegraph, 213, 214, 439
Ovid, N.Y., xix, 158, 167
Owens Lake, Valley, River, 534, 535, 539
Owensville, 537, 539
Owls, 96, 106, 130, 521

Pacheco (Contra Costa County), 197, 262
Pacheco Pass, 150, 275, 286, 288, 506, 508, 511, *505*
Pacheco Peak, 149, 150, 288

Pacific Railroad Survey and *Reports*, 75, 87, 89, 99, 335, 344, *323*
Pajaro Valley, 152
Palo Scrito, 110, 156
Panama, 5, 6
Panoche River, Plain, 136, 137, 139, 144
Panther, *see* Lion, California
Pass Creek, 390
Paul Pry (steamer), 355, 356, 357, 358, 359, 360, 361
Pecaya, volcano, Guatemala, 565
Pelicans, 130, 348, 495
Pence's Ranch, 340, 344
Penitentia Canyon, 353
Perry, Mr. ——, 311
Pescadero Ranch, 102, 104, 105, 114, 129, 131, *104, 108*
Petaluma, 257, 259, 348
Piety Hill, 299
Pilgrim Camp, 472
Pilot Peak (Feather River), 455, 463
Pine Mountain, 227
Pine nuts, 339, 536, 540, 548
Pines, 93, 136, 144, 153, 155, 166, 311, 324, 339, 398, 409, 410, 413, 520, 526, 527, 536, 540, 548
Pioneer Mine, 228, 234
Pit River, 304, 322, 463, 467, 468, 488
Pitt, Mount, 476, 484, 488
Placerville, 552
Placerville Road, 439, 440, 552
Plato Mine, 558
Plumas County, *451*
Pluton River, Canyon, 228, 230, 231
Pluto's Cave, 473, 474
Point Reyes, *see* Punta de los Reyes
Poker Flat, 455
Politics, 48, 175, 176, 177, 178, 179, 354, 426, 427, 448, 452, 477, 499
Pomeroy, Mr. ——, 492
Pony Express, 9, 102, 123, 200
Population, 497

Porcupine Flat, 407
Posé Flat, 394
Potosi, 455
Prices, 40, 144, 175, 214, 329, 330, 418, 420, 447, 448, 456, 480, 495, 509, 539, 554
Puerto, Cañada del, 279, 280, 289
Pumpelly, Raphael, 196, 197, 198, 212, 213, 236
Punta de los Reyes (Point Reyes), 349, 496
Puta [Putas, Putah] Creek, 294, 307; Mount, 226
Putnam, Katie (Mrs. Katherine Putnam Hooker), 355, 356, 357, 360, 361
Putnam, Samuel Osgood, 175, 189, 355, 359, 360; Mrs. ——, 426
Pyramid Peak, 265, 441, 463

Quicksilver, 111, 142, 159, 160, 202, 226, 228, 229, 248, 327, 328, 371, 376, 454, 483, 557, 562; *see also* New Almaden, New Idria, Enriquita, and Guadalupe mines
Quincy, 456

Raccoon, 193
Races, mixture of, 13, 69, 70, 128, 181, 215, 395
Rag Canyon, 294, 307, *293*
Ragged Peak, 415, 427
Railroads, 6, 33, 44, 451, 499, 551, 563; street, 498
Rainfall, 20, 25, 121, 196, 241, 242, 243, 250, 253, 262, 368, 392, 393, 471
Rattlesnakes, 75, 79, 93, 110, 180, 189, 190, 206, 279, 383, 386, 400, 457, 524, 530, 531, 533
Rawhide, 48, 78
Red Bluff, 295, 307, 333, 334, 336, *294*
Redwoods (*Sequoia sempervirens*), 153, 154, 166, 225, 350, 490, 491, 492, 495
Reed, Mr. ——, 162

Reeves, Mr. ——, 480
Rémond, Auguste, 261, 273, 303, 323, 325
Reptiles, 96; *see also* Rattlesnakes
Ribbon Fall, 406, 413
Richthofen, Baron Friedrich von, 359, 361, 561
Ripley, Mount, 264
Roaring River, 522, 524, 532
Roberts, Captain ——, 567
Roberts, Mr. ——, 392
Robinson, Todd, 477
Rockland District, 485
Rockville, 293, *293*
Rocky Bar Ledge, 454
Rowland Flat, 455
Rúbio, Father José María González, 57, 58, 72
Rudesill's Landing, 257, 273
Russell, Major ——, 185, 212
Russian River, 224, 226, 230, 259
Russian ship, 496

Sacramento, 187, 241, 242, 248, 249, 295, 451, 453, 551, 562, *182, 187*
Sacramento River, Valley, 187, 242, 243, 244, 248, 264, 294, 295, 296, 298, 303, 304, 305, 307, 316, 332, 333, 334, 336, 464, 564, *294*
Sage, sagebrush, 209, 212, 389, 535, 536, 538
Sailor Diggings (Waldo), 482, 483, 484
St. Helena, 224, 225, 235
St. Helena, Mount, 224, 226, 230, 234, 238, 264, 462
St. John Mountain, 224, 225, 238
St. John, Mr. ——, 401
St. Johns, Mr. ——, 549
Salinas, 129, 131
Salinas Valley, 92, 95, 97, 98, 100, 110, 129, 155
Salmon Mountains, 462, 463
San Antonio River, 94, 97, 100, *93*
San Benito River, Valley, 125, 135, 289

San Bernardino County, 197; peaks, 13, 38, 236
San Buenaventura, 49, 50, 51, 53
San Buenaventura Mission, 49, 50, 53
San Carlos (Owens Valley), 537
San Carlos Mine (New Idria), 138, 139
San Carlos Mission, *see* Carmel Mission
San Emidio Ranch, 385, 386
San Fernando Mission, 45
San Fernando Valley, 44, 45, *43*
San Francisco, 9, 117, 118, 119, 127, 182, 235, 243, 250, 253, 258, 264, 289, 294, 347, 496, 497, 498, 499, 500, 505, 548, 564, *117, 191, 221, 241, 242, 243, 244, 245, 250, 284, 294, 333, 337, 347, 353, 354, 365, 367, 371, 483, 489, 496, 545*
San Francisco Bay, 230, 247, 255, 264, 294, 347, 462
San Gabriel, *33*
San Gabriel Canyon, River, 24, 25, *20, 29, 35*
San Gabriel Mission, 21, 22, 23, 27
San Gabriel Range (Sierra Madre), 22, 23, 27, 38
San Joaquin, Valley, plain, River, 151, 202, 203, 205, 220, 242, 264, 268, 269, 275, 276, 278, 286, 287, 294, 333, 367, 377, 379, 395, 510, 511, 512, 540, 541, 544, 545, 548, *375, 533, 541, 544*
San Jose, 120, 156, 158, 247, 248, 353, 365, 367, 371, 499, 506, *169, 172*
San Jose Mission, 179, 181, 189, 182, 352, *179*
San Juan, 117, 118, 128, 131, 147, 288, 367, *122, 128, 143*
San Juan del Norte, Nicaragua, 569
San Juan del Sur, Nicaragua, 565, 566
San Juan Mission, 127, 128, 132
San Juan North, 455

San Juan Plain, 150, 289
San Juan River, Nicaragua, 568
San Leandro, 182, 351
San Lorenzo River, 154, 289
San Lucas, Cape, 565
San Luis de Gonzaga, 286, 288, 509
San Luis Obispo, port, 11; town, 82, 83, *79, 82;* buttes near, 82, 83, 84, 89; pass, 92
San Pablo Bay, 224, 259
San Pedro, 12
San Rafael, 255, 256, 257, 350
San Ramon Valley, 184, 185, 202, 213, *210*
Santa Ana Mountains, 33, 34, 36, 37; River, 43, *31*
Santa Barbara, 12, 55, 56, 57, 58, 59, 72, *55, 60*
Santa Barbara Mission, 56, 57, 58, 69, 72
Santa Clara, 120, 156, 173
Santa Clara College, 368, 373
Santa Clara River, 49
Santa Clara Valley, 118, 156, 158, 169, 506
Santa Cruz, 152, 154, *149*
Santa Cruz Mountains, 153, 154, 155, 375
Santa Inez Mission and College, 73, 75, 76, 77, 89, *73*
Santa Inez River, 66, 75
Santa Lucia Mountains, 84, 92, 99, 100
Santa Margarita Mission (*Asistencia*), 92, 93, 114
Santa Margarita Valley, 100
Santa Maria River, 78
Santa Monica, Sierra, 13, 45, *15*
Santa Rosa Valley, 230
Santa Susana Mountains, 45, 52
Sargent Ranch, 118
Sausalito, 256, 257, 273
Savage Mine, 556
Scandinavian Canyon, 433
Schmidt, —— (cook), 261, 279, 303, 323, 325, 332, 347, 453

Sciad Creek, 480
Scott River, Valley, 477
Scott's Bar, 477, 478
Seals and sea lions, 11, 52, 130, 496
Sebastopol, 221, 237, *217*
Secession, Secessionists, *see* Civil War
Sequoia gigantea, see Big Trees; *sempervirens, see* Redwoods
Shasta Balley, 331, 332
Shasta (City), 297, 298, 323, 325, 332, 464, *299*
Shasta, Mount, 296, 298, 304, 305, 306, 309, 310, 311, 312, 313, 314, 315, 316, 317, 318, 319, 323, 324, 336, 337, 344, 460, 462, 463, 464, 467, 471, 473, 474, 525, 527, *302, 309*
Shasta Springs, 306
Shasta Valley, River, 307, 473
Sheep, 40, 74, 99, 100, 136, 277, 280, 288, 506, 509; mountain, "bighorn," 474, 487
Shell Mountain, 518, 532
Sierra Madre, *see* San Gabriel Range
Sierra Nevada, 140, 260, 265, 266, 276, 283, 284, 285, 298, 341, 376, 377, 380, 381, 382, 407, 409, 441, 463, 465, 510, 515, 518, 525, 535, 537, 538, 552, 558
Silliman, Mount, 523
Silliman, Professor Benjamin, Jr., xviii, 523, 561
Silver and silver mines, 24, 104, 129, 431, 432, 444, 536, 553, 554, 556, 557, 562
Silver City, Nevada, 553, 554
Silver Lake, 437
Silver Mountain (peak and town), 431, 432, 433
Silver Valley, 430, 431
Siskiyou Mountains, 316, 463, 476, 480, 482
Sisson, 324
Skunks, 181, 223

Slate Range, 392
Slippery Ford, 439, 440
Snow, 23, 37, 314, 315, 316, 317, 318, 410, 411, 434, 435, 461, 462, 465, 519, 525, 540, 552
Snowshoe Thompson, 449
Snowshoes, 435, 449
Soldiers, 513, 519, 527, 528, 531, 532, 536, 539, 540, 541, 542, 545
Soledad Mission, 99, 101
Sonoma, 259
Sonoma County, 347
Sonora, 241, 242, 400, 401, 425
Sonora Pass, 423
Southern, Sim, 306, 310
Spanish Californians, 13, 18, 51, 70, 71, 74, 75, 77, 128, 157, 160, 161
Spanish grants, 99, 160, 169, 170, 222, 257, 258, 292, 293, 349, 384, 387
Speculation, xv, xx, 35, 257, 291, 292, 354, 395, 422, 432, 452, 457, 511, 554
Spratt, —— (soldier), 532, 541, 542, 543
Springs, hot and mineral, 35, 39, 55, 59, 60, 61, 169, 175, 179, 193, 225, 231, 232, 233, 297, 306, 315, 407, 408, 417, 418, 422, 427, 459, 460, 538, 561, 562
Squaw Valley, 446
Stanislaus River, 397, 398, 400, 423, 424, *417*
Steamboat Springs (near Mount Lassen), 459
Steamboat Springs, Nevada, 561
Stockton, 259, 260, 261, 275, 548
Strawberry Valley (base of Mount Shasta), 310, 311, 319, *302, 309*
Sugarloaf Creek, 522, 532
Suisun, 289, 394
Suisun Creek, 293
Sulphur Mountain, 230
Summit Lake, 437
Suñol, 351
Suscol, 216, 235, *214*

582 UP AND DOWN CALIFORNIA

Swans, 495
Sweet Briar Ranch, 320
Sycamore Canyon, *43*

Table Mountain, 397
Table mountains, 340, 341, 342, 343, 344, 395, 424, 436
Tahoe, Lake, 265, 437, 438, 441, 442, 443, 444, 449, 450, 553, *436*
Tailholt, 394, 396
Tamalpais, Mount, 230, 256, 273
Tarantulas, 97, 203, 209, 210
Tehachapi Pass, Valley, 389
Tehama, 296, 297
Tejon Pass, Indian Reservation, 386, 396; *see also* Fort Tejon
Telegraph, 213, 214, 233, 243, 244, 250, 507; Indian, 539
Temescal, 34, 39, 44
Tenaya, Lake, 407
Terry, Judge [David S.], 426
Thomas' Mill, 514, 516, 528, *512*
Three Devils, 480
Timilick Valley, 444
Tin mines, 35, 39, 40
Tioga Pass, 409, 413
Tomales, 348
Tomales Bay, 230, 348, 349, 350
Tompkins, Edward, 102, 114, 129, 263, 265
Towers, 325, 331
Tragedy Springs, 436, 449
Trees, 93, 153, 155, 225, 227, 277, 312, 328, 350, 359, 360, 380, 394, 398, 403, 409, 410, 435; *see also* Cedars, Cypress, Douglas fir, Firs, Laurel, Madroña, Nutmeg tree, Oaks, Pines
Tres Pinos, 135
Trinity County, Mountains, River, 325, 326, 329
Triunfo Ranch, 47, 49, 53, *45*
Trout, 443, 459, 529, 530, 533, 544
Truckee River, 445, 446, 563, *443;* Pass, 563
Tulare, Lake, 203, 379; Plain, *508*

Tule River, 381, 395
Tules, 219, 220
Tuolumne County, 241, 242, 342
Tuolumne Grove (Big Trees), 403, 413
Tuolumne River, Meadows, Soda Springs, 401, 407, 408, 409, 410, 412, 549, *406*
Tuscan Springs, 297, 336
Twenty-one Mile House, 169, 506
Tyndall, Mount, 527, 532

Unicorn Peak, 412
Union, the, *see* Civil War
University of California, xxv, 283, 355, 368, 370, 373
Uvas, Cañada de las, 383, 396

Vallecito Canyon (San Benito region), 137, 144
Vallejo, 216, 292
Ventura, *see* San Buenaventura
Vernal Fall, 405
Views, 15, 22, 26, 34, 37, 38, 84, 100, 109, 110, 124, 139, 141, 150, 151, 155, 170, 174, 175, 185, 192, 202, 217, 219, 224, 229, 234, 248, 256, 257, 259, 262, 264, 265, 266, 268, 278, 296, 305, 306, 311, 314, 316, 317, 329, 335, 336, 384, 391, 394, 401, 403, 405, 407, 408, 409, 411, 416, 423, 431, 441, 446, 455, 460, 471, 476, 484, 486, 518, 523, 524, 525, 545, 558
Vigilance Committee, 499, 500
Virgin Bay, City, Nicaragua, 566, 567
Virginia City, 559, 561, 562, 563, *551*
Visalia, 380, 381, 395, 512, 513, 519, 528, *378*
Vivaparoa, 63
Volcanic phenomena, 8, 30, 31, 82, 90, 150, 206, 312, 313, 315, 318, 334, 335, 336, 339, 340, 341, 342, 408, 416, 418, 424, 431, 433, 434, 436, 437, 459, 463, 464, 465, 467, 473

Volcano (town), 435, 436, *429*

Warm Springs, 506
Waldo, *see* Sailor Diggings
Walker River, 422
Walker's Basin, 390
Walker's Diggings, 420
Walker's Pass, 391, 392
Walkinshaw, the Misses, 157, 160, 469
Walkinshaw, Robert, 167
Walnut Creek, 289
Walsh, Judge R. J., 459, 460, 470
Ward, H. A., 300
Washington and Jefferson College, xix
Washoe, 173, 184, 189, 439, 440, 444, 446, 552, 559
Wasps, 209, 210
Watkins, C. E., 406, 413
Watsonville, 152
Wattles, Mr. ——, 228
Wawona (Clark's Ranch), 546, 547, 549
Weaver Balley, 328, 329
Weaver Creek, 326
Weaverville, 325, 326, 329, 330, 331
Wentworth, Mr. ——, 387
Whales, 63, 105, 130; fossil, 212
Wheelock, Mr. ——, 300
Whirlwinds, 511
Whiskey (mining town), 325
Whiskey Diggings, 455
White River, 381, 394, 396
Whitman, Mr. ——, 214
Whitney, Eleanor (Nora), 430
Whitney, Josiah Dwight, xvi, xxiv, xxviii, xxix, 3, 11, 15, 18, 24, 36, 63, 119, 128, 129, 157, 158, 173, 184, 187, 189, 191, 192, 200, 201, 213, 225, 245, 246, 261, 262, 263, 266, 303, 325, 398, 399, 409, 410, 413, 425, 426, 429, 430, 435, 453
Whitney, Mrs. J. D., 157, 263, 266, 398, 429, 496
Whitney, Mount, 525, 526, 527, 532
Wildcats, 386
Willey, Rev. S. H., 368, 373
Williamson, Mount, 532
Williamson, R. S., 309, 323
Willow Lake, 459
Wilson, Benjamin Davis (Benito), 15, 21, 23, 26, 27
Wilson [Captain John], 87
Wilson, Mr. ——, (Englishman), 162, 163
Wilson, Mr. ——, (formerly of the Navy), 457
Wolf, 412
Wragg Canyon, *see* Rag Canyon

Yale University, xvi, xviii, xix, xxiii, xxv, xxvii, 159, 167, 196, 429, 430, 451, 468, 469, 548
Yosemite Valley, 403, 404, 405, 406, 530, 532, 547, *397;* Falls, 404, 405, 531, 547
Yosemite (Steamer), 355
Young, John, 157, 167
Yount, George, 221, 222, 223, 237, 238
Yreka, 474, 475, 477
Yuba River, 455
Yucca, 388, 389, 392, 396

Zimmermann's Mountain House, 204, 211, 275

31901059448714

CPSIA information can be obtained at www.ICGtesting.com
Printed in the USA
LVOW11s0003240516

489519LV00002B/8/P